Advanced Applications of Multi-Criteria Decision-Making Methods in Operational Research

Advanced Applications of Multi-Criteria Decision-Making Methods in Operational Research

Guest Editors

Marcio Basilio
Valdecy Pereira
Marcos Dos Santos

MDPI

Basel • Beijing • Wuhan • Barcelona • Belgrade • Novi Sad • Cluj • Manchester

Guest Editors
Marcio Basilio
Military Police of the Rio de Janeiro
Rio de Janeiro
Brazil

Valdecy Pereira
Department of Production Engineering
Federal Fluminense University
Niteroi
Brazil

Marcos Dos Santos
Military Institute of Engineering
Rio de Janeiro
Brazil

Editorial Office
MDPI AG
Grosspeteranlage 5
4052 Basel, Switzerland

This is a reprint of the Special Issue, published open access by the journal *Mathematics* (ISSN 2227-7390), freely accessible at: https://www.mdpi.com/si/mathematics/902HY38776.

For citation purposes, cite each article independently as indicated on the article page online and as indicated below:

Lastname, A.A.; Lastname, B.B. Article Title. *Journal Name* **Year**, *Volume Number*, Page Range.

ISBN 978-3-7258-5775-3 (Hbk)
ISBN 978-3-7258-5776-0 (PDF)
https://doi.org/10.3390/books978-3-7258-5776-0

Contents

About the Editors

Marcio Basilio

Marcio Basilio is a Professor at the Military Police College, Colonel PM Jorge da Silva of the State of Rio de Janeiro (ESPM/PMERJ), Brazil. He holds a bachelor's degree and a master's degree in management from Getulio Vargas Foundation (FGV), as well as a PhD in Production Engineering, obtained at Fluminense Federal University (UFF). His research interests lie in decision support methods for complex environments, operational research, multicriteria methods, topic modeling, and text mining. Dr. Basilio is actively involved in projects focusing on decision support models related to the application of law by government agencies in the fight against local crime, as well as in advancing existing multicriteria methods. He has an extensive publication record in both national and international journals, indexed in the leading databases, with significant contributions to multicriteria decision support methods. His scientific work has been recognized with several awards, including the Order of Military Police Merit of Rio de Janeiro, the Dom João VI Merit Medal, and the Tiradentes Medal—the highest honor granted by the Legislative Power of the State of Rio de Janeiro.

Valdecy Pereira

Valdecy Pereira is Adjunct Professor in the Department of Production Engineering at the Federal Fluminense University (UFF), Brazil. He holds a bachelor's degree, master's degree, and PhD in Production Engineering from UFF. His research interests span operations research, production planning and control, multivariate data analysis, multicriteria decision methods, and the full spectrum of Artificial Intelligence, from Machine Learning to Deep Learning. Dr. Pereira is actively involved in projects that bridge advanced quantitative methodologies with real-world industrial challenges, ranging from optimizing production systems and developing data-driven decision-support tools to applying AI to complex operational problems. He has published widely in leading national and international journals, with significant contributions to multicriteria decision-making, data analytics, and intelligent systems. His scientific work has been recognized with several awards, such as the Order of Military Police Merit of Rio de Janeiro and the Highly Commended Award from the Emerald Literati Network.

Marcos Dos Santos

Marcos Dos Santos holds two postdoctoral degrees, one of them in Space Sciences and Technologies from the Instituto Tecnológico de Aeronáutica (ITA - Aeronautics Institute of Technology). He obtained his master's degree in Industrial Engineering from COPPE UFRJ and his Ph.D. in Industrial Engineering from UFF. He currently serves as an Operations Research professor at the Instituto Militar de Engenharia (IME - Military Institute of Engineering), working in the fields of Computer Engineering, Transportation Engineering, and Defense Engineering. Additionally, he teaches in the MBA in Data Science and Analytics program at the Universidade de São Paulo (USP). With extensive teaching experience, Prof. Santos has taught Operations Research to over 10,000 students both in Brazil and internationally. He was a board directors of the Sociedade Brasileira de Pesquisa Operacional (SOBRAPO—Brazilian Operations Research Society) from 2019 to 2021. In the field of decision-making, with an H-index 35 in the Scopus database, he has presented and published more than 900 studies in over 20 countries, including the United States of America, China, Italy, England, France, Serbia, India, Nepal, Chile, Portugal, and Spain, among others. His global impact was recognized by Stanford University, which listed him among the top 2% of the world's most influential scientists.

Preface

It is with great pleasure that we present this Reprint, a Special Issue that brings together nine articles focused on "Advanced Applications of Multi-Criteria Decision-Making Methods in Operational Research". The scope of this collection is comprehensive and deeply detailed, exploring everything from the development of innovative tools for decision analysis, such as an online tool and a new R package, to advancements in complex scenarios like bi-matrix games with fuzzy payoffs and parametric equilibria. The articles also delve into hybrid methodologies for criteria weighting, such as a new method combining ENTROPY and CRITIC with PROMETHEE, and the integration of Data Envelopment Analysis (DEA) with other techniques, such as Projection Pursuit Regression (DEA-PPR), to enhance performance measurement and prediction. Furthermore, this Special Issue explores multi-threat assessment methods with heterogeneous information, procedures for increasing the consistency of pairwise comparison matrices in AHP, and resource allocation schemes under uncertainty using possibilistic logic and fuzzy sets. Additionally, efficiency evaluations of postgraduate activities in Brazilian higher education institutions are presented through dynamic network DEA models, as well as the application of MCDM in human resource management to prioritize innovative HRM practices. Our primary objective is to provide a comprehensive and up-to-date overview of the challenges and innovative solutions arising from the application of these powerful analytical tools in increasingly complex and dynamic decision-making contexts.

The motivation for compiling this Reprint lies in the growing need to address decision problems involving multiple and often conflicting criteria, a common reality in the current global landscape. We believe that this scientific work will be an invaluable resource for researchers, academics, postgraduate students, and professionals seeking to enhance their skills in complex decision problem-solving, serving as a source of inspiration for new research and practical applications.

This Special Issue is the result of a collaborative effort. We would like to express our profound gratitude to all the authors involved for their invaluable contributions and for sharing their innovative and highly significant research. Special thanks are extended to the reviewers, whose diligent and insightful work was crucial in ensuring the high quality and scientific rigor of the articles presented herein. We also extend our appreciation to the entire editorial team for their unwavering support, dedication, and professionalism, which were fundamental to the successful completion of this reprint. We hope it will help shape future directions of research and practice within the field of operational research.

We trust that reading this Special Issue will prove enriching and beneficial for all readers.

Marcio Basilio, Valdecy Pereira, and Marcos Dos Santos
Guest Editors

mathematics

MDPI

Editorial

Preface to the Special Issue "Advanced Applications of Multi-Criteria Decision-Making Methods in Operational Research"

Marcio Pereira Basilio [1,*], Valdecy Pereira [2] and Marcos dos Santos [3]

1 Military Police of the Rio de Janeiro, Rio de Janeiro 21941-901, Brazil
2 Department of Production Engineering, Faculty of Engineering, Praia Vermelha Campus, Federal Fluminense University, Niteroi 24210-240, Brazil; valdecy.pereira@gmail.com
3 Systems and Computing Department, Military Institute of Engineering, Rio de Janeiro 22290-270, Brazil; marcosdossantos@ime.eb.br
* Correspondence: marciopbasilio@gmail.com

Decision-making is a consistent part of the daily activities of individuals and organizations. All decisions are based on the evaluation of individual decision options, usually grounded in the preferences, experience, and other data of the decision-maker. Some decisions are relatively simple, especially when the consequences of a wrong choice are small, while others are highly complex and have significant effects. Frequently, real-life problem-solving involves several competing viewpoints that need to be considered to arrive at a reasonable decision. Formally, a decision can be defined as a choice made based on available information or a course of action intended to solve a specific decision problem. Multi-criteria decision analysis (MCDA) involves assessing various courses of action or options, ultimately selecting the most preferable alternative or ranking them from best to worst. In our daily lives, the use of MCDA is crucial to indicate the best rational alternative to the decision-maker, allowing for the allocation of finite resources among competing and alternative interests, whether in an organizational or household environment. Recognizing the importance and advancement of this field, this Special Issue, entitled "Advanced Applications of Multi-criteria Decision-Making Methods in Operational Research," presents nine articles selected from the 24 submissions received. These articles, which successfully passed the peer-review process and were published between February 2023 and April 2025, bring original research ideas that significantly contribute to operational research, with a strong emphasis on developing and applying decision-support methods.

In the first article, Barbara et al. (Contribution 1) present waspasWEB, an online decision-making tool based on the WASPAS method, and an R package available on CRAN. The tool facilitates the application of multi-criteria decision analysis by providing an intuitive solution. The article details the platform and validates its application through a case study. Li et al. in (Contribution 2) advance the study of multi-objective bi-matrix games by incorporating fuzzy payoffs (MBGFP), addressing the challenge of imprecise information in game theory. The main innovations include establishing the conditions for a fuzzy Pareto–Nash equilibrium and developing a parametric bilinear programming method to calculate this equilibrium. In addition, the article introduces the concept of fuzzy weighted Pareto–Nash equilibrium, providing the existence conditions and a calculation method, thus offering new tools for analyzing games with fuzzy uncertainties. Basilio et al. in (Contribution 3) present a new hybrid method, EC-PROMETHEE, for weighting criteria in decision-making processes. This method's innovation uses a weight range per criterion, combining the ENTROPY and CRITIC methods with the PROMETHEE method. This approach generates multiple sets of weights, allowing for multiple final rankings and providing decision-makers with a more robust analysis. The EC-PROMETHEE method aims to reduce uncertainty and improve the quality of decisions by considering a range of weights rather than a single weight per criterion. In the fourth article, Yu and Lou (Contribution 4) present a new approach, integrating Data Envelopment Analysis (DEA) with Projection Pursuit Regression (PPR) to improve performance measurement and prediction. This DEA-PPR combined model addresses the limitations of traditional DEA models, particularly their inability to forecast future efficiency, and outperforms other combined models like DEA-BPNN and DEA-SVR, especially with small and non-normal distribution samples. The model demonstrates superior global optimization, convergence, accuracy, and robustness, offering a more reliable efficiency analysis and prediction tool. Gao and Lyu in (Contribution 5) propose a new three-target multiple threat assessment method designed to deal with heterogeneous information and assign relevance in complex battlefield environments. The method innovatively uses heterogeneous forms to represent dynamic assessment information and employs heterogeneous CRITIC to calculate attribute weights. It also adaptively determines risk avoidance coefficients and uses the weighted Heronian mean operator to construct comprehensive loss function matrices. In the sixth article, Salomon and Gomes (Contribution 6) present a powerful and efficient procedure for increasing the consistency of AHP pairwise comparison matrices. Utilizing means and standard deviations, the method addresses stalled decisions by deriving a more consistent matrix with minimal alterations. Ekel et al. in (Contribution 7) address the problem of resource allocation with various objectives, developing a decision-making scheme for uncertain conditions. The methodology employs a possibilistic approach with fuzzy set theory to handle uncertainty and integrate quantitative and qualitative data through transformation functions. Innovations include the uncertainty scheme and combining fuzzy sets and transformation functions for robust solutions. In the eighth article, Torres and Ramos (Contribution 8) evaluate the efficiency of postgraduate activities in Brazilian Higher Education Institutions (HEIs) using a two-stage dynamic network DEA model. It introduces a novel approach that considers graduate programs' formative and scientific production stages and incorporates shared resources. The study also presents an efficiency decomposition method and a bi-dimensional representation of the efficiency frontier, offering new insights into evaluating HEI performance. Finally, Mirčetić et al. in (Contribution 9) address the application of MCDM methods in HRM, identifying a gap in understanding how these methods prioritize innovative HRM practices and classify companies. The study proposes an innovative MCDM approach using CRITIC and PIPRECIA-S to prioritize HRM practices and COBRA to assess companies.

The Guest Editors sincerely thank all authors for their valuable contributions to this Special Issue. We are also profoundly grateful to the anonymous reviewers for their insightful and professional evaluation reports, which have significantly enhanced the quality of the submitted manuscripts. Furthermore, we acknowledge the excellent collaboration with the publisher, the constant assistance provided by the MDPI associate editors in bringing this project to an end, and the excellent support of the Managing Editor of this Special Issue, Ms. Kelly Su.

Conflicts of Interest: The authors declare there are no conflicts of interest.

List of Contributions:

1. Barbara, F.; dos Santos, M.; Silva, A.S.; Moreira, M.Â.L.; Fávero, L.P.; Pereira Júnior, E.L.; dos Anjos Carvalho, W.; Muradas, F.M.; de Moura Pereira, D.A.; Portella, A.G. Interactive Internet

Framework Proposal of WASPAS Method: A Computational Contribution for Decision-Making Analysis. *Mathematics* **2023**, *11*, 3375. https://doi.org/10.3390/math11153375.
2. Li, W.; Li, D.; Feng, Y.; Zou, D. Fuzzy Weighted Pareto–Nash Equilibria of Multi-Objective Bi-Matrix Games with Fuzzy Payoffs and Their Applications. *Mathematics* **2023**, *11*, 4266. https://doi.org/10.3390/math11204266.
3. Basilio, M.P.; Pereira, V.; Yigit, F. New Hybrid EC-Promethee Method with Multiple Iterations of Random Weight Ranges: Applied to the Choice of Policing Strategies. *Mathematics* **2023**, *11*, 4432. https://doi.org/10.3390/math11214432.
4. Yu, X.; Lou, W. An Exploration of Prediction Performance Based on Projection Pursuit Regression in Conjunction with Data Envelopment Analysis: A Comparison with Artificial Neural Networks and Support Vector Regression. *Mathematics* **2023**, *11*, 4775. https://doi.org/10.3390/math11234775.
5. Gao, Y.; Lyu, N. A New Multi-Target Three-Way Threat Assessment Method with Heterogeneous Information and Attribute Relevance. *Mathematics* **2024**, *12*, 691. https://doi.org/10.3390/math12050691.
6. Salomon, V.A.P.; Gomes, L.F.A.M. Consistency Improvement in the Analytic Hierarchy Process. *Mathematics* **2024**, *12*, 828. https://doi.org/10.3390/math12060828.
7. Ekel, P.I.; Libório, M.P.; Ribeiro, L.C.; Ferreira, M.A.D.d.O.; Pereira Junior, J.G. Multi-Criteria Decision under Uncertainty as Applied to Resource Allocation and Its Computing Implementation. *Mathematics* **2024**, *12*, 868. https://doi.org/10.3390/math12060868.
8. Torres, L.M.L.d.S.; Ramos, F.S. Are Brazilian Higher Education Institutions Efficient in Their Graduate Activities? A Two-Stage Dynamic Data-Envelopment-Analysis Cooperative Approach. *Mathematics* **2024**, *12*, 884. https://doi.org/10.3390/math12060884.
9. Mirčetić, V.; Popović, G.; Vukotić, S.; Mihić, M.; Kovačević, I.; Đoković, A.; Slavković, M. Navigating the Complexity of HRM Practice: A Multiple-Criteria Decision-Making Framework. *Mathematics* **2024**, *12*, 3769. https://doi.org/10.3390/math12233769.

mathematics

MDPI

Article

Interactive Internet Framework Proposal of WASPAS Method: A Computational Contribution for Decision-Making Analysis

Flavio Barbara [1], Marcos dos Santos [2], Antônio Sergio Silva [1], Miguel Ângelo Lellis Moreira [3,*], Luiz Paulo Fávero [1], Enderson Luiz Pereira Júnior [3], Wagner dos Anjos Carvalho [4], Fernando Martins Muradas [5], Daniel Augusto de Moura Pereira [3] and Anderson Gonçalves Portella [6]

1 School of Economics, University of São Paulo, São Paulo 05508-010, Brazil; flavio.barbara@gmail.com (F.B.); antoniosergio@preventsenior.com.br (A.S.S.); lpfavero@usp.br (L.P.F.)
2 Systems and Computing Department, Military Institute of Engineering, Rio de Janeiro 22290-270, Brazil; marcosdossantos@ime.eb.br
3 Production Engineering Department, Federal Fluminense University, Rio de Janeiro 24210-240, Brazil; endersonlpj@id.uff.br (E.L.P.J.); danielmoura@ufcg.edu.br (D.A.d.M.P.)
4 Administration Department, Federal University of Rio de Janeiro, Rio de Janeiro 21941-630, Brazil; wagner.acarvalho@gmail.com
5 Naval Systems Analysis Centre, Operational Research Department, Rio de Janeiro 20091-000, Brazil; fernando.muradas@marinha.mil.br
6 Production Engineering Department, Veiga de Almeida University, Rio de Janeiro 20271-020, Brazil; andersonportella@yahoo.com.br
* Correspondence: miguellellis@hotmail.com

Abstract: Concerning the development of computational tools and solutions as a decision-making aid, this paper presents the results of the waspasWEB project, which strives to provide decision-makers with a readily accessible mechanism to employ the weighted aggregated sum product assessment (WASPAS) method. The social contribution of the project encompasses the development of a user-friendly and publicly accessible internet tool, as well as a package launched on the Comprehensive R Archive Network (CRAN) to serve the community of users of the R language. The use of operational research methodologies is crucial to justify decisions, and this effort seeks to advance the adoption of such methodologies, offering managers, researchers, and the general public an intuitive and easily accessible multi-criteria decision-making tool. In this way, we present the technical specifications, usability, and interactivity of the user with the computational platform, being validated its viability through a hypothetical case study. At the end of the research, it exposes the limitations and feasibility of the proposed computational model along with future research.

Keywords: CRAN; decision theory; operational research; R language; shiny

MSC: 90-04

1. Introduction

The scientific community has been actively involved in the exploration and dissemination of methodologies, procedures, and algorithms aimed at enhancing the field of decision-making [1]. Decision-making, a fundamental aspect of human society since ancient times, holds profound implications for both individuals and organizations. As elucidated in [2], the discipline of multi-criteria decision analysis (MCDA) [3] is currently experiencing accelerated growth within the realm of operational research (OR), manifesting in a proliferation of diverse methods and their practical implementations [4].

The study [5] reflects the paramount importance of OR in the realm of decision-making, tracing its significant role back to the aftermath of the Second World War. Technological advancements have ushered in transformative changes in the business landscape, introducing elements of uncertainty and complexity [6]. Consequently, decision-making processes have become increasingly intricate. Organizations have devised strategies to identify, evaluate, mitigate, and monitor events and conditions that exert influence on their operational frameworks [7]. These strategies heavily rely on decision-making procedures that encompass multiple criteria, often derived from extensive multidimensional data sources [8].

Drawing inspiration from the field of OR, this research proposes a solution to the challenges posed by MCDA problematic, employing a range of analytical techniques such as AHP [9,10], ANP [11], PROMETHEE [12–14], THOR [15], SAPEVO [16], TOPSIS [17], and WASPAS [18], among others. In this environment, some new studies of areas have been proposed, integrating consensus reaching for ordinal classification-based group decision-making with heterogeneous preference information, where a group of decision-makers with different preferences and heterogeneous information aims to reach a consensus on the ranking or classification of alternatives based on ordinal data [19].

In MCDA, criteria weights reflect the importance or priority assigned to each criterion in the decision-making process. The weights are typically determined based on the decision-maker's preferences, and they influence how the alternatives are evaluated and compared [20,21]. Strategic weight manipulation refers to a strategy employed in MCDA where decision-makers strategically manipulate the weights assigned to criteria to influence the overall decision outcome or ranking of alternatives. This strategy involves adjusting the relative importance of criteria to achieve a desired result, often driven by personal biases or preferences [22].

Regarding the popular literature in MCDA, the WASPAS method may have limited available literature, but it exhibits promising potential for both academic research and practical applications in the public and private management environment [23]. As expressed in [24,25], the credibility of WASPAS concerns the integration of two prominent MCDA approaches, namely, the weighted sum model and the weighted product model.

This method enhances the analytical depth by evaluating the sensitivity of each underlying approach in response to the criteria weighting system, thereby incorporating various perspectives for decision-makers [26]. The practical efficacy of the aggregate method is demonstrated in [24] through its application as an effective MCDA tool to address eight decision-making problems in industrial manufacturing processes [11]. The proposed methodology has made numerous practical contributions, such as [27] utilizing the method for single and multiple response optimization in non-traditional machining processes [28]. These processes are employed in industries such as aerospace, nuclear, missile, turbine, automobile, and tool-and-die manufacturing, which impose stringent requirements [29].

The method's applicability extends to all multi-criteria decision processes. For instance, in the realm of healthcare [30], the study employs the WASPAS method to prioritize patient care in the Ghanaian health system, where population growth surpasses the availability of medical resources, leading to constraints that often result in treatment delays and increased probabilities of complications and mortality. In a distinct context, a study conducted in India [31] utilizes WASPAS to propose an integrated weighting approach for essential factors affecting client satisfaction with the care experience, aiming to enhance their overall level of satisfaction. The study employed real data collected from the largest health service provider in Calcutta and addressed the demands arising from the sector's economic growth and increased competition in the private healthcare domain in the region.

To substantiate the implementation of the proposed approach, a recent publication [18] serves as an illustrative example, addressing a critical public security issue in Rio de Janeiro: determining the optimal choice for the acquisition of a helicopter by the State Military Police. This study presents a highly intricate decision problem characterized by various constraints, including the high cost and advanced nature of the equipment, the requirement for operational versatility and precision, and the necessity to adhere to stringent safety criteria. The research provides a comprehensive investigation and rigorous application of the proposed method, thus serving as an invaluable resource for information and validation of the implemented algorithms and developed systems [32].

In this scenario, as a motivational character, in the search to enable the dissemination of knowledge within the scope of the MCDA, the study aims to provide tangible products to the community by offering a publicly accessible mechanism on the internet that empowers decision-makers to utilize the WASPAS method as a supporting mechanism. The mechanism is user-friendly and intuitive and abstracts the computational intricacies involved in the calculation algorithms from the user, thereby eliminating the need for programming or mathematical expertise.

Embedded within the context, the modest contribution of this study aims to concretize and offer a tangible product to the academic community. The product takes the form of a publicly accessible mechanism on the internet, ensuring unrestricted access [33]. The intention is to empower decision-makers with the capability to utilize the WASPAS method as a supporting mechanism without requiring programming or mathematical expertise [34]. The computational intricacies involved in the calculation algorithms are abstracted from the user, who only needs to input the relevant information pertaining to the problem through a user-friendly and intuitive graphical interface.

This paper is structured into six sections. After the contextualization in the introduction section, the second sections describe the concepts of the WASPAS methodology and computational development through material and methods. The third section approaches the technological framework proposal. Exploring the feasibility of the computational model, a case study is presented in Section 4, exposing the main concepts of the technological proposal. Section 5 presents the discussion within the limitations and gains of the framework. Finally, Section 6 concludes the study along with future study proposals.

2. Materials and Methods

This section is divided into three subtopics: "The WASPAS Method", "Used Infrastructure", and "Delivered Results". These subtopics serve as an organizational framework for presenting the key aspects of the research. Notably, the emphasis is placed on the topic of "The WASPAS Method", as it holds significant importance within this study. While references to individual publications related to the WASPAS theme exist in this paper, direct quotations from these works will be avoided to ensure a clear and coherent presentation.

2.1. The WASPAS Method

In the context of intricate decision-making processes involving extensive sets of alternatives and criteria, the application of multi-criteria decision-making (MCDA) systems has proven to be effective. It has been established that combining multiple methods yields higher accuracy compared to applying each method individually. the weighted aggregated sum product assessment (WASPAS) method implements this principle by aggregating the well-known weighted sum model (WSM) and weighted product model (WPM) methods. WSM is widely recognized and extensively used in MCDA for addressing problems of this nature, while WPM is a variation that replaces the sum of multiplications (rating x weight) with the exponentiation of product weights [35].

It is important to note that the WSM and WPM methods are applicable exclusively to quantitative data. It is advisable to refrain from employing criteria with qualitative ratings. If the inclusion of qualitative data is deemed necessary, it is crucial to employ appropriate methods capable of converting qualitative information into numerical rating without introducing arbitrary weighting, whether directly or indirectly.

The underlying steps of WASPAS, namely WSM and WPM, share initial procedures. The first step involves constructing the decision matrix, as MCDA problems are defined by sets of m alternatives and m criteria. Consequently, a matrix is created, containing a known rating of the m criteria for each of the n alternatives, as illustrated in matrix b (1).

$$X_{ij} = \begin{bmatrix} c_j a_i & a_1 & a_2 & \dots & a_n \\ c_1 & x_{11} & x_{21} & \dots & x_{1n} \\ c_2 & x_{21} & x_{22} & \dots & x_{2n} \\ \dots & \dots & \dots & \dots & \dots \\ c_m & x_{m1} & x_{m2} & \dots & x_{mn} \end{bmatrix} \quad (1)$$

where x is the algebraic matrix formed by the rating of the set of criteria associated with each of the alternatives under analysis in the study of the MCDA problem, where $a_i = 1, \dots, n$, and $c_i = 1, \dots, m$. The variable x_{ij} represents the performance of alternative a_i bbbbin the criterion c_j.

In the second step, the matrix rating is normalized due to their tendency to be highly disparate. It is common for one criterion to vary by thousands while another varies by units, resulting in difficulty when comparing and evaluating alternatives. Criteria can also be either monotonic of cost or benefit. For example, in the above case, price is a cost criterion, where lower ratings are preferred, and warranty is a benefit criterion, where higher ratings are desired. Thus, a distinct normalization formula is employed for each criterion type.

For benefit criteria, the normalization process involves dividing the rating of each alternative by the maximum rating of the set of ratings for that specific criterion. The performance rating of alternatives with respect to each criterion is normalized such that, for the criterion vector of rating. x_{ij}. Equation (2) below illustrates the normalization function for monotonic benefit criteria.

$$x_{ij} = \frac{x_{ij}}{max_i(x_{ij})} \quad (2)$$

where x_{ij}: represents the normalized rating for a specific tuple (alternative, criterion); x_{ij} is the original rating that needs to be normalized; $max(x_{ij})$ denotes the largest rating within the set of rating for a specific monotonic benefit criterion across all alternatives.

For monotonic cost criteria, the vector of performance rating associated with the specific criterion is normalized by dividing each rating obtained for the criterion by the smallest rating within the set of ratings. In this case, the normalization of the alternatives' performance rating with respect to the criterion involves applying a function to the vector of rating. x_{ij} for the criterion. This function divides the minimum rating of the criterion's rating vector by the rating of the ith alternative. The formula for this normalization process is represented by Equation (3).

$$x_{ij} = \frac{min_i(x_{ij})}{x_{ij}} \quad (3)$$

where x_{ij}: is the normalized rating for a specific tuple (alternative, criterion); x_{ij} is the original rating to be normalized; $min(x_{ij})$ represents the smallest rating within the set of rating for a specific monotonic cost criterion across all alternatives, and the index i ranges from 1 to m, representing the number of alternatives.

In the subsequent step, the criteria are assigned weights based on their relative importance in the decision-making process, with these weights being determined by the decision-maker. The WSM and WPM methods differ in their approach to determining the best alternatives based on the weighting function.

Some MCDA methods have as objective the construction of ranking based on alternative performance in multiple criteria. This ranking provides decision support and serves as a means of communication to stakeholders affected by the decision-making process. The classification process in the WSM method is as follows: w_j represents the relative importance (weight) assigned to the criterion, and IR_i denotes the calculated relative importance of the alternative. The relative importance rating is obtained by summing the normalized rating of the set of criteria assigned to the alternative being evaluated.

Since there are m criteria involved, the formula for calculating the relative importance (IR_i) according to the WSM method is as follows in Equation (4).

$$IR_i = \sum_{j=1}^{m} x_{ij} w_j \tag{4}$$

where IR_i: is the relative importance of alternative i obtained by the sum of normalized rating x_{ij} weighted by the arbitrated weight of criterion w_j ranging from 1 to m, where m is the number of criteria in the problem.

In the case of the WPM method, we follow a similar approach by obtaining the normalized rating of the set of criteria assigned to the alternatives (x_{ij}), where i ranges from 1 to n. These normalized ratings are then raised to the power of the weight assigned to the relative importance of the j criterion, as indicated in the weights vector. Equation (5) represents the classification function used in the WPM method.

$$IR_i = \prod_{j=1}^{n} (x_{ij})^{W_j} \tag{5}$$

where IR_i: is the relative importance of alternative i obtained by the product of the normalized rating x_{ij} raised to the arbitrated weight of the criterion w_j ranging from 1 to m, and m is the number of criteria in the problem.

The WASPAS method incorporates the relative importance derived from the *WSM* and *WPM* methods to assess the sensitivity of the alternatives and criteria. To achieve this, a lambda (λ) parameter, ranging from 0 to 1, is introduced and applied to the alternatives versus criteria set. The objective is to determine the total relative importance by combining the weighted relative importance of *WSM* and *WPM* based on lambda. This weighting is obtained by multiplying the relative importance obtained from the *WSM* method by lambda and adding it to the relative importance obtained from the *WPM* method multiplied by the complement of lambda (1 − λ). This approach allows for different emphasis on the *WSM* and *WPM* relative importance based on the rating of lambda. When lambda is set to 1, the *WSM* relative importance is fully utilized, while the *WPM* relative importance is disregarded. Conversely, when lambda is set to 0, only the *WPM* relative importance is considered for the determination of the total relative importance. For lambda rating of 0.5, the total relative importance is computed as the arithmetic mean of the *WSM* and *WPM* relative importance. In a simplified explanation, the total relative importance of the criterion (IRT_j) can be calculated as the sum of lambda multiplied by the relative importance from the *WSM* method (*WSM*) and the complement of lambda (1 − λ) multiplied by the relative importance from the *WPM* method (*WPM*):

$$IRT_j = \lambda \times IR_j(WSM) + (1-\lambda) \times IR_j(WPM) \tag{6}$$

where IRT_j is the total relative importance of alternative i, obtained by the WASPAS method; $IR_j(WSM)$ is its relative importance obtained by the *WSM* method, and $IR_j(WPM)$ is its relative importance obtained by the *WPM* method; λ (lambda) is a rating ranging from 0 to 1.

By substituting the weighted sum (*WSM*) and weighted product (*WPM*) formulas, we arrive at Equation (7), which is commonly encountered in the relevant technical literature:

$$IRT_j = \lambda \times \sum_{j=1}^{m} \overline{X}_{ij} W_j + (1-\lambda) \prod_{j=1}^{n} \left(\overline{X}_{ij}\right)^{W_j}, \lambda_j \in [0,1] \tag{7}$$

where IRT_j is the total relative importance of alternative i, obtained through the WASPAS method, is calculated using the formula above, where the lambda factor multiplies the sum indicated in Equation (4), and its complement (1 − λ) multiplies the product indicated in Equation (5).

The high-level software (IDEs and web portals) used in the production of the work are listed and briefly described in Table 1, more detailed explanations of their functionality and use are presented throughout this topic.

Table 1. List of IDEs and web portals used.

System	Description
RStudio 2022.07.2	RStudio is a graphical development environment that provides productivity tools for systems development in the R language. It is distributed by Posit Software company, PBC, and is licensed under version 3 of the GNU General Public License.
posit.cloud	Posit Cloud is a cloud-based solution or web service that offers a browser-based experience similar to RStudio. It serves as an alternative IDE for R users and developers.
Shiny	Shiny is a free and open-source R package used for developing web applications (Apps). It is integrated with RStudio and posit.cloud, allowing for enhanced productivity in application prototyping.
shinyapps.io	shinyapps.io is a web portal that provides free services for hosting and publishing applications developed in the R language on the internet. It is part of the suite of solutions offered by Posit Software.
GitHub	GitHub is a file repository hosting platform that is integrated with the Git version control system. It can be used with various IDEs, including Rstudio, offering a graphical interface for interacting with the web service.
CRAN	CRAN (Comprehensive R Archive Network) is the central repository of packages for R language development. Supported by the R Foundation, it includes package source codes and precompiled binary files for Windows and macOS. CRAN was created in 1997 by Kurt Hornik and Friedrich Leisch.

The R language [36] serves as the foundation for all the development in this work. It was initially created in 1993 by Robert Gentleman and Ross Ihaka, statisticians from the University of Auckland in New Zealand. R was specifically designed to be a high-performance language for statistical analysis, data mining, machine learning, and database exploration to identify patterns. Being an open-source language, it benefits from numerous packages available primarily through the CRAN repository. The extensive collection of free packages enables R to be widely used across various domains beyond statistics and data science. R is recognized as one of the most popular languages for statistical analysis, statistical graphing, and data science projects. Moreover, it has been gaining popularity in general terms as well.

The prominence of R is attributed not only to its extensibility, robustness, and versatility but also to the active support from a large community of volunteers who contribute to frequent updates of the language. For instance, the development of this work was carried out using version 4.2.2 of R, released on 31 October 2022. As of the time of writing, the current version is 4.2.3 ("Shortstop Beagle"), released on 15 March 2023, with version 4.3.0 ("Already Tomorrow") scheduled for release on 21 April 2023. The progress of R is driven by a core group of developers supported by contributions from the community.

This community is primarily manifested through the "R Foundation," which holds the copyright and oversees the management of the R software and documentation. Established as a non-profit public interest organization, the foundation was founded by members of the core development team with the goal of providing support for the R Project (www.r-project.org, accessed on 16 January 2023) and fostering innovation in statistical computing. With R having reached a high level of maturity, the R Foundation strives to ensure its ongoing development through continuous advancements in statistical and computational research software. The foundation also serves as a reference point for individuals, institutions, and companies seeking to support or engage with the R development community, including organizing meetings and conferences in the field of statistical computing.

The R Foundation serves as the maintainer of the CRAN package repository, which currently hosts 19,312 packages (source: Contributed Packages) as of the time of writing. The R Package developed within the scope of this work was accepted and added to CRAN on 9 March 2023, making it globally available for use by the entire R community.

2.2. *RStudio and Posit Cloud IDEs*

In this study, the RStudio integrated development environment (IDE) played a crucial role. Developed using Java, C++, and JavaScript, the IDE is compatible with Linux, macOS, and Windows operating systems. It is distributed under the GNU Affero General Public License v3 by Posit Software PBC. This organization holds significant prominence within the R community, offering essential resources widely utilized in the field of data science. Therefore, it is pertinent to provide an overview of this organization, as their free products played an exceptional role in the completion of this work [37].

Originally established in 2009 as RStudio, Inc., the organization began distributing free and open-source products. In February 2011, they released the IDE bearing the same name. Over time, they continued to introduce significant contributions to the R community, including the launch of Shiny in 2012, RStudio and Shiny SERVER PRO versions in 2014, and the Spark and tidyverse packages in 2016. In 2020, RStudio, Inc. transformed into Posit PBC, expanding its business to include the Python community with the release of Shiny for Python. Posit PBC is classified as a public benefit corporation (PBC) and holds B-Corp certification. PBC companies are profit-oriented entities with a defined social mission. They are legally structured to prioritize societal well-being alongside shareholder rating maximization. This framework enables companies to focus on both profits and social benefits while also necessitating transparency in demonstrating how the public benefit purpose is served and how member interests are promoted [38].

Since March 2018, several US states have enacted legislation to support PBC companies. These companies must demonstrate a commitment to social good, conduct activities responsibly and transparently without generating adverse environmental, social, or economic impacts, and empower management to make decisions in the public interest, even if they may affect profitability. While pursuing profit is an objective, PBC companies are not required to obtain B-Corp certification. However, Posit Software PBC has earned this certification from B-Lab, which independently assesses the company's social and environmental performance, responsibility, and transparency based on rigorous standards.

Posit Software PBC fulfills a significant public purpose by creating and distributing high-quality open-source software for data scientists while also providing various free resources to the data science community. In the course of this study, posit.cloud (https://posit.cloud/, accessed on 16 January 2023) was initially employed on an experimental basis to facilitate seamless integration with the web platform where the web application was hosted. Initial development and initial deployments were carried out using this tool. However, posit.cloud aims to provide an online environment nearly identical to the RStudio IDE, eliminating the need for downloads, installations, and configurations. Once the integration with the shinyapps.io web platform was complete, the standalone IDE was reverted to for usage. It is worth noting that the free version of the web IDE has certain usage limitations, including a maximum of 50 projects, 25 h of computation per month, 1 GB of RAM, and no support. To utilize posit.cloud, users are required to create an account, thereby gaining access to their designated work area.

2.3. *shinyapps.io the Web Hosting Platform*

The web service developed in this study is made accessible to the community through the shinyapps.io portal (www.shinyapps.io, accessed on 16 January 2023). This portal offers a free membership plan with certain limitations, including the ability to host up to 5 apps and a maximum of 25 h of availability per month. If the allotted hours of usage are exceeded, the application becomes temporarily unavailable under the free plan. However, paid plans offer extended availability, and if usage exceeds the allocated hours, additional charges may apply, but the service remains accessible [39].

To utilize the shinyapps.io portal, users are required to create an account, granting them access to the application hosting platform. This platform offers all the necessary resources for hosting a web application developed in R using the shiny package. Upon accessing the service, users are presented with a dashboard (Figure 1) that provides convenient hyperlinks to the application's management.

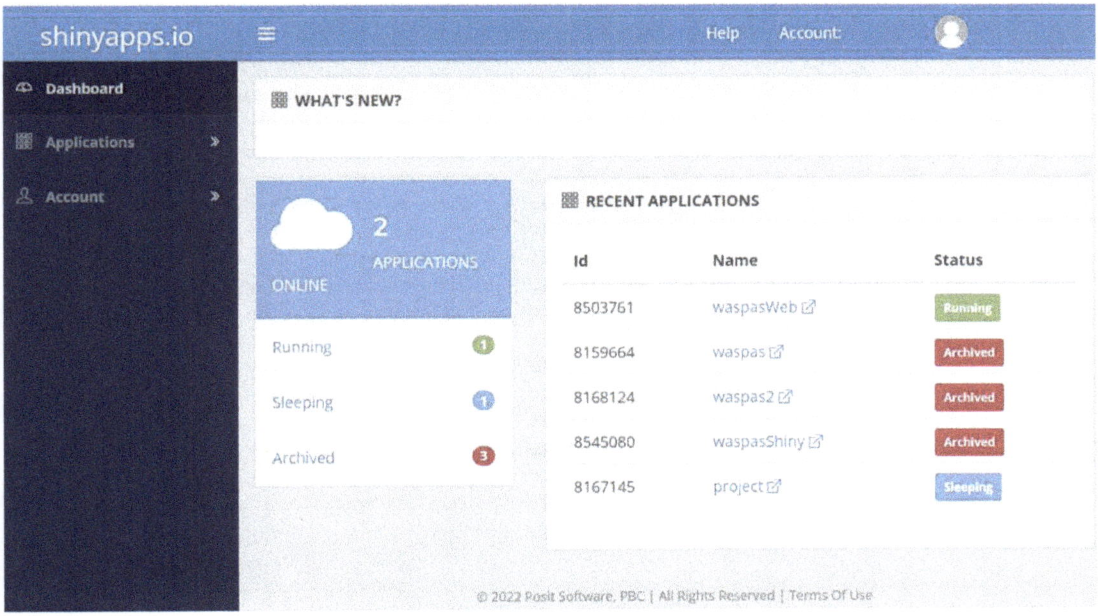

Figure 1. "Dashboard" screen, which showcases a list of web applications hosted on the shinyapps.io platform. This screen serves as a centralized hub for managing and monitoring the various applications.

The process of publishing the developed web application involves utilizing the available functionalities within the IDEs (posit.cloud or Rstudio), which offer a streamlined publishing option specifically designed for Shiny applications. This option is automatically displayed when the project is created as a Shiny. Throughout the development process, the application can be run in a browser or within the IDE's runtime viewer, which also provides the capability to command the publication of the application on the web.

2.4. GitHub

GitHub, Inc. is a prominent internet company that exemplifies the success story of young visionary founders starting in a garage and experiencing exponential growth to become a technology giant. Originally established in 2008 as Logical Awesome LLC (Limited Liability Company), it introduced a collaborative software version control platform based on Git. Presently, the company boasts a revenue of approximately USD one billion and a workforce of 2500 employees as of 2022. In 2012, Microsoft became a significant user of GitHub's services, and in October 2018, it acquired the company, assuming its current ownership. As highlighted on its official website, GitHub offers a comprehensive and scalable platform that enables development teams to securely create and deploy their products. It presently serves over 100 million developers across more than 4 million organizations and hosts over 330 million repositories [40].

The utilization of GitHub in this context is motivated by its integration with RStudio and its widespread adoption within the Information Technology community, encompassing both academic and commercial spheres. This choice allows for the broad accessibility of the software developed in this project to these diverse communities. The entire material developed as an R library (package) is publicly available in a repository on GitHub [41].

The waspasR package can be obtained directly from CRAN or through GitHub through the link www.github.com/flavio-barbara/waspasR (accessed on 16 January 2023), and the application code can be retrieved through the link www.github.com/flavio-barbara/waspasWeb (accessed on 16 January 2023).

2.5. CRAN, Package Building, and Submission Process

Building a software package offers significant advantages, including componentization, code reuse, context isolation, improved code readability, and standardized design. Additionally, it facilitates sharing functions with other developers, fostering an engaged community. In the case of this study, the R language was chosen for implementation, and the package was promoted through CRAN [42].

CRAN is a vast repository of R packages supported by a global network of FTP servers or mirrors. These mirrors store updated versions of component packages, providing sophisticated resources for R development. CRAN serves as the primary instrument for the R Project, which aims to support the continuous development of the R language and explore new methodologies in statistical computing and data science. The R Project is maintained by the R Foundation, located at the Institute of Statistics and Mathematics of the University of Economics and Business in Vienna, Austria. The CRAN network consists of 94 servers, with the main server, 0-Cloud, automatically routing to the other servers worldwide. Rstudio organization maintains the 0-Cloud server.

All packages available on CRAN undergo a rigorous certification process to ensure compliance with strict standards. CRAN has a set of policies that must be adhered to for package submission. The repository emphasizes hosting quality packages and requires contributors to make relevant contributions. Compliance with legal requirements for code and documentation distribution is also essential, considering CRAN operates in multiple jurisdictions. The policies aim to ensure that mirror server distributors fulfill their legal obligations without overloading their work. The CRAN Repository Policy page provides submission instructions, an online form for package submission, and a checklist to aid contributors in meeting submission requirements.

The development process of the package followed a prototyping approach, which is widely recognized in software engineering. As an individual project, communication was ad hoc, based on the needs of the CBT project and the availability of the advisor and student. The process involved analyzing requirements, designing the package structure, implementing the code, thorough testing, creating comprehensive documentation, submitting the package to CRAN, addressing feedback, and continuously improving the package. It is important to note that activities overlapped and proceeded in parallel during the development process.

1. Agreement between the authors on the topic to be developed.
2. Initial guidance provided to the development team regarding objectives, deliverables, and deadlines.
3. Study of the WASPAS method based on recorded classes taught by one of the co-authors.
4. Implementation of functions:
    a. Selection of a validation database.
    b. Construction of functions.
    c. Validation of results.
    d. Correction of defects.

   e. Iteration between steps 4c and 4d until optimization is achieved.
5. Development of the Shiny application.
   a. Debugging process following the steps outlined in 4.
6. Publication of the application on shinyapps.io.
7. Software registration with the INPI.
8. Structuring of the package for submission to CRAN:
   a. Re-engineering of functions to meet the required requirements.
   b. Application of verification programs.
   c. Adjustments to meet the required standards.
   d. Iteration between steps 8b and 8c until optimization is achieved.
9. Submission of the package to CRAN.
   a. Re-submission with necessary cosmetic corrections until accepted.
10. Re-engineering of the Shiny application:
   a. Integration of the waspasR package.
   b. Replacement of functions with calls to package functions.
   c. Deletion of the original functions.

3. Interactive Framework Proposal

The R package, which includes functions for implementing various solutions based on MCDM (multiple criteria decision-making), also makes a small contribution to the R community. It has been accepted in CRAN and is readily available worldwide through the simple installation command "install.packages("waspasR")" in any R environment. The package is also publicly available on GitHub.

As the main product, the waspasWEB project is an academic scientific research project that proposes to implement a decision-making support tool using the WASPAS method, proposed by Zavadskas [35]. The implementation was performed in the R language using the Shiny package for internet publishing and the shinyapps.io hosting service.

The "WASPAS for Dummies" service is a tool to support multi-criteria decision-making, or MCDM, which stands for "Multi-Criteria Decision Making". This type of problem involves a set of alternatives, from which one wants to select the best choice, and a set of evaluation criteria, weighted according to the relative importance that the decision maker considers to be applicable in the decision-making process. There are many methods developed to solve MCDM problems. The study [18] cites more than 25 different methods, including AHP, MACBETH, ELECTRE, MELCHIOR, PAMSSEM, EVAMIX, QUALIFLEX, SAPEVO-M, WASPAS, and several others. Rani et al. [2] point out that multi-criteria decision-making processes are one of the areas of OR that grows the most, both in terms of method diversification and their application in the market [43].

Considering an example of real application, if one decision maker wants to choose the best automobile that will satisfy his needs; based on these needs, a set of criteria is established: economy, power, transport capacity, comfort items, safety items, price, etc., once the criteria are defined, a degree of importance must be assigned to each one of them, that is, the price may be a more relevant criterion than engine power, but this may be considered more relevant than autonomy. For the mathematical calculations of the method, it is also essential to determine whether the criterion regards cost or benefit, that is, if the higher is better or if the lower is better. The algorithms need to know whether the maximization or minimization of rating is intended, for example, price is a cost criterion, that is, the lower, the better, and autonomy is a benefit criterion, the higher, the better. With this information in hand, it is possible to apply mathematical models that classify the most appropriate alternatives for any problem that has this structure.

The software waspasWEB can be accessed by [44]. The first page is divided into two areas, as Figure 2 exposes.

Figure 2. The first page of waspasWEB.

On the right side, there is a presentation column with important information and links to supporting files (e.g., this operation manual). To the left of this display column is the workspace with four tabs, such as: Input, containing the field for uploading the database to be analyzed; View, with visualization of input data and button to command the calculation; Output, presenting the list of alternatives properly ranked; and Radar Chart, exposing radar-type graphic with classification.

The first step is to download the spreadsheet with the database model. To do so, just click on the Database Template link: WASPAS_Data_Template.xlsx (Figure 2), follow the well-known dialog for choosing the folder (directory) where you want to save the file and click on save.

Once the spreadsheet with the data model has been downloaded, edit the file using MS Excel or LibreOffice Calc, for example, and save it with a name you deem appropriate. We will explain later how to fill in the worksheet with data from the multi-criteria decision-making problem. The system will validate the format of the loaded data and show a brief report of what was imported. If there is no formatting error, the screen that will be displayed is the one shown below (Figure 3).

There are scroll bars on the right and bottom that allow you to view all the loaded data. After applying the WASPAS method to the database, the "Output" tab opens automatically and shows the classification made by the WASPAS method and by the two underlying methods, WSM and WPM. There is a "Slider" object on the screen that allows you to very quickly change the lambda rating that weigh the relative rating of each of the underlying methods. The closer to 0 (zero) the lambda rating is, the greater the weight of the WPM method, and the closer to 1, the greater the relative weight of the ranking obtained by WSM. The screen shown in the Output tab contains the rankings by the WSM (summation) and WPM (product) methods and the WASPAS ranking, which changes dynamically according to the change in the lambda rating in the "slider" object, as Figure 4 presents.

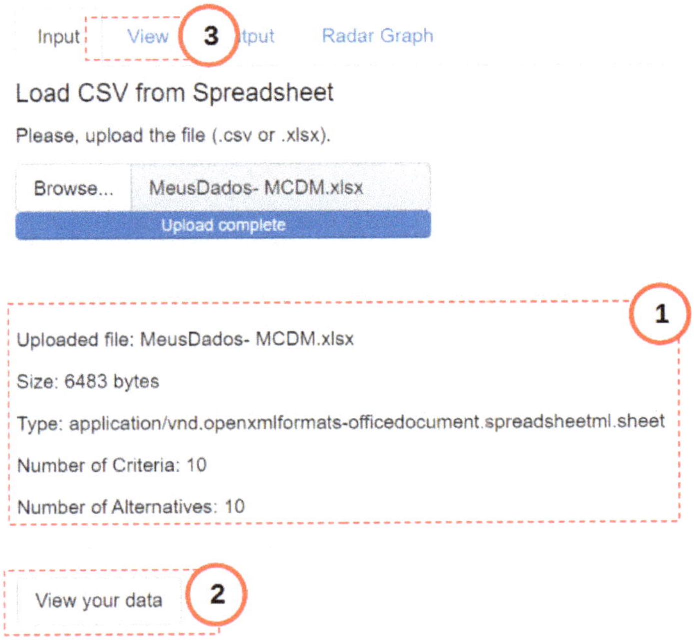

Figure 3. Database load result presentation screen, where: (1) Report on the loading process: file name and size, number of criteria, and database alternatives; (2) A button that directs to the visualization tab of the imported data (One can click directly on the "View" (3) tab to view the data).

Alternative	WSM_Rank	WPM_Rank	WASPAS_Rank
Alternative 4	0.638	0.605	0.622
Alternative 1	0.719	0.498	0.609
Alternative 2	0.617	0.469	0.543
Alternative 3	0.569	0.452	0.511
Alternative 5	0.526	0.429	0.478

Figure 4. Slider that allows dynamic visualization of the lambda.

The same result is displayed on the "Radar Chart" tab in a graphic format also known as spider web chart, Kiviat diagram, and other names. In this tab, there is also a "Slider" object that allows changing the lambda rating dynamically (Figure 5). The Radar Chart slider is synchronized with the Output tab object. When modifying the lambda rating in this component, the slider on the "Output" tab is also changed to the same rating, updating the WASPAS ranking.

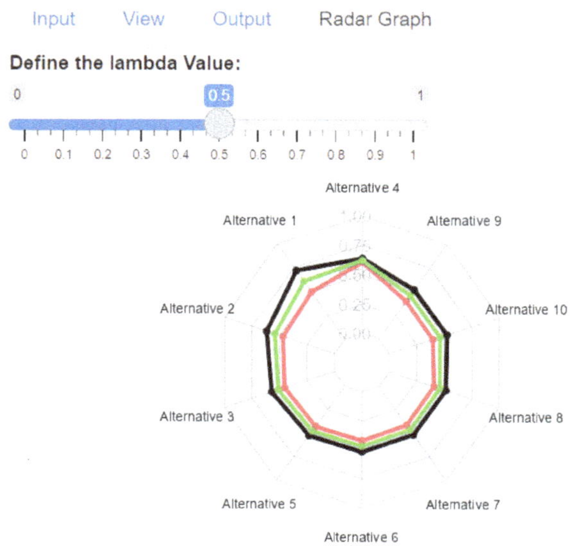

Figure 5. Slider and radar chart (spider web).

After downloading the spreadsheet (WASPAS_Data_Template.xlsx), open it in MS Excel, LibreOffice Calc, or the application of your choice. The worksheet will be the one shown in (Figure 6), but without any color, the colors are merely for didactic purposes.

F	Cost	Benefit	Cost	Benefit	Cost	Benefit	Cost	Benefit	Cost	Cost
W	0,1	0,1	0,1	0,1	0,1	0,1	0,1	0,1	0,1	0,1
C	Criterion 1	Criterion 2	Criterion 3	Criterion 4	Criterion 5	Criterion 6	Criterion 7	Criterion 8	Criterion 9	Criterion 10
Alternative 1	1	2	3	4	5	6	7	8	9	10
Alternative 2	2	3	4	5	6	7	8	9	10	11
Alternative 3	3	4	5	6	7	8	9	10	11	12
Alternative 4	4	123	6	7	8	9	10	11	12	13
Alternative 5	5	6	7	8	9	10	11	12	13	14
Alternative 6	6	7	8	9	10	11	12	13	14	15
Alternative 7	7	8	9	10	11	12	13	14	15	16
Alternative 8	8	9	10	11	12	13	14	15	16	17
Alternative 9	9	10	11	12	13	14	15	16	17	18
Alternative 10	10	11	12	13	14	15	16	17	18	19

Figure 6. Template spreadsheet for structuring input data.

The spreadsheet that structures the database of the multi-criteria decision-making problem that will be submitted to the WASPAS calculation provided by the "WASPAS for Dummies" page must respect the above structure. Separated by colors, there are six areas in the worksheet: indicators, flags, weights, criteria, alternatives, and alternative criterion rating.

Detailing each of them, we have in cells (1,1), (2,1), and (3,1) the indicators (Figure 6) of lines 1, 2, and 3 are informed, which must be "F" for "Flags", "W" for weights, and "C" for "Criteria". That is, "F" means the Cost or Benefit flag, "W" is the weight (importance of each criterion), this information (metadata) allows the user of the "WASPAS for Dummies" service to inform the data of criteria, weights, and cost–benefit in the line that suits you, the service will process each line according to the indicator in the cell.

With the indicator properly defined, it is necessary that the data in the referred rows are appropriate. The "F" indicator line should contain only "Flags" that indicate whether the criterion is cost or benefit. For this, the cells must contain words starting with "C" (cost) or "B" (Benefit).

The row whose indicator is the "W" (weight) must contain the weights arbitrated by the decision maker in relation to the relative importance of each of the criteria. The sum of the weights in the "W" row must add up to 1 (100%). And the indicator line "C" should contain the problem criteria. These are brief descriptions such as price, weight, size, capacity, etc.

The area in red is the part where the alternatives to the problem are introduced. There are no limits to the number of alternatives, just as there are no limits to the number of criteria. We suggest that in these cells (column 1, rows 4 to n, last alternative), brief descriptions of the alternatives be introduced, as well as in row "C" (criteria).

And the most important part is the one that has the rating. The ratings are, in general, obtained in the market and refer to the performance of that alternative in relation to the criterion. For example, the price rating of product X, the maximum speed of alternative Y, and the boiling temperature of element Z. The gray hatched ratings in Figure 6 are those that will be submitted to the decision support algorithm that is determined by criteria alternatives.

For a better understanding of the interactive internet-based model, Figure 7 presents a flowchart exposing the main steps of analysis using the computational framework.

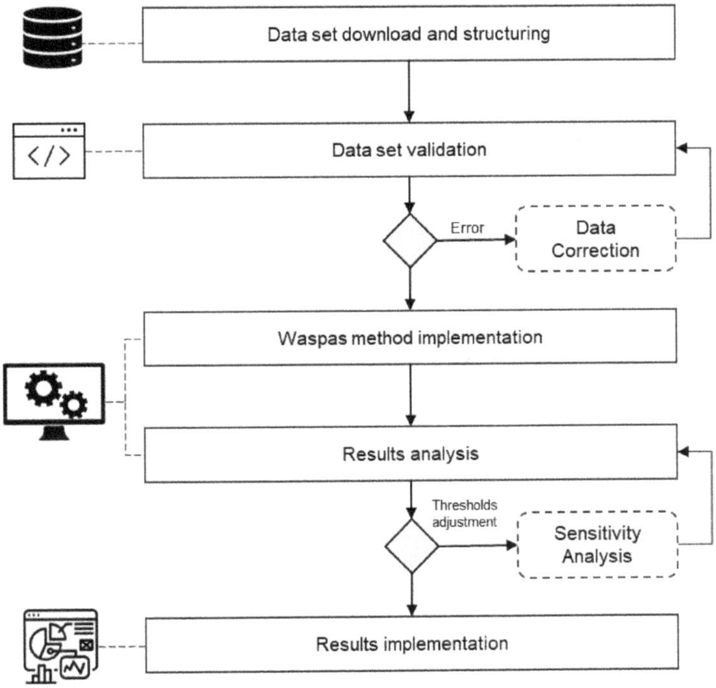

Figure 7. Computational analysis flowchart.

4. Case Study

After validating the algorithm implemented by comparing all the results obtained (partial and final) by the work [18], several other exercises were applied. We will report one of them. The most important thing is that the public tool derived from this work can be used for any MCDM study based on the methods presented here (WSM, WPM, and WASPAS), which can be accessed through the link www.flaviob.shinyapps.io/waspasWeb, accessed on 16 January 2023.

For this case study, we used the interactive software proposal using a step-by-step guide that serves as a drive for any further study that uses the tool.

First, we download the database template and edit it using the application of your choice. Then, we download a public database available on Kaggle [45] with various models of cell phones presenting technical specifications and prices in USD. Regarding the base used and the computational processing capacity of the mathematical and computational model, a limit between the number of criteria and alternatives is not identified. However, it should be noted that the use of heterogeneous and non-redundant criteria becomes validated, allowing a more accurate assessment in the application of a given real case.

In the sequence, the analysis and data preparation were conducted on the CSV file obtained in the previous step. This process involved data scrubbing, which entailed removing non-numeric entries and unnecessary columns that were not relevant to the MCDM process. The original dataset has 22 columns, as can be seen in Figure 8. The first cleansing operation involved removing non-numeric columns (highlighted in yellow). Next, column "A" was removed as it served only as an indexer (highlighted in gray). Then, it was observed that column "B" was a combination of columns "C" and "D" (highlighted in green), hence they were also eliminated.

A	B	C	D	E	F	G	H	I	J	K	L	M	N	O	P	Q	R	S	T	U	V
	Name	Brand	Model	Battery capacity (mAh)	Screen size (inches)	Touchscreen	Resolution x	Resolution y	Processor	RAM (MB)	Internal storage (GB)	Rear camera	Front camera	Operating system	Wi-Fi	Bluetooth	GPS	Number of SIMs	3G	4G/ LTE	Price
0	OnePlus	OnePlus	7T Pro M	4085	6.67	Yes	1440	3120	8	12000	256.0	48.0	16.0	Android	Yes	Yes	Yes	2	Yes	Yes	58998
1	Realme	Realme	X2 Pro	4000	6.5	Yes	1080	2400	8	6000	64.0	64.0	16.0	Android	Yes	Yes	Yes	2	Yes	Yes	27999
2	iPhone 1	Apple	iPhone 1	3969	6.5	Yes	1242	2688	6	4000	64.0	12.0	12.0	iOS	Yes	Yes	Yes	2	Yes	Yes	106900
3	iPhone 1	Apple	iPhone 1	3110	6.1	Yes	828	1792	6	4000	64.0	12.0	12.0	iOS	Yes	Yes	Yes	2	Yes	Yes	62900
4	LG G8X	LG	G8X Thi	4000	6.4	Yes	1080	2340	8	6000	128.0	12.0	32.0	Android	Yes	Yes	Yes	1	No	No	49990
5	OnePlus	OnePlus	7T	3800	6.55	Yes	1080	2400	8	8000	128.0	48.0	16.0	Android	Yes	Yes	No	2	Yes	Yes	34930
6	OnePlus	OnePlus	7T Pro	4085	6.67	Yes	1440	3120	8	8000	256.0	48.0	16.0	Android	Yes	Yes	Yes	2	Yes	Yes	52990
7	Samsun	Samsun	Galaxy N	4300	6.8	Yes	1440	3040	8	12000	256.0	12.0	10.0	Android	Yes	Yes	Yes	2	Yes	Yes	79699
8	Asus RO	Asus	ROG Ph	6000	6.59	Yes	1080	2340	8	8000	128.0	48.0	24.0	Android	Yes	Yes	Yes	1	Yes	Yes	37999
9	Xiaomi R	Xiaomi	Redmi K	4000	6.39	Yes	1080	2340	8	6000	128.0	48.0	20.0	Android	Yes	Yes	Yes	2	No	No	23190
10	Oppo K3	Oppo	K3	3765	6.5	Yes	1080	2340	8	6000	64.0	16.0	16.0	Android	Yes	Yes	Yes	2	Yes	Yes	23990
11	Realme	Realme	X	3765	6.53	Yes	1080	2340	8	4000	128.0	48.0	16.0	Android	Yes	Yes	Yes	2	Yes	Yes	14999
12	Xiaomi R	Xiaomi	Redmi K	4000	6.39	Yes	1080	2340	8	6000	64.0	48.0	20.0	Android	Yes	Yes	Yes	2	Yes	Yes	19282
13	OnePlus	OnePlus	7 Pro	4000	6.67	Yes	1440	3120	8	6000	128.0	48.0	16.0	Android	Yes	Yes	Yes	2	Yes	Yes	39995
14	Oppo Re	Oppo	Reno 10	4065	6.6	Yes	1080	2340	8	6000	128.0	48.0	16.0	Android	Yes	Yes	Yes	2	Yes	Yes	36990

Figure 8. The original dataset was loaded in a spreadsheet tool.

So, the prepared dataset at the end has the following list of criteria: battery capacity (mAh), screen size (inches), resolution x, resolution y, processor, RAM (MB), internal storage (GB), rear camera, front camera, number of SIMs, and price.

Assign weights to the criteria. At this point, the decision maker is required to assign weights to each criterion in such a manner that the summation of weights is equal to 1 (or 100%). Since there are eleven criteria, and the main one is the price, some exercises of criteria importance powering can be easily performed. For example, a weight of 0.2 or 20% can be assigned to the price criterion, and the remaining ten criteria can be equally divided into a weight of 0.08 or 8% each (Figure 9). Similarly, if a weight of 50% is assigned to the price criterion, the other ten criteria will have a weight of 0.05 or 5% to fit the sum of 100% (Figure 10). Alternatively, a specific weight rating can be assigned for each criterion, provided that the sum of the weights equals 1. These types of exercises have been performed and will be described in more detail later on.

Assign and specify for each criterion whether it is a cost or benefit criterion. It is important to bear in mind that such a definition can be subjective. For instance, a specific criterion may be considered a benefit by one decision-maker, while it may be viewed as a cost by another. In the case being studied, the screen size, which is typically regarded as a benefit by most individuals, will be considered a cost criterion by someone seeking an exceptionally small phone for whatever reason. In other words, the smaller the screen, the better it is perceived as a cost criterion in their case. This classification is also performed in the spreadsheet editor and can be seen in Figure 11.

	A	B	C	D	E	F	G	H	I	J	K	L
1	**Weights**	**0.08**	**0.08**	**0.08**	**0.08**	**0.08**	**0.08**	**0.08**	**0.08**	**0.08**	**0.08**	**0.20**
2	**Criteria**	Battery capacity (mAh)	Screen size (inches)	Resolution x	Resolution y	Processor	RAM (MB)	Internal storage (GB)	Rear camera	Front camera	Number of SIMs	Price
3	OnePlus 7T P	4085	6.67	1440	3120	8	12000	256.0	48.0	16.0	2	58998
4	Realme X2 Pr	4000	6.5	1080	2400	8	6000	64.0	64.0	16.0	2	27999
5	iPhone 11 Pro	3969	6.5	1242	2688	6	4000	64.0	12.0	12.0	2	106900
6	iPhone 11	3110	6.1	828	1792	6	4000	64.0	12.0	12.0	2	62900

Figure 9. Weight assignments (highlighted in light yellow) were used in the first exercise.

	A	B	C	D	E	F	G	H	I	J	K	L
1	**Weights**	**0.05**	**0.05**	**0.05**	**0.05**	**0.05**	**0.05**	**0.05**	**0.05**	**0.05**	**0.05**	**0.50**
2	**Criteria**	Battery capacity (mAh)	Screen size (inches)	Resolution x	Resolution y	Processor	RAM (MB)	Internal storage (GB)	Rear camera	Front camera	Number of SIMs	Price
3	OnePlus 7T P	4085	6.67	1440	3120	8	12000	256.0	48.0	16.0	2	58998
4	Realme X2 Pr	4000	6.5	1080	2400	8	6000	64.0	64.0	16.0	2	27999
5	iPhone 11 Pro	3969	6.5	1242	2688	6	4000	64.0	12.0	12.0	2	106900
6	iPhone 11	3110	6.1	828	1792	6	4000	64.0	12.0	12.0	2	62900

Figure 10. Weight assignments are used in the second example mentioned.

	A	B	C	D	E	F	G	H	I	J	K	L
1	Flags	Benefit	Benefit	Benefit	Benefit	Benefit	Benefit	Benefit	Benefit	Benefit	Benefit	Cost
2	**Weights**	**0.08**	**0.08**	**0.08**	**0.08**	**0.08**	**0.08**	**0.08**	**0.08**	**0.08**	**0.08**	**0.20**
3	**Criteria**	Battery capacity (mAh)	Screen size (inches)	Resolution x	Resolution y	Processor	RAM (MB)	Internal storage (GB)	Rear camera	Front camera	Number of SIMs	Price
4	OnePlus 7T P	4085	6.67	1440	3120	8	12000	256.0	48.0	16.0	2	58998
5	Realme X2 Pr	4000	6.5	1080	2400	8	6000	64.0	64.0	16.0	2	27999
6	iPhone 11 Pro	3969	6.5	1242	2688	6	4000	64.0	12.0	12.0	2	106900
7	iPhone 11	3110	6.1	828	1792	6	4000	64.0	12.0	12.0	2	62900

Figure 11. Flags cost–benefit defined. Screen size as benefit.

The import of the data and the result of the upload is then displayed as shown in Figure 12, Encircled with a dotted orange dashed line. Since it is not possible to edit the data after it is loaded, the criterion weighting exercises should be performed using a spreadsheet editing tool (such as Excel, LibreOffice, etc.), and reloading the data, which means going back to the previous Step 4 and repeating the exercises. It is recommended to refresh the page by clicking the "Reload this Page" button in the browser whenever a new load is performed. Additionally, it is important to remember to save the spreadsheet after making any changes.

In the sequence of software implementation, we visualize the data and submit it to the WASPAS algorithm. After loading the data, you can click on the "View your data" button or the "View" tab for a visualization of the imported dataset's contents, as shown highlighted in red in Figure 13.

Once the data loading results have been reviewed and confirmed to be successful, click on the "Calculate WASPAS" button to apply the method's algorithms to the imported database. The user will be automatically directed to the "Output" tab screen (as shown in Figure 14), where only the top 20 ranked items, with the application of lambda = 0.5, will be displayed. Limiting the number of observations to 20 is very useful in this case, given the thousands of options. Once we surpass a dozen of alternatives, it is better to create a shortlist and work with it. This is due to the presence of a set of "losers" that do not deserve attention and would burden the computational effort required for applying the algorithms.

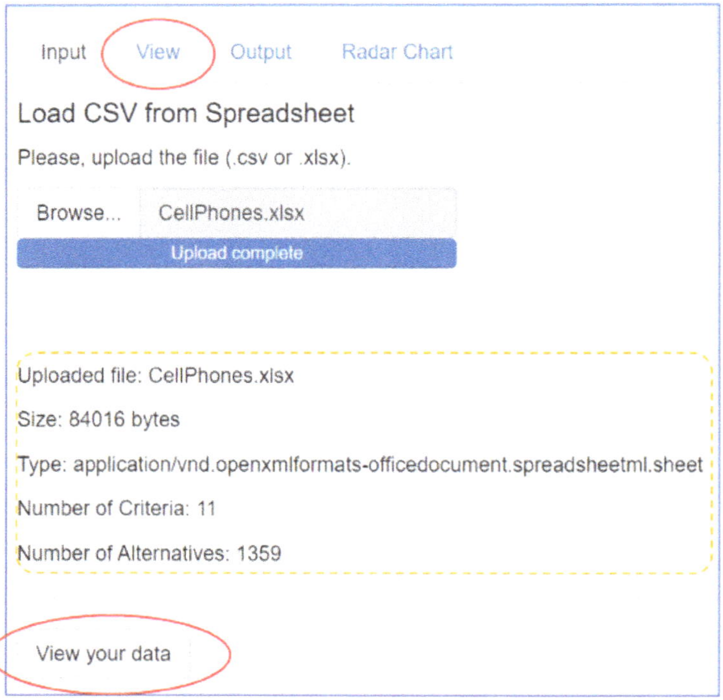

Figure 12. A dataset with 1359 cell phones with 11 criteria was successfully uploaded.

Input View Output Radar Chart

Calculate WASPAS

...1	...2	...3	...4	...5	...6	...7	...8	...9	...10	...11	...12
Flags	Benefit	Benefit	Benefit	Benefit	Benefit	Benefit	Benefit	Benefit	Benefit	Benefit	Cost
Weights	0.08	0.08	0.08	0.08	0.08	0.08	0.08	0.08	0.08	0.08	0.2
Criteria	Battery capacity (mAh)	Screen size (inches)	Resolution x	Resolution y	Processor	RAM (MB)	Internal storage (GB)	Rear camera	Front camera	Number of SIMs	Price
OnePlus 7T Pro McLaren Edition	4085	6.67	1440	3120	8	12000	256.0	48.0	16.0	2	58998
Realme X2 Pro	4000	6.5	1080	2400	8	6000	64.0	64.0	16.0	2	27999
iPhone 11 Pro Max	3969	6.5	1242	2688	6	4000	64.0	12.0	12.0	2	106900
iPhone 11	3110	6.1	828	1792	6	4000	64.0	12.0	12.0	2	62900

Figure 13. A dataset with 1359 cell phones with 11 criteria was successfully uploaded.

Dynamically different lambda ratings to observe the ranking change. The slider object allows the dynamic application of the lambda rating and the immediate visualization of the sensitivity of each of the underlying methods (WSM and WPM) to the dataset under study. It is very interesting to observe that a small push of lambda from zero to 2.5 already produces drastic changes, indicating how the database under study is sensitive to the weighting between the WSM and WPM methods, as can be observed in Figure 15.

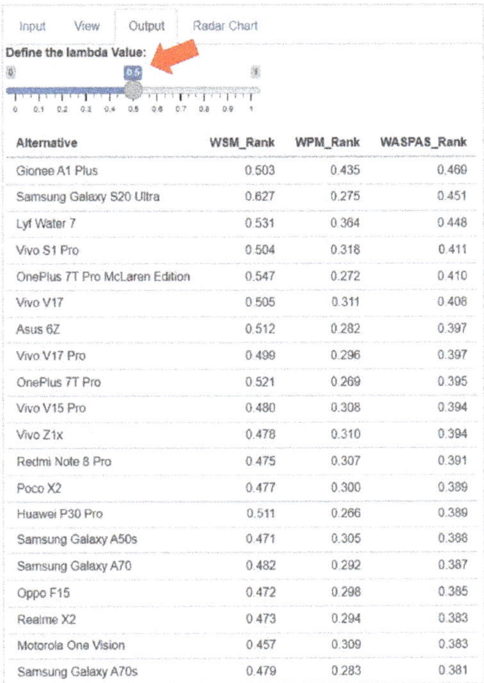

Alternative	WSM_Rank	WPM_Rank	WASPAS_Rank
Gionee A1 Plus	0.503	0.435	0.469
Samsung Galaxy S20 Ultra	0.627	0.275	0.451
Lyf Water 7	0.531	0.364	0.448
Vivo S1 Pro	0.504	0.318	0.411
OnePlus 7T Pro McLaren Edition	0.547	0.272	0.410
Vivo V17	0.505	0.311	0.408
Asus 6Z	0.512	0.282	0.397
Vivo V17 Pro	0.499	0.296	0.397
OnePlus 7T Pro	0.521	0.269	0.395
Vivo V15 Pro	0.480	0.308	0.394
Vivo Z1x	0.478	0.310	0.394
Redmi Note 8 Pro	0.475	0.307	0.391
Poco X2	0.477	0.300	0.389
Huawei P30 Pro	0.511	0.266	0.389
Samsung Galaxy A50s	0.471	0.305	0.388
Samsung Galaxy A70	0.482	0.292	0.387
Oppo F15	0.472	0.298	0.385
Realme X2	0.473	0.294	0.383
Motorola One Vision	0.457	0.309	0.383
Samsung Galaxy A70s	0.479	0.283	0.381

Figure 14. Ranking of alternatives in a standard analysis.

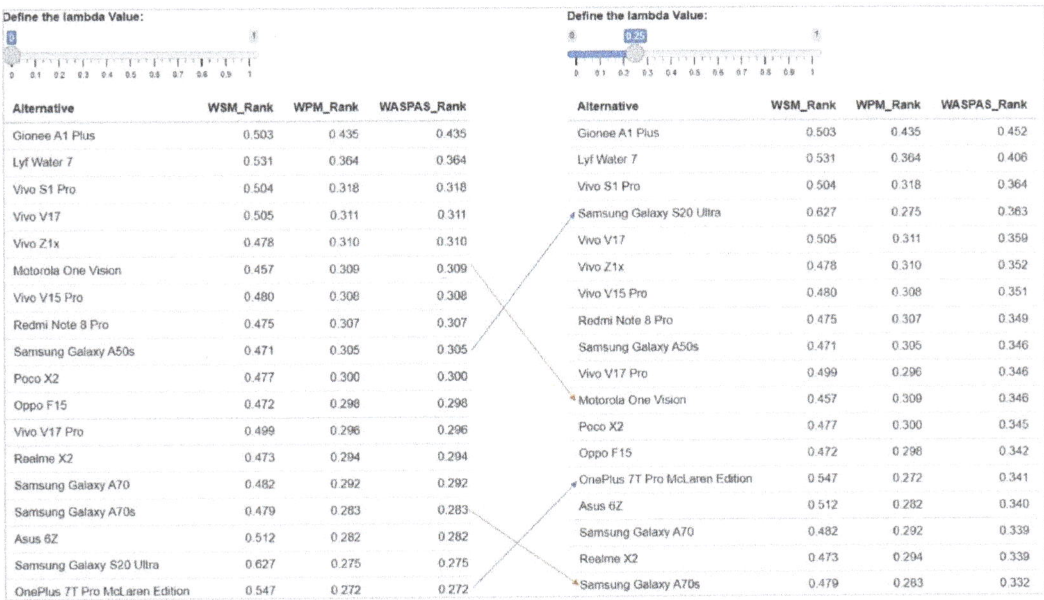

Alternative	WSM_Rank	WPM_Rank	WASPAS_Rank
Gionee A1 Plus	0.503	0.435	0.435
Lyf Water 7	0.531	0.364	0.364
Vivo S1 Pro	0.504	0.318	0.318
Vivo V17	0.505	0.311	0.311
Vivo Z1x	0.478	0.310	0.310
Motorola One Vision	0.457	0.309	0.309
Vivo V15 Pro	0.480	0.308	0.308
Redmi Note 8 Pro	0.475	0.307	0.307
Samsung Galaxy A50s	0.471	0.305	0.305
Poco X2	0.477	0.300	0.300
Oppo F15	0.472	0.298	0.298
Vivo V17 Pro	0.499	0.296	0.296
Realme X2	0.473	0.294	0.294
Samsung Galaxy A70	0.482	0.292	0.292
Samsung Galaxy A70s	0.479	0.283	0.283
Asus 6Z	0.512	0.282	0.282
Samsung Galaxy S20 Ultra	0.627	0.275	0.275
OnePlus 7T Pro McLaren Edition	0.547	0.272	0.272

Alternative	WSM_Rank	WPM_Rank	WASPAS_Rank
Gionee A1 Plus	0.503	0.435	0.452
Lyf Water 7	0.531	0.364	0.406
Vivo S1 Pro	0.504	0.318	0.364
Samsung Galaxy S20 Ultra	0.627	0.275	0.363
Vivo V17	0.505	0.311	0.359
Vivo Z1x	0.478	0.310	0.352
Vivo V15 Pro	0.480	0.308	0.351
Redmi Note 8 Pro	0.475	0.307	0.349
Samsung Galaxy A50s	0.471	0.305	0.346
Vivo V17 Pro	0.499	0.296	0.346
Motorola One Vision	0.457	0.309	0.346
Poco X2	0.477	0.300	0.345
Oppo F15	0.472	0.298	0.342
OnePlus 7T Pro McLaren Edition	0.547	0.272	0.341
Asus 6Z	0.512	0.282	0.340
Samsung Galaxy A70	0.482	0.292	0.339
Realme X2	0.473	0.294	0.339
Samsung Galaxy A70s	0.479	0.283	0.332

Figure 15. Major changes in the ranking, changing lambda from 0 to 0.25.

Again, changing it to 0.75, since we started with 0.5 and did not need to repeat, we can observe some radical changes. Notice how the direction arrows of the changes become more aggressive; observe Figure 16. By pushing lambda further towards making WASPAS exactly like a WPM, signifying making lambda equal to 1.0, we can observe a few more changes, as depicted in Figure 17. Note that the dashed arrows indicate alternatives that did not undergo any changes, while the solid arrows indicate new changes.

Alternative	WSM_Rank	WPM_Rank	WASPAS_Rank
Gionee A1 Plus	0.503	0.435	0.435
Lyf Water 7	0.531	0.364	0.364
Vivo S1 Pro	0.504	0.318	0.318
Vivo V17	0.505	0.311	0.311
Vivo Z1x	0.478	0.310	0.310
Motorola One Vision	0.457	0.309	0.309
Vivo V15 Pro	0.480	0.308	0.308
Redmi Note 8 Pro	0.475	0.307	0.307
Samsung Galaxy A50s	0.471	0.305	0.305
Poco X2	0.477	0.300	0.300
Oppo F15	0.472	0.298	0.298
Vivo V17 Pro	0.499	0.296	0.296
Realme X2	0.473	0.294	0.294
Samsung Galaxy A70	0.482	0.292	0.292
Samsung Galaxy A70s	0.479	0.283	0.283
Asus 6Z	0.512	0.282	0.282
Samsung Galaxy S20 Ultra	0.627	0.275	0.275
OnePlus 7T Pro McLaren Edition	0.547	0.272	0.272

Alternative	WSM_Rank	WPM_Rank	WASPAS_Rank
Samsung Galaxy S20 Ultra	0.627	0.275	0.539
Lyf Water 7	0.531	0.364	0.490
Gionee A1 Plus	0.503	0.435	0.486
OnePlus 7T Pro McLaren Edition	0.547	0.272	0.478
OnePlus 7T Pro	0.521	0.269	0.458
Vivo S1 Pro	0.504	0.318	0.457
Vivo V17	0.505	0.311	0.456
Asus 6Z	0.512	0.282	0.455
Huawei P30 Pro	0.511	0.266	0.450
Vivo V17 Pro	0.499	0.296	0.448
Vivo V15 Pro	0.480	0.308	0.437
Vivo Z1x	0.478	0.310	0.436
Samsung Galaxy A70	0.482	0.292	0.435
Redmi Note 8 Pro	0.475	0.307	0.433
Poco X2	0.477	0.300	0.433
Samsung Galaxy A70s	0.479	0.283	0.430
Samsung Galaxy A50s	0.471	0.305	0.429
Oppo F15	0.472	0.298	0.428
Realme X2	0.473	0.294	0.428
Motorola One Vision	0.457	0.309	0.420

Figure 16. Extreme changes were observed in the ranking, changing lambda from 0 to 0.75.

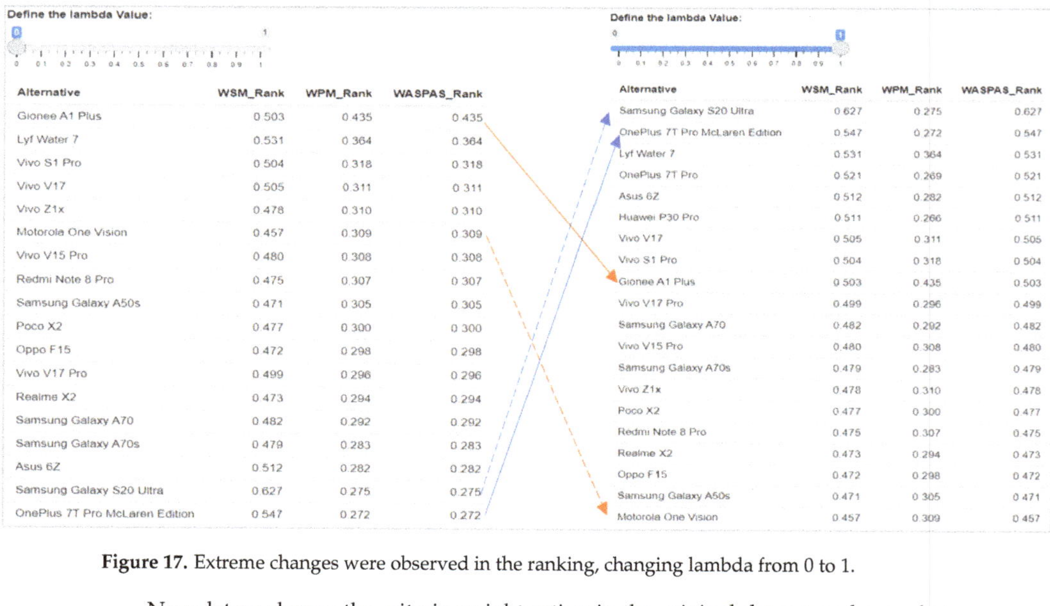

Alternative	WSM_Rank	WPM_Rank	WASPAS_Rank
Gionee A1 Plus	0.503	0.435	0.435
Lyf Water 7	0.531	0.364	0.364
Vivo S1 Pro	0.504	0.318	0.318
Vivo V17	0.505	0.311	0.311
Vivo Z1x	0.478	0.310	0.310
Motorola One Vision	0.457	0.309	0.309
Vivo V15 Pro	0.480	0.308	0.308
Redmi Note 8 Pro	0.475	0.307	0.307
Samsung Galaxy A50s	0.471	0.305	0.305
Poco X2	0.477	0.300	0.300
Oppo F15	0.472	0.298	0.298
Vivo V17 Pro	0.499	0.296	0.296
Realme X2	0.473	0.294	0.294
Samsung Galaxy A70	0.482	0.292	0.292
Samsung Galaxy A70s	0.479	0.283	0.283
Asus 6Z	0.512	0.282	0.282
Samsung Galaxy S20 Ultra	0.627	0.275	0.275
OnePlus 7T Pro McLaren Edition	0.547	0.272	0.272

Alternative	WSM_Rank	WPM_Rank	WASPAS_Rank
Samsung Galaxy S20 Ultra	0.627	0.275	0.627
OnePlus 7T Pro McLaren Edition	0.547	0.272	0.547
Lyf Water 7	0.531	0.364	0.531
OnePlus 7T Pro	0.521	0.269	0.521
Asus 6Z	0.512	0.282	0.512
Huawei P30 Pro	0.511	0.266	0.511
Vivo V17	0.505	0.311	0.505
Vivo S1 Pro	0.504	0.318	0.504
Gionee A1 Plus	0.503	0.435	0.503
Vivo V17 Pro	0.499	0.296	0.499
Samsung Galaxy A70	0.482	0.292	0.482
Vivo V15 Pro	0.480	0.308	0.480
Samsung Galaxy A70s	0.479	0.283	0.479
Vivo Z1x	0.478	0.310	0.478
Poco X2	0.477	0.300	0.477
Redmi Note 8 Pro	0.475	0.307	0.475
Realme X2	0.473	0.294	0.473
Oppo F15	0.472	0.298	0.472
Samsung Galaxy A50s	0.471	0.305	0.471
Motorola One Vision	0.457	0.309	0.457

Figure 17. Extreme changes were observed in the ranking, changing lambda from 0 to 1.

Now, let us change the criteria weight rating in the original dataset and see what happens. These first exercises demonstrate the ease of use of the waspasWEB tool (WASPAS for Dummies website). Now let us apply a more "human" weighting to the criteria since the technical specifications generally have different relative importance among themselves, and the weight of the price criterion may, in fact, not have the oversized importance that we used previously (in many cases, the price is less important than other performance criteria). The weighted set of criteria is shown in Table 2.

Table 2. Weighting criteria that would likely make more sense to a decision-maker.

Flags	Weights	Criteria
Benefit	0.10	Battery capacity (mAh)
Cost	0.07	Screen size
Benefit	0.09	Resolution X
Benefit	0.09	Resolution Y
Benefit	0.06	Processors
Benefit	0.10	RAM (MB)
Benefit	0.12	Internal storage (GB)
Benefit	0.09	Rear camera
Benefit	0.09	Front camera
Benefit	0.05	Number of SIMs
Cost	0.14	Price

One could say that the set of criteria now assigned in the case study scope is not only more rigorous but also has a more human aspect. For instance, even though cost remains the most important criterion in the selection process of the best option, it no longer presents such a significant difference compared to other criteria. As a result, those devices whose only advantage is a low price but have poor technical characteristics will not be artificially overrated.

Another significant change made for the upcoming exercises was the inversion of the "Screen size" criterion from a monotonically increasing benefit criterion to a monotonically increasing cost criterion. This change makes sense within the context that a subjective criterion related to the desires, tastes, and personal preferences of the decision-maker is the most important aspect of the entire process. Thus, what is negative for some may be positive for others. Mathematical methods, algorithms, formulas, and all the technical, scientific tools serve as support so that, from a human perspective, whether individual or as a group, the best decision can be made. The tool used should apply computational effort to support decisions without ever neglecting human interests. The mechanism that ensures that human desire overrides the coldness of calculations is the subjective imposition of the relative importance weighting of the criteria associated with the direction of maximizing or minimizing the performance rating of each option under analysis determined based on the evaluated criteria.

In this new set of weights, in addition to the decrease in the relative importance of the price of the mobile device, the decision-maker also considers the battery performance, memory, and data storage capacity as important. The quantity of processors and SIM cards, on the other hand, is considered less important. A reasonable explanation for these decisions could be, for example, the lack of intention to use the smartphone for gaming, making higher processing power less relevant. As for SIM cards, there may not be an immediate intention to use two (or more) phone numbers, but since it could become a future necessity, the criterion, although considered of low relative importance, should not be excluded from the set of criteria.

Now, simply execute some steps again using the new metadata configuration. The first observation, as evidenced by Figure 18, is the elimination of some options from the shortlist and changes in the ranking, which is now much less sensitive to price compared to other technical criteria.

Now, just as it was performed previously, we will alter the rating of lambda to observe the changes in the ranking derived from the sensitivity of the set of options in relation to the percentage weight assigned to the underlying WSM and WPM methods, as proposed by the WASPAS method.

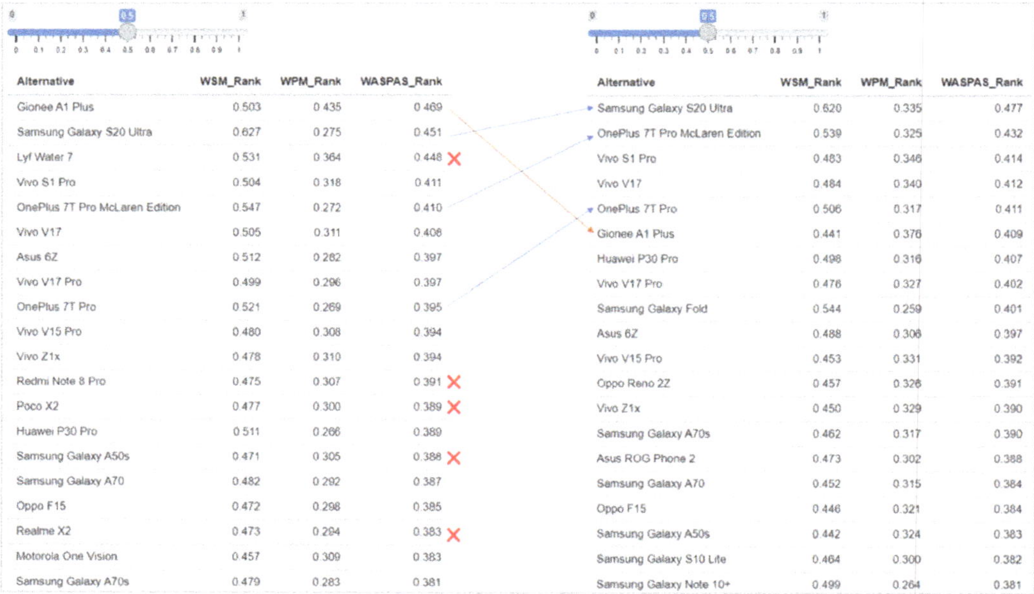

Alternative	WSM_Rank	WPM_Rank	WASPAS_Rank
Gionee A1 Plus	0.503	0.435	0.469
Samsung Galaxy S20 Ultra	0.627	0.275	0.451
Lyf Water 7	0.531	0.364	0.448
Vivo S1 Pro	0.504	0.318	0.411
OnePlus 7T Pro McLaren Edition	0.547	0.272	0.410
Vivo V17	0.505	0.311	0.408
Asus 6Z	0.512	0.282	0.397
Vivo V17 Pro	0.499	0.296	0.397
OnePlus 7T Pro	0.521	0.269	0.395
Vivo V15 Pro	0.480	0.308	0.394
Vivo Z1x	0.478	0.310	0.394
Redmi Note 8 Pro	0.475	0.307	0.391
Poco X2	0.477	0.300	0.389
Huawei P30 Pro	0.511	0.266	0.389
Samsung Galaxy A50s	0.471	0.305	0.388
Samsung Galaxy A70	0.482	0.292	0.387
Oppo F15	0.472	0.298	0.385
Realme X2	0.473	0.294	0.383
Motorola One Vision	0.457	0.309	0.383
Samsung Galaxy A70s	0.479	0.283	0.381

Alternative	WSM_Rank	WPM_Rank	WASPAS_Rank
Samsung Galaxy S20 Ultra	0.620	0.335	0.477
OnePlus 7T Pro McLaren Edition	0.539	0.325	0.432
Vivo S1 Pro	0.483	0.346	0.414
Vivo V17	0.484	0.340	0.412
OnePlus 7T Pro	0.506	0.317	0.411
Gionee A1 Plus	0.441	0.376	0.409
Huawei P30 Pro	0.498	0.316	0.407
Vivo V17 Pro	0.476	0.327	0.402
Samsung Galaxy Fold	0.544	0.259	0.401
Asus 6Z	0.488	0.306	0.397
Vivo V15 Pro	0.453	0.331	0.392
Oppo Reno 2Z	0.457	0.326	0.391
Vivo Z1x	0.450	0.329	0.390
Samsung Galaxy A70s	0.462	0.317	0.390
Asus ROG Phone 2	0.473	0.302	0.388
Samsung Galaxy A70	0.452	0.315	0.384
Oppo F15	0.446	0.321	0.384
Samsung Galaxy A50s	0.442	0.324	0.383
Samsung Galaxy S10 Lite	0.464	0.300	0.382
Samsung Galaxy Note 10+	0.499	0.264	0.381

Figure 18. Extreme changes were observed in the ranking, fixing the lambda in 0.5.

Once again, we observe a noticeable alteration due to the weighting of the underlying methods, as demonstrated in Figure 19. This leads us to believe that the prices of the products composing the set of alternatives exhibit a significant internal discrepancy, suggesting a parallel analysis of this criterion. We then conducted a parallel analysis and found the following results: The maximum price rating is 35,423% higher than the minimum, the average is 11,466, and the standard deviation is high at 13,852.

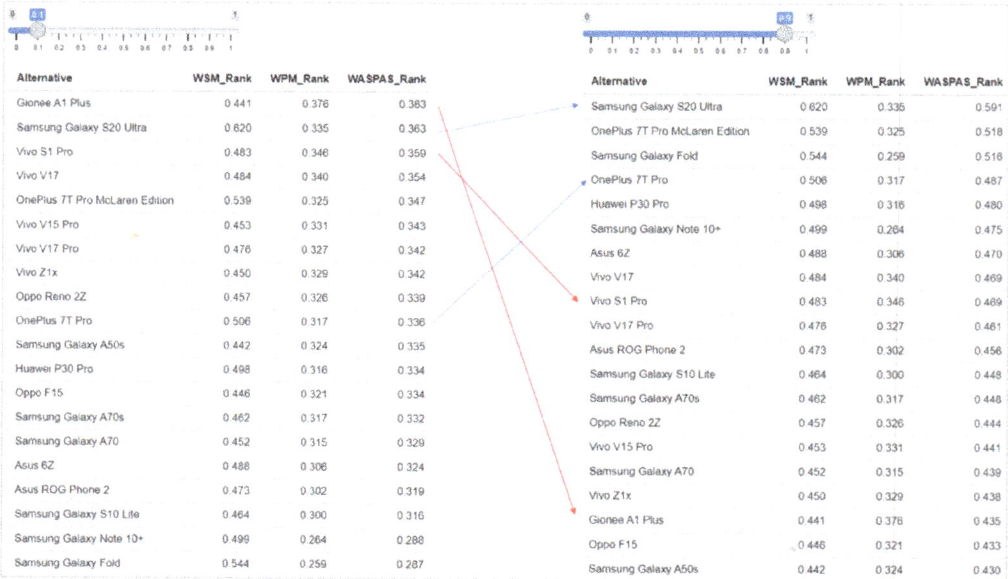

Alternative	WSM_Rank	WPM_Rank	WASPAS_Rank
Gionee A1 Plus	0.441	0.376	0.383
Samsung Galaxy S20 Ultra	0.620	0.335	0.363
Vivo S1 Pro	0.483	0.346	0.359
Vivo V17	0.484	0.340	0.354
OnePlus 7T Pro McLaren Edition	0.539	0.325	0.347
Vivo V15 Pro	0.453	0.331	0.343
Vivo V17 Pro	0.476	0.327	0.342
Vivo Z1x	0.450	0.329	0.342
Oppo Reno 2Z	0.457	0.326	0.339
OnePlus 7T Pro	0.506	0.317	0.336
Samsung Galaxy A50s	0.442	0.324	0.335
Huawei P30 Pro	0.498	0.316	0.334
Oppo F15	0.446	0.321	0.334
Samsung Galaxy A70s	0.462	0.317	0.332
Samsung Galaxy A70	0.452	0.315	0.329
Asus 6Z	0.488	0.306	0.324
Asus ROG Phone 2	0.473	0.302	0.319
Samsung Galaxy S10 Lite	0.464	0.300	0.316
Samsung Galaxy Note 10+	0.499	0.264	0.288
Samsung Galaxy Fold	0.544	0.259	0.287

Alternative	WSM_Rank	WPM_Rank	WASPAS_Rank
Samsung Galaxy S20 Ultra	0.620	0.335	0.591
OnePlus 7T Pro McLaren Edition	0.539	0.325	0.518
Samsung Galaxy Fold	0.544	0.259	0.516
OnePlus 7T Pro	0.506	0.317	0.487
Huawei P30 Pro	0.498	0.316	0.480
Samsung Galaxy Note 10+	0.499	0.264	0.475
Asus 6Z	0.488	0.306	0.470
Vivo V17	0.484	0.340	0.469
Vivo S1 Pro	0.483	0.346	0.469
Vivo V17 Pro	0.476	0.327	0.461
Asus ROG Phone 2	0.473	0.302	0.456
Samsung Galaxy S10 Lite	0.464	0.300	0.448
Samsung Galaxy A70s	0.462	0.317	0.448
Oppo Reno 2Z	0.457	0.326	0.444
Vivo V15 Pro	0.453	0.331	0.441
Samsung Galaxy A70	0.452	0.315	0.439
Vivo Z1x	0.450	0.329	0.438
Gionee A1 Plus	0.441	0.376	0.435
Oppo F15	0.446	0.321	0.433
Samsung Galaxy A50s	0.442	0.324	0.430

Figure 19. Extreme changes were observed in the ranking, fixing the lambda in 0.5.

We also observed that the options with extreme ratings are the ones that appear most frequently on the shortlist. This certainly occurs because these options with very high prices have extremely high technical criteria, resulting in a set of options that are radically opposite. There is no technical issue in this aspect, and it is still very important that the purpose of the presented case study is to demonstrate the power of the tool offered to the public, which has already been demonstrated at this point.

Therefore, the subsequent operations serve only a more didactic purpose in relation to the discipline of OR. Then, we performed a summary cut of the devices with prices above 10,000 and below 5000 monetary units. As a result, the original set of 1359 alternatives (see Figure 19) was reduced to a subset of 524 alternatives, representing a volume reduction of over 60%.

So, a flat cut-off of devices with prices above 10,000 and below 5000 monetary units was made. As a result, the original set of 1359 alternatives was reduced to a subset of 524 alternatives, representing a volume reduction of over 60%.

This seems like a fair and more meticulous contention. It is important to remember that the imposition of constraints is often one of the steps in OR processes. To enhance the ongoing decision-making process, we have implemented a revision in the weighting of criteria, and the new setup can be seen in Table 3.

Table 3. Weighting criteria that drastically reduce the importance of the price criterion.

Flags	Weights	Criteria
Benefit	0.10	Battery capacity (mAh)
Cost	0.05	Screen size
Benefit	0.10	Resolution X
Benefit	0.10	Resolution Y
Benefit	0.07	Processors
Benefit	0.14	RAM (MB)
Benefit	0.14	Internal storage (GB)
Benefit	0.10	Rear camera
Benefit	0.10	Front camera
Benefit	0.05	Number of SIMs
Cost	0.05	Price

In this configuration, the weight of the price criterion has been reduced, and it has been redistributed to other criteria.

So, we applied a price restriction based on a spending ceiling and a purchasing budget theory, and also considering that very cheap devices probably will not have technically advanced features with good performance and could compromise the shortlist due to their extremely low price. It would be more appropriate to apply restrictions to each of the criteria, but the most important within the scope of this work, as mentioned earlier, is to present the public decision support tool based on WASPAS, as well as the development process and the contributions to society derived from this research.

Now we have a winner. It can be observed in Figure 20 that even when moving lambda between its extreme rating, the top-ranked option remains unchanged.

It is also observed that for lambda ratings between a weight of 90% for the underlying WPM method (lambda = 0.1) and an equal weighting between the two methods (WSM and WPM), the top four rankings remain unchanged. The alteration of this "elite group" is only observed when we apply a lambda rating close to 1 (in the example of Figure 20, lambda = 0.9).

The waspasWEB public service also offers a radar graph view, in which the lambda rating is also dynamically applied, and immediate visualization of the WASPAS (green), WSM (black), and WPM (red) ranking lines is obtained. The radar charts, also known as spider web charts or Kiviat diagrams, are shown in Figure 21.

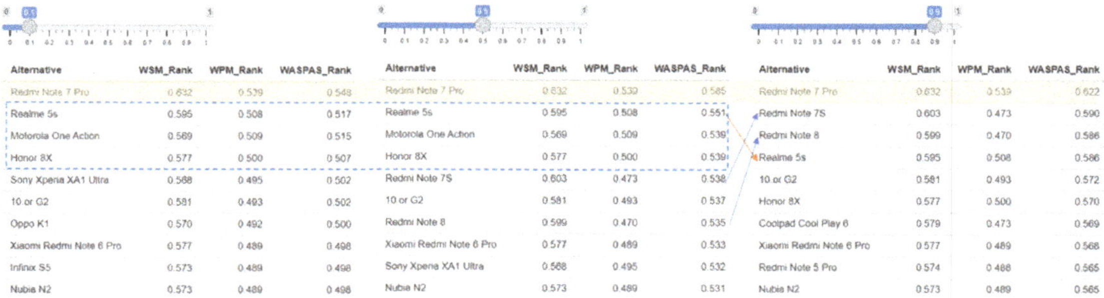

Alternative	WSM_Rank	WPM_Rank	WASPAS_Rank
Redmi Note 7 Pro	0.632	0.539	0.548
Realme 5s	0.595	0.508	0.517
Motorola One Action	0.569	0.509	0.515
Honor 8X	0.577	0.500	0.507
Sony Xperia XA1 Ultra	0.568	0.495	0.502
10.or G2	0.581	0.493	0.502
Oppo K1	0.570	0.492	0.500
Xiaomi Redmi Note 6 Pro	0.577	0.489	0.498
Infinix S5	0.573	0.489	0.498
Nubia N2	0.573	0.489	0.498

Alternative	WSM_Rank	WPM_Rank	WASPAS_Rank
Redmi Note 7 Pro	0.632	0.539	0.585
Realme 5s	0.595	0.508	0.551
Motorola One Action	0.569	0.509	0.539
Honor 8X	0.577	0.500	0.539
Redmi Note 7S	0.603	0.473	0.538
10 or G2	0.581	0.493	0.537
Redmi Note 8	0.599	0.470	0.535
Xiaomi Redmi Note 6 Pro	0.577	0.489	0.533
Sony Xperia XA1 Ultra	0.568	0.495	0.532
Nubia N2	0.573	0.489	0.531

Alternative	WSM_Rank	WPM_Rank	WASPAS_Rank
Redmi Note 7 Pro	0.632	0.539	0.622
Redmi Note 7S	0.603	0.473	0.590
Redmi Note 8	0.599	0.470	0.586
Realme 5s	0.595	0.508	0.586
10.or G2	0.581	0.493	0.572
Honor 8X	0.577	0.500	0.570
Coolpad Cool Play 6	0.579	0.473	0.569
Xiaomi Redmi Note 6 Pro	0.577	0.489	0.568
Redmi Note 5 Pro	0.574	0.488	0.565
Nubia N2	0.573	0.489	0.565

Figure 20. The winning option remains unchanged.

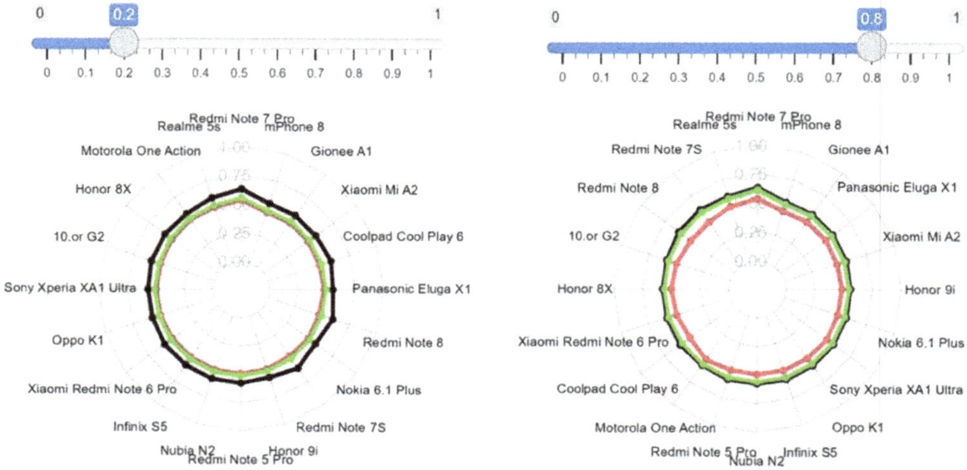

Figure 21. Spider chart is a functionality available in the waspasWEB service.

The spider chart available in the WASPASWEB service presents a green line representing the ranking distribution by the WASPAS method, a black line shows the distribution by the WSM (sum) method, and a red line represents the ranking by the WPM (product) method. By moving the slider and dynamically applying a different rating for the lambda, the WASPAS result (green line) can be observed moving between complete overlap with the red line (WPM) when lambda is equal to zero and expanding until it completely overlaps with the black line (WSM) for lambda equal to 1.

In Figure 21, it is evident that the green line is very close to the red line for a lambda rating of 0.2, and it is very close to the black line for a lambda equal to 0.8. In Figure 22, the line derived from WASPAS for equal weighting between WSM and WPM (lambda = 0.5) is positioned in the middle of the two lines.

In Figure 22, it is evident that the green line is very close to the red line for a lambda rating of 0.2, and it is very close to the black line for a lambda equal to 0.8.

As presented, the proposed computational model works as an aid in the implementation of WASPAS methodology, performing the aggregation of numerical preferences through numerical and graphical resources, helping the clarity of results transparently. The numerical example used in this study works just as an aid in the understanding of the interactivity of the internet-based platform, and the implementation of different case studies with different levels of complexity in operational, tactical, and strategical environments is possible.

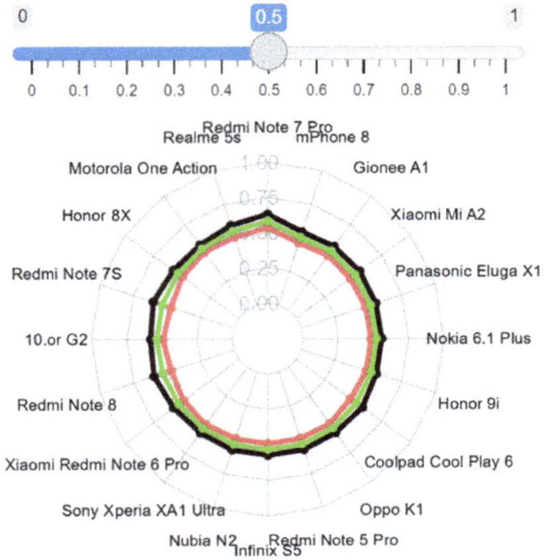

Figure 22. Spider chart is a functionality available in the waspasWEB service.

5. Discussion

The primary focus of this work was to obtain practical results and apply the authors' knowledge in the development of tangible products [46]. Although the contributions may be modest, they serve as a valuable resource for the community [47]. The website created during this work serves as a useful tool for professionals in the field of OR and individuals seeking to make informed decisions based on reliable mathematical models without the need for complex calculations or software implementations [48].

As questions of limitations to the proposed model, we identify the need for an axiomatic understanding of the mathematical model to enable its correct and satisfactory application. The software is limited to the implementation of the WASPAS multi-criteria method.

Regarding the source code complexity perspective, all the necessity of computational programming is transcribed into an internet-based platform, where there is no necessity for coding by the user, as is presented in some computational models [37,49], being just necessary for the alignment of the problematic situation to the WASPAS methodology and basic knowledge to interactive computational platforms, where on the website, is possible to understand all software functionalities through a manual guide to support the users.

A computational model needs to be constantly updated and technically adapted. For future research, we seek to increase the computational model and practical application in different case studies, clarifying the limitations of the mathematical model, thus continuing the research regarding the development of new axiomatic techniques that can incorporate the base method and thus provide improvements and new possibilities to the present computational model.

6. Conclusions

The present study was based on presenting a computational interactive web model as support in the decision-making process through the implementation of the WASPAS method, built under the multi-criteria decision support approach.

The WASPAS method is a flexible approach and can be adapted to different types of problems and scenarios. However, it is important to remember that it depends on the accuracy of the weights and ratings assigned to the criteria and alternatives, which can be difficult to obtain in some situations. Furthermore, the WASPAS methodology can be mathematically demanding for problems with many alternatives or criteria being necessary for the computation support.

The computation proposed framework presents an interactive approach concerning the user, enabling the implementation of the mathematical model along with the performance of sensitivity analysis in the changing of the weights and thresholds of the methodology. As a form of future studies, we search for the integration of a module for open format exportation of the provided calculations and their results, along with the chart exportation by vectorial graphics, with high-quality images. Also, we consider the implementation of the model in other case studies framed in the specifications of the method, providing not only the resolution of these but also the identification of improvement points for greater robustness in the method and computational model.

Author Contributions: Conceptualization, F.B., M.d.S., A.S.S. and M.Â.L.M.; Methodology, F.B., A.S.S. and M.Â.L.M.; Software, F.B. and M.Â.L.M.; Validation, M.d.S. and L.P.F.; Formal analysis, F.B. and M.Â.L.M.; Investigation, F.B., A.S.S., E.L.P.J. and D.A.d.M.P.; Resources, D.A.d.M.P. and W.d.A.C.; Data curation, M.d.S. and M.Â.L.M.; Writing—original draft, F.B. and M.Â.L.M.; Supervision, M.d.S. and L.P.F.; Project administration, M.d.S., F.M.M. and A.G.P.; Funding acquisition, W.d.A.C., F.M.M. and A.G.P. All authors have read and agreed to the published version of the manuscript.

Funding: This research received no external funding.

Data Availability Statement: Not applicable.

Conflicts of Interest: The authors declare no conflict of interest.

References

1. Sun, A.Y.; Scanlon, B.R. How Can Big Data and Machine Learning Benefit Environment and Water Management: A Survey of Methods, Applications, and Future Directions. *Environ. Res. Lett.* **2019**, *14*, 73001. [CrossRef]
2. Rani, P.; Mishra, A.R.; Pardasani, K.R. A Novel WASPAS Approach for Multi-Criteria Physician Selection Problem with Intuitionistic Fuzzy Type-2 Sets. *Soft Comput.* **2020**, *24*, 2355–2367. [CrossRef]
3. de Almeida, I.D.P.; Costa, I.P.A.; Costa, A.P.A.; Corriça, J.V.P.; Moreira, M.Â.L.; Gomes, C.F.S.; Santos, M. A Multicriteria Decision-Making Approach to Classify Military Bases for the Brazilian Navy. *Procedia Comput. Sci.* **2022**, *199*, 79–86. [CrossRef]
4. Drumond, P.; Costa, I.P.A.; Moreira, M.Â.L.; Santos, M.; Gomes, C.F.S.; Maêda, S.M.N. Strategy Study to Prioritize Marketing Criteria: An Approach in the Light of the DEMATEL Method. *Procedia Comput. Sci.* **2022**, *199*, 448–455. [CrossRef]
5. Payton, E.; Khubchandani, J.; Thompson, A.; Price, J.H. Parents' Expectations of High Schools in Firearm Violence Prevention. *J. Community Health* **2017**, *42*, 1118–1126. [CrossRef]
6. Chen, C.J. Developing a Model for Supply Chain Agility and Innovativeness to Enhance Firms' Competitive Advantage. *Manag. Decis.* **2019**, *57*, 1511–1534. [CrossRef]
7. Patil, S.; Nemade, V.; Soni, P.K. Predictive Modelling for Credit Card Fraud Detection Using Data Analytics. *Procedia Comput. Sci.* **2018**, *132*, 385–395. [CrossRef]
8. Jiang, W.; Liu, J.; Dong, Y.; Wang, L. Assignment of Duplicate Storage Locations in Distribution Centres to Minimise Walking Distance in Order Picking. *Int. J. Prod. Res.* **2021**, *59*, 4457–4471. [CrossRef]
9. Saaty, T.L. A Scaling Method for Priorities in Hierarchical Structures. *J. Math. Psychol.* **1977**, *15*, 234–281. [CrossRef]
10. Chan, F.T.S. Performance Measurement in a Supply Chain. *Int. J. Adv. Manuf. Technol.* **2003**, *21*, 534–548. [CrossRef]
11. Maghrabie, H.F.; Beauregard, Y.; Schiffauerova, A. Grey-Based Multi-Criteria Decision Analysis Approach: Addressing Uncertainty at Complex Decision Problems. *Technol. Forecast. Soc. Chang.* **2019**, *146*, 366–379. [CrossRef]
12. Brans, J.P.; Vincke, P.; Mareschal, B. *Prométhée: A New Family of Outranking Methods in Multicriteria Analysis*; ULB Institutional Repository; Universite Libre de Bruxelles: Brussels, Belgium, 1984.
13. Brans, J.-P.; Vincke, P. A Preference Ranking Organisation Method: The PROMETHEE Method for Multiple Criteria Decision-Making. *Manag. Sci.* **1985**, *31*, 47–56. [CrossRef]
14. Brans, J.P.; Mareschal, B. The PROMCALC & GAIA Decision Support System for Multicriteria Decision Aid. *Decis. Support Syst.* **1994**, *12*, 297–310. [CrossRef]
15. Tenorio, F.M.; Dos Santos, M.; Gomes, C.F.S.; Araujo, J.D.C.; De Almeida, G.P. THOR 2 Method: An Efficient Instrument in Situations Where There Is Uncertainty or Lack of Data. *IEEE Access* **2021**, *9*, 161794–161805. [CrossRef]
16. Moreira, M.Â.L.; Silva, F.C.A.; Costa, I.P.A.; Gomes, C.F.S.; Santos, M. SAPEVO-H^2 a Multi-Criteria Systematic Based on a Hierarchical Structure: Decision-Making Analysis for Assessing Anti-RPAS Strategies in Sensing Environments. *Processes* **2023**, *11*, 352. [CrossRef]
17. Hwang, C.; Yoon, K. *Multiple Attribute Decision Making: Methods and Application*, 1st ed.; Springer: New York, NY, USA, 1981.

18. de Assis, G.S.; dos Santos, M.; Basilio, M.P. Use of the WASPAS Method to Select Suitable Helicopters for Aerial Activity Carried Out by the Military Police of the State of Rio de Janeiro. *Axioms* **2023**, *12*, 77. [CrossRef]
19. Li, Z.; Zhang, Z.; Yu, W. Consensus Reaching for Ordinal Classification-Based Group Decision Making with Heterogeneous Preference Information. *J. Oper. Res. Soc.* **2023**, *1*, 1–22. [CrossRef]
20. Dong, Y.; Liu, Y.; Liang, H.; Chiclana, F.; Herrera-Viedma, E. Strategic Weight Manipulation in Multiple Attribute Decision Making. *Omega* **2018**, *75*, 1339–1351. [CrossRef]
21. Maêda, S.M.d.N.; Basílio, M.P.; Costa, I.P.d.A.; Moreira, M.Â.L.; dos Santos, M.; Gomes, C.F.S. The SAPEVO-M-NC Method. *Front. Artif. Intell. Appl.* **2021**, *341*, 89–95. [CrossRef]
22. Liu, Y.; Li, Y.; Zhang, Z.; Xu, Y.; Dong, Y. Classification-Based Strategic Weight Manipulation in Multiple Attribute Decision Making. *Expert Syst. Appl.* **2022**, *197*, 116781. [CrossRef]
23. Greco, S.; Ehrgott, M.; Figueira, J.R. *Multiple Criteria Decision Analysis State of the Art Surveys*, 2nd ed.; Springer: New York, NY, USA, 2016; ISBN 038723067X.
24. Chakraborty, S.; Zavadskas, E.K. Application of WASPAS Method in Manufacturing Decision Making. *Informatica* **2014**, *25*, 1–20. [CrossRef]
25. Chakraborty, S.; Saha, A.K. A Framework of LR Fuzzy AHP and Fuzzy WASPAS for Health Care Waste Recycling Technology. *Appl. Soft Comput.* **2022**, *127*, 109388. [CrossRef]
26. Pamucar, D.; Yazdani, M.; Montero-Simo, M.J.; Araque-Padilla, R.A.; Mohammed, A. Multi-Criteria Decision Analysis towards Robust Service Quality Measurement. *Expert Syst. Appl.* **2021**, *170*, 114508. [CrossRef]
27. Chakraborty, S.; Zavadskas, E.K.; Antucheviciene, J. Applications of WASPAS Method as a Multi-Criteria Decision-Making Tool. *Econ. Comput. Econ. Cybern. Stud. Res.* **2015**, *49*, 5–22.
28. Parreiras, R.O.; Ekel, P.Y.; Martini, J.S.C.C.; Palhares, R.M. A Flexible Consensus Scheme for Multicriteria Group Decision Making under Linguistic Assessments. *Inf. Sci.* **2010**, *180*, 1075–1089. [CrossRef]
29. Rocha, C.d.S., Jr.; Moreira, M.Â.L.; Santos, M. Selection of Interns for Startups: An Approach Based on the AHP-TOPSIS-2N Method and the 3DM Computational Platform. *Procedia Comput. Sci.* **2022**, *199*, 984–991. [CrossRef]
30. Darko, A.P.; Liang, D. Probabilistic Linguistic WASPAS Method for Patients' Prioritization by Developing Prioritized Maclaurin Symmetric Mean Aggregation Operators. *Appl. Intell.* **2022**, *52*, 9537–9555. [CrossRef]
31. Singh, S.P.; Kundu, T.; Adhikari, A.; Basu, S. An Integrated Weighting-Based Modified WASPAS Methodology for Assessing Patient Satisfaction. In Proceedings of the 2020 International Conference on Decision Aid Sciences and Application (DASA), Sakheer, Bahrain, 8–9 November 2020; pp. 592–596.
32. Simić, V.; Lazarević, D.; Dobrodolac, M. Picture Fuzzy WASPAS Method for Selecting Last-Mile Delivery Mode: A Case Study of Belgrade. *Eur. Transp. Res. Rev.* **2021**, *13*, 43. [CrossRef]
33. Zhang, Y.; Ma, S.; Yang, H.; Lv, J.; Liu, Y. A Big Data Driven Analytical Framework for Energy-Intensive Manufacturing Industries. *J. Clean. Prod.* **2018**, *197*, 57–72. [CrossRef]
34. Do Nascimento Maêda, S.M.; Basílio, M.P.; De Araújo Costa, I.P.; Lellis Moreira, M.Â.; Dos Santos, M.; Gomes, C.F.S.; Pereire de Almeida, I.D.; de Araujo Costa, A.P. Investments in Times of Pandemics: An Approach by the SAPEVO-M-NC Method. In Proceedings of the 2nd Conference on Modern Management Based on Big Data, MMBD, Quanzhou, China, 8–11 November 2021; and Proceedings of the 3rd Conference on Machine Learning and Intelligent Systems, MLIS, Xiamen, China, 8–11 November 2021. pp. 162–168.
35. Zavadskas, E.K.; Turskis, Z.; Antucheviciene, J.; Zakarevicius, A. Optimization of Weighted Aggregated Sum Product Assessment. *Elektron. Elektrotechnika* **2012**, *122*, 3–6. [CrossRef]
36. R Core Team. *R: A Language and Environment for Statistical Computing*; R-Project: Vienna, Austria, 2021.
37. Campbell, M.; Campbell, M. Rstudio Projects. In *Learn RStudio IDE: Quick, Effective, and Productive Data Science*; Apress: Pune, India, 2019; pp. 39–48.
38. Corsi, A.; de Souza, F.F.; Pagani, R.N.; Kovaleski, J.L. Big Data Analytics as a Tool for Fighting Pandemics: A Systematic Review of Literature. *J. Ambient Intell. Humaniz. Comput.* **2021**, *12*, 9163–9180. [CrossRef]
39. Sievert, C. *Interactive Web-Based Data Visualization with R, Plotly, and Shiny*; CRC Press: Boca Raton, FL, USA, 2020.
40. Cosentino, V.; Izquierdo, J.L.C.; Cabot, J. A Systematic Mapping Study of Software Development with GitHub. *IEEE Access* **2017**, *5*, 7173–7192. [CrossRef]
41. Jardim, R.; dos Santos, M.; Neto, E.; Muradas, F.M.; Santiago, B.; Moreira, M. Design of a Framework of Military Defense System for Governance of Geoinformation. *Procedia Comput. Sci.* **2022**, *199*, 174–181. [CrossRef]
42. Zeileis, A.; Bivand, R.; Eddelbuettel, D.; Hornik, K.; Vialaneix, N. CRAN Task Views: The Next Generation. *arXiv* **2023**, arXiv:2305.17573.
43. Diaz-Balteiro, L.; González-Pachón, J.; Romero, C. Measuring Systems Sustainability with Multi-Criteria Methods: A Critical Review. *Eur. J. Oper. Res.* **2017**, *258*, 607–616. [CrossRef]
44. Barbara, F. WaspasWeb. Available online: https://flaviob.shinyapps.io/waspasWeb/ (accessed on 16 January 2023).
45. Garai, P. Mobile Phone Specifications and Prices. Available online: https://www.kaggle.com/datasets/pratikgarai/mobile-phone-specifications-and-prices (accessed on 16 January 2023).
46. Cinelli, M.; Kadziński, M.; Miebs, G.; Gonzalez, M.; Słowiński, R. Recommending Multiple Criteria Decision Analysis Methods with a New Taxonomy-Based Decision Support System. *Eur. J. Oper. Res.* **2022**, *302*, 633–651. [CrossRef]

47. Cinelli, M.; Kadziński, M.; Gonzalez, M.; Słowiński, R. How to Support the Application of Multiple Criteria Decision Analysis? Let Us Start with a Comprehensive Taxonomy. *Omega* **2020**, *96*, 102261. [CrossRef]
48. dos Santos, F.B.; dos Santos, M. Choice of Armored Vehicles on Wheels for the Brazilian Marine Corps Using PrOPPAGA. *Procedia Comput. Sci.* **2022**, *199*, 301–308. [CrossRef]
49. Wątróbski, J.; Bączkiewicz, A.; Sałabun, W. Pyrepo-Mcda—Reference Objects Based MCDA Software Package. *SoftwareX* **2022**, *19*, 101107. [CrossRef]

mathematics MDPI

Article

Fuzzy Weighted Pareto–Nash Equilibria of Multi-Objective Bi-Matrix Games with Fuzzy Payoffs and Their Applications

Wen Li, Deyi Li †, Yuqiang Feng *,† and Du Zou †

School of Science, Wuhan University of Science and Technology, Wuhan 430065, China; liwenmath@wust.edu.cn (W.L.); lideyi@wust.edu.cn (D.L.); zoudumath@wust.edu.cn (D.Z.)
* Correspondence: fengyuqiang@wust.edu.cn
† These authors contributed equally to this work.

Abstract: Based on our previous research, this paper further discusses the multi-objective bi-matrix game with fuzzy payoffs (MBGFP), which is a special case of the fuzzy constrained multi-objective game with fuzzy payoffs. We first prove that any bi-matrix game with interval payoffs (BGIP) has at least one Pareto–Nash equilibrium. Then, with the help of BGIP, we obtain the necessary and sufficient conditions for the existence of fuzzy Pareto–Nash equilibrium of MBGFP. Secondly, based on the bilinear programming method for calculating Nash equilibrium in crisp bi-matrix games, we established a bilinear programming method with parameters for calculating fuzzy Pareto–Nash equilibrium. By considering the importance of each objective to the players, MBGFP is transformed into a bi-matrix game with fuzzy payoffs (BGFP). Furthermore, we obtained the necessary and sufficient conditions for the existence of fuzzy weighted Pareto–Nash equilibrium and its calculation method. Finally, a practical example is used to illustrate the effectiveness of our proposed calculation method.

Keywords: multi-objective bi-matrix game; fuzzy payoffs; fuzzy weighted Pareto–Nash equilibrium; bi-matrix game with interval payoffs; bilinear programming model with parameters

MSC: 91A86; 47H10

1. Introduction

Matrix games are widely used in economics, management, sociology, political science, military science, and other fields. In many real-world situations, the information or the payoffs are imprecise, uncertain, or incomplete. In such cases, using the traditional matrix game becomes difficult as it relies on precise and complete information. Since fuzzy numbers can better describe the uncertainty of payoffs, matrix games with fuzzy payoffs are usually considered. Nowadays, the study of matrix games with fuzzy payoffs is an essential branch of game theory that helps in making the analysis and decision-making in various fields more practical and efficient.

Zadeh [1] first introduced the fuzzy set theory. Subsequently, Butnariu [2] first introduced fuzzy sets in non-cooperative games. Dubois and Prade [3] studied the two-player zero-sum game with fuzzy payoffs, and regarded the payoffs in the game as fuzzy numbers, which is an earlier paper on fuzzy matrix games. Campos [4] proposed a fuzzy linear programming method to solve the two-player zero-sum fuzzy game. This method is more suitable for solving the problem that the payoff is a triangular fuzzy number. Maeda [5] discussed the bi-matrix game with fuzzy payoffs. Based on the literature [6,7], Clemente [8] introduced the standard fuzzy orders to compare the fuzzy payoffs, and then studied the Pareto-optimal security strategies of zero-sum matrix games with fuzzy payoffs. Li [9] introduced an approach to computing fuzzy values of matrix games with single objectives and triangular fuzzy payoffs. Chandra and Aggarwal [10] wrote a note on the work of Li [9] for solving the two-player zero-sum games with payoffs of triangular fuzzy numbers and proposed a new methodology for solving such games. For more literature on fuzzy matrix games, see, e.g., Refs. [11–18].

Matrix games with multiple non-comparable objectives are called multi-objective matrix games. Zeleny [19] introduced a parameter vector and a vector with weighting coefficients, and analyzed the multi-objective two-player zero-sum game through parameter changes. Buckley [20] used the decision principles of Bellman and Zadeh [21] in fuzzy environments to formulate multi-objective non-cooperative games under uncertainty. Sakawa and Nishizaki [22] consider multi-objective two-player zero-sum matrix games with fuzzy payoffs and fuzzy goals. Utilizing a degree of attainment of the fuzzy goal, the corresponding max-min strategy of this game is obtained. Fernandez and Puerto [23] showed that a multi-objective zero-sum matrix game corresponds to a multi-objective linear programming problem, and verified that the effective solution set of the linear programming problem is consistent with the Pareto optimal security strategy set for one of the players in the original game. Based on Ref. [22], Nishizaki and Sakawa [24] examined fuzzy bi-matrix games incorporating fuzzy goals in single and multiple objective environments. Bigdeli and Hassanpour [25] discussed the multi-objective zero-sum matrix game with triangular fuzzy numbers. The game is converted to several multi-objective matrix games with interval payoffs by using the α-cuts of fuzzy payoffs. For more literature on multi-objective fuzzy matrix games, see, e.g., Refs. [20,22–25] and references therein. In addition, multi-objective games in which the payoff function is a fuzzy vector-valued function are more general game models. The definition of equilibrium, as well as the existence and stability of equilibrium, hold significant importance in this game. Based on the partial order of fuzzy vectors, Li et al. [26] proposed the concept of fuzzy Pareto–Nash equilibrium in fuzzy constrained multi-objective games with fuzzy payoffs. Furthermore, the existence and stability of fuzzy Pareto–Nash equilibrium are researched.

Moore [27] introduced the concept of interval analysis about interval numbers and functions with interval coefficients. Subsequently, many scholars further developed the theory of interval arithmetic and interval-valued functions, see Refs. [28–32] and references therein. Fei and Li [33] developed an effective bilinear programming method for solving bi-matrix games with interval payoffs. The current application of interval analysis in game theory is mainly to discuss the existence of equilibria for matrix games with interval payoffs and their calculation methods, see, e.g., Refs. [33–36].

As far as we know, there are few theoretical studies and applications on multi-objective bi-matrix games with fuzzy payoffs (MBGFP). Different from the classic (multi-objective) bi-matrix game, the player's expected payoff is a (fuzzy vector) fuzzy number in MBGFP. The order on the fuzzy number set is also different from the natural order of real numbers. It is a partial order. There are two existing research ideas on MBGFP. One research idea is to introduce fuzzy goals, construct two attainment degree functions as two players' payoff functions, and then transform the MBGFP into a two-player crisp game. Another research idea is to use the cut sets of fuzzy numbers to transform the MBGFP into some multi-objective bi-matrix games with interval payoffs, and then use the interval optimization method to solve the interval value of the game. However, there is no literature starting from the partial order on the fuzzy number (vector) set and directly establishing the equilibrium of the MBGFP under this partial order. Based on this, and based on our recent research in Ref. [26], this paper proposes the concept of fuzzy (weighted) Pareto–Nash equilibrium for MBGFP, and obtains calculation methods for these two types of equilibria.

This article is structured as follows. In Section 2, we review some basic terminology and related conclusions. In Section 3, we study the fuzzy Pareto–Nash equilibrium and fuzzy weighted Pareto–Nash equilibrium of MBGFP and show the relationship between them. First, the necessary and sufficient condition for the existence of fuzzy weighted Pareto–Nash equilibrium is given using the existence of Pareto–Nash equilibrium of BGIP. Second, the calculation method of fuzzy weighted Pareto–Nash equilibrium is obtained through the optimal solution of the bilinear programming problem with parameters. Based on Section 3, we specifically discuss the two-type two-company competition problem with triangular fuzzy payoffs in Section 4 and use Lingo software (v.19) to calculate the fuzzy weighted Pareto–Nash equilibrium of the game under a given grade of membership. Finally, the conclusion is presented in Section 5.

2. Preliminaries and Terminology

Throughout this paper, we write $\mathbb{R}$ for the set of all real numbers, $\mathcal{I}(\mathbb{R})$ for the set of all closed intervals in $\mathbb{R}$, and $\mathcal{F}(\mathbb{R})$ for the set of all fuzzy numbers in $\mathbb{R}$. We first review basic terminology on fuzzy numbers, fuzzy vectors, and some related conclusions. In order to study MBGFP, we will propose the concept of Pareto–Nash equilibrium of a bi-matrix game with interval payoffs (BGIP), and prove that for any BGIP there is at least one Pareto–Nash equilibria in the sense of mixed strategies.

Let X denote a universal set. A fuzzy subset $\widetilde{a}$ of X is defined by its membership function $\mu_{\widetilde{a}} : X \to [0,1]$, which assigns to each element $x \in X$ a real number $\mu_{\widetilde{a}}(x)$ in the interval $[0,1]$. Especially, $\mu_{\widetilde{a}}(x)$ is the grade of membership of x in the set $\widetilde{a}$. The α-cut of the fuzzy set $\widetilde{a}$, denoted by $\widetilde{a}^{\alpha}$, is a set defined by

$$\widetilde{a}^{\alpha} = \{x \in X \mid \mu_{\widetilde{a}}(x) \geq \alpha\},$$

when $\alpha \in (0,1]$. And $\widetilde{a}^{0} = cl\{x \in X \mid \mu_{\widetilde{a}}(x) > 0\}$, where cl denotes the closure of sets. For more about the properties of fuzzy sets, please refer to Refs. [1,3].

Definition 1 (See Ref. [37]). *A fuzzy number $\widetilde{A}$ is a fuzzy set on $\mathbb{R}$, whose membership function $\mu_{\widetilde{A}}(\cdot) : \mathbb{R} \to [0,1]$ satisfies the following conditions:*

1. $\mu_{\widetilde{A}}(x) = 0$ *for all* $x \in (-\infty, c]$,
2. $\mu_{\widetilde{A}}(\cdot)$ *is strictly increasing and continuous on* $[c,a]$,
3. $\mu_{\widetilde{A}}(x) = 1$ *for all* $x \in [a,b]$,
4. $\mu_{\widetilde{A}}(\cdot)$ *is strictly decreasing and continuous on* $[b,d]$,
5. $\mu_{\widetilde{A}}(x) = 0$ *for all* $x \in [d,+\infty)$,

where $-\infty < c \leq a \leq b \leq d < +\infty$.

For $\widetilde{A} \in \mathcal{F}(\mathbb{R})$, the α-cut of $\widetilde{A}$ is a closed interval, i.e., $\widetilde{A}^{\alpha} = [\widetilde{A}_L(\alpha), \widetilde{A}_R(\alpha)]$, where $\widetilde{A}_L(\alpha) = \inf\{x \in \mathbb{R} : \mu_{\widetilde{A}}(x) \geq \alpha\}$ and $\widetilde{A}_R(\alpha) = \sup\{x \in \mathbb{R} : \mu_{\widetilde{A}}(x) \geq \alpha\}$.

Suppose that the membership functions of fuzzy numbers $\widetilde{A}$ and $\widetilde{B}$ are represented as follows:

$$\mu_{\widetilde{A}}(x) = \begin{cases} \mu^L_{\widetilde{A}}(x), & x \in [a_1, a_2] \\ 1, & x \in [a_2, a_3] \\ \mu^R_{\widetilde{A}}(x), & x \in [a_3, a_4] \\ 0, & \text{otherwise} \end{cases}$$

and

$$\mu_{\widetilde{B}}(x) = \begin{cases} \mu^L_{\widetilde{B}}(x), & x \in [b_1, b_2] \\ 1, & x \in [b_2, b_3] \\ \mu^R_{\widetilde{B}}(x), & x \in [b_3, b_4] \\ 0, & \text{otherwise} \end{cases}$$

where $-\infty < a_1 \leq a_2 \leq a_3 \leq a_4 < +\infty$, $-\infty < b_1 \leq b_2 \leq b_3 \leq b_4 < +\infty$. Then, the sum of $\widetilde{A}$ and $\widetilde{B}$ is denoted by $\widetilde{A} + \widetilde{B}$, whose membership function is defined by

$$\mu_{\widetilde{A}+\widetilde{B}}(x) = \begin{cases} \mu^L_{\widetilde{A}+\widetilde{B}}(x), & x \in [a_1 + b_1, a_2 + b_2] \\ 1, & x \in [a_2 + b_2, a_3 + b_3] \\ \mu^R_{\widetilde{A}+\widetilde{B}}(x), & x \in [a_3 + b_3, a_4 + b_4] \\ 0, & \text{otherwise} \end{cases}$$

where $(\mu^L_{\widetilde{A}+\widetilde{B}})^{-1}(\alpha) = (\mu^L_{\widetilde{A}})^{-1}(\alpha) + (\mu^L_{\widetilde{B}})^{-1}(\alpha)$ and $(\mu^R_{\widetilde{A}+\widetilde{B}})^{-1}(\alpha) = (\mu^R_{\widetilde{A}})^{-1}(\alpha) + (\mu^R_{\widetilde{B}})^{-1}(\alpha)$, for all $\alpha \in [0,1]$.

The product of a scalar k and a fuzzy number $\widetilde{A}$, denoted by $k\widetilde{A}$, is defined as follows:
Case 1: $k \geq 0$, then

$$\mu_{k\widetilde{A}}(x) = \begin{cases} \mu^L_{k\widetilde{A}}(x), & x \in [ka_1, ka_2] \\ 1, & x \in [ka_2, ka_3] \\ \mu^R_{k\widetilde{A}}(x), & x \in [ka_3, ka_4] \\ 0, & \text{otherwise} \end{cases}$$

where $(\mu^L_{k\widetilde{A}})^{-1}(\alpha) = k \cdot (\mu^L_{\widetilde{A}})^{-1}(\alpha)$, $(\mu^R_{k\widetilde{A}})^{-1}(\alpha) = k \cdot (\mu^R_{\widetilde{A}})^{-1}(\alpha)$, for all $\alpha \in [0,1]$.
Case 2: $k < 0$, then

$$\mu_{k\widetilde{A}}(x) = \begin{cases} \mu^L_{k\widetilde{A}}(x), & x \in [ka_4, ka_3] \\ 1, & x \in [ka_3, ka_2] \\ \mu^R_{k\widetilde{A}}(x), & x \in [ka_2, ka_1] \\ 0, & \text{otherwise} \end{cases}$$

where $(\mu^L_{k\widetilde{A}})^{-1}(\alpha) = k \cdot (\mu^R_{\widetilde{A}})^{-1}(\alpha)$, $(\mu^R_{k\widetilde{A}})^{-1}(\alpha) = k \cdot (\mu^L_{\widetilde{A}})^{-1}(\alpha)$, for all $\alpha \in [0,1]$.

From the above definitions, we can see that $(\widetilde{A} + \widetilde{B})^\alpha = \widetilde{A}^\alpha + \widetilde{B}^\alpha$ and $(k\widetilde{A})^\alpha = k\widetilde{A}^\alpha$, for all $\alpha \in [0,1]$. Moreover, the set $\mathcal{F}(\mathbb{R})$ is closed under addition and scalar multiplication.

Write $\mathcal{F}(\mathbb{R})^d$ for the family of all d-dimensional fuzzy vectors. For $\widetilde{\mathbf{A}} = (\widetilde{A}_1, \ldots, \widetilde{A}_d) \in \mathcal{F}(\mathbb{R})^d$ and $\alpha \in [0,1]$. Then the α-cut of $\widetilde{\mathbf{A}}$ is an interval vector, that is,

$$\widetilde{\mathbf{A}}^\alpha = (\widetilde{A}_1^\alpha, \ldots, \widetilde{A}_d^\alpha) \in \mathcal{I}(\mathbb{R})^d,$$

where $\mathcal{I}(\mathbb{R})^d$ is the set of all d-dimensional interval vector. It is worth noting that every $\widetilde{\mathbf{A}}^\alpha$ corresponds to two d-dimensional vectors. They are recorded as

$$\widetilde{\mathbf{A}}_L(\alpha) = ((\widetilde{A}_1)_L(\alpha), \ldots, (\widetilde{A}_d)_L(\alpha))$$

and

$$\widetilde{\mathbf{A}}_R(\alpha) = ((\widetilde{A}_1)_R(\alpha), \ldots, (\widetilde{A}_d)_R(\alpha)).$$

Similarly, the addition and scalar multiplication of fuzzy vectors are defined for $\widetilde{\mathbf{A}} = (\widetilde{A}_1, \ldots, \widetilde{A}_d), \widetilde{\mathbf{B}} = (\widetilde{B}_1, \ldots, \widetilde{B}_d) \in \mathcal{F}(\mathbb{R})^d, k \in \mathbb{R}$ by

$$\widetilde{\mathbf{A}} + \widetilde{\mathbf{B}} = (\widetilde{A}_1 + \widetilde{B}_1, \ldots, \widetilde{A}_d + \widetilde{B}_d)$$

and

$$k \cdot \widetilde{\mathbf{A}} = (k \cdot \widetilde{A}_1, \ldots, k \cdot \widetilde{A}_d).$$

Definition 2 (See Ref. [26]). *Let $\widetilde{\mathbf{A}}$ and $\widetilde{\mathbf{B}}$ be two elements of $\mathcal{F}(\mathbb{R})^d$.*

1. *$\widetilde{\mathbf{A}}$ is said to be dominated by $\widetilde{\mathbf{B}}$ from below if $\widetilde{\mathbf{B}}_L(\alpha) \leq \widetilde{\mathbf{A}}_L(\alpha)$ and $\widetilde{\mathbf{B}}_R(\alpha) \leq \widetilde{\mathbf{A}}_R(\alpha)$ for all $\alpha \in [0,1]$, and we rewrite this property as $\widetilde{\mathbf{B}} \preceq \widetilde{\mathbf{A}}$. Otherwise, we write $\widetilde{\mathbf{B}} \npreceq \widetilde{\mathbf{A}}$.*
2. *$\widetilde{\mathbf{A}}$ is said to be strictly dominated by $\widetilde{\mathbf{B}}$ from below if $\widetilde{\mathbf{B}}_L(\alpha) < \widetilde{\mathbf{A}}_L(\alpha)$ and $\widetilde{\mathbf{B}}_R(\alpha) < \widetilde{\mathbf{A}}_R(\alpha)$ for all $\alpha \in [0,1]$, and we rewrite this property as $\widetilde{\mathbf{B}} \prec \widetilde{\mathbf{A}}$. Otherwise, we write $\widetilde{\mathbf{B}} \nprec \widetilde{\mathbf{A}}$.*

In particular, when $d = 1$, the partial order $\preceq$ in Definition 2 is precisely equivalent to the fuzzy maximum order in Ref. [7].

Next, consider a bi-matrix games with interval payoffs (BGIP)

$$(\mathcal{A},\mathcal{B}) = \begin{pmatrix} (a_{11}, b_{11}) & (a_{12}, b_{12}) & \cdots & (a_{1n}, b_{1n}) \\ (a_{21}, b_{21}) & (a_{22}, b_{22}) & \cdots & (a_{2n}, b_{2n}) \\ \cdots & \cdots & \cdots & \cdots \\ (a_{m1}, b_{m1}) & (a_{m2}, b_{m2}) & \cdots & (a_{mn}, b_{mn}) \end{pmatrix}$$

where $a_{ij} = [\underline{a}_{ij}, \overline{a}_{ij}], b_{ij} = [\underline{b}_{ij}, \overline{b}_{ij}]$ $(i = 1, \ldots, m; j = 1, \cdots, n)$. Let $\underline{\mathcal{A}} = (\underline{a}_{ij})_{n\times m}$, $\overline{\mathcal{A}} = (\overline{a}_{ij})_{n\times m}$, $\underline{\mathcal{B}} = (\underline{b}_{ij})_{n\times m}$ and $\overline{\mathcal{B}} = (\overline{b}_{ij})_{n\times m}$. The sets of all mixed strategies for Players I and II are defined as $\mathcal{S}^m = \{(x_1, \ldots, x_m) \in \mathbb{R}^m \mid \sum_{i=1}^m x_i = 1, x_i \geq 0, i = 1, \ldots, m\}$ and $\mathcal{S}^n = \{(y_1, \ldots, y_n) \in \mathbb{R}^n \mid \sum_{j=1}^n y_j = 1, y_j \geq 0, j = 1, \ldots, n\}$. The interval-valued expected payoffs for Players I and II are defined as $x^T\mathcal{A}y = [x^T\underline{\mathcal{A}}y, x^T\overline{\mathcal{A}}y]$ and $x^T\mathcal{B}y = [x^T\underline{\mathcal{B}}y, x^T\overline{\mathcal{B}}y]$ for $(x, y) \in \mathcal{S}^m \times \mathcal{S}^n$, where $x^T\underline{\mathcal{A}}y = \sum_{i=1}^m \sum_{j=1}^n \underline{a}_{ij}x_iy_j$, $x^T\overline{\mathcal{A}}y = \sum_{i=1}^m \sum_{j=1}^n \overline{a}_{ij}x_iy_j$, $x^T\underline{\mathcal{B}}y = \sum_{i=1}^m \sum_{j=1}^n \underline{b}_{ij}x_iy_j$, and $x^T\overline{\mathcal{B}}y = \sum_{i=1}^m \sum_{j=1}^n \overline{b}_{ij}x_iy_j$.

For $a = [\underline{a}, \overline{a}]$, $b = [\underline{b}, \overline{b}]$, if $\underline{a} \leq \underline{b}$ and $\overline{a} \leq \overline{b}$, then we write $a \preceq b$. If $\underline{a} < \underline{b}$ and $\overline{a} < \overline{b}$, then we write $a \prec b$. The relation $\preceq$ is a partial order on $\mathcal{I}(\mathbb{R})$. However, two intervals with true inclusion relations are incomparable under the partial order $\preceq$. For example, $[3,4] \subset [2,5]$, but $[3,4] \npreceq [2,5]$ and $[2,5] \npreceq [3,4]$. So, $[3,4]$ and $[2,5]$ are incomparable.

Now we will introduce the equilibrium of bi-matrix games with interval payoffs $(\mathcal{A},\mathcal{B})$.

Definition 3. *A strategy profile $(x^*, y^*) \in \mathcal{S}^m \times \mathcal{S}^n$ is called a Pareto–Nash equilibrium of $(\mathcal{A},\mathcal{B})$, if for each $(x,y) \in \mathcal{S}^m \times \mathcal{S}^n$ such that*

$$x^{*T}\underline{\mathcal{A}}y^* \geq x^T\underline{\mathcal{A}}y^* \ (\text{or } x^{*T}\overline{\mathcal{A}}y^* \geq x^T\overline{\mathcal{A}}y^*)$$

and

$$x^{*T}\underline{\mathcal{B}}y^* \geq x^{*T}\underline{\mathcal{B}}y \ (\text{or } x^{*T}\overline{\mathcal{B}}y^* \geq x^{*T}\overline{\mathcal{B}}y).$$

Lemma 1. *Assume that $(x^*, y^*) \in \mathcal{S}^m \times \mathcal{S}^n$ is a Nash equilibrium to one of the four crisp bi-matrix games $(\underline{\mathcal{A}},\underline{\mathcal{B}})$, $(\underline{\mathcal{A}},\overline{\mathcal{B}})$, $(\overline{\mathcal{A}},\underline{\mathcal{B}})$ and $(\overline{\mathcal{A}},\overline{\mathcal{B}})$. Then, (x^*, y^*) is also a Pareto–Nash equilibrium of BGIP $(\mathcal{A},\mathcal{B})$.*

Proof. Without loss of generality, assume that $(x^*, y^*) \in \mathcal{S}^m \times \mathcal{S}^n$ is a Nash equilibrium of the crisp bi-matrix game $(\underline{\mathcal{A}},\underline{\mathcal{B}})$. Then

$$x^{*T}\underline{\mathcal{A}}y^* = \max_{x\in\mathcal{S}^m} x^T\underline{\mathcal{A}}y^* \text{ and } x^{*T}\underline{\mathcal{B}}y^* = \max_{y\in\mathcal{S}^n} x^{*T}\underline{\mathcal{B}}y.$$

That is, for each $(x,y) \in \mathcal{S}^m \times \mathcal{S}^n$, we have $x^{*T}\underline{\mathcal{A}}y^* \geq x^T\underline{\mathcal{A}}y^*$ and $x^{*T}\underline{\mathcal{B}}y^* \geq x^{*T}\underline{\mathcal{B}}y$. From Definition 3, it follows that (x^*, y^*) is a Pareto–Nash equilibrium of BGIP $(\mathcal{A},\mathcal{B})$. Similarly, the other three cases can be verified. □

In the sense of mixed strategies, any crisp bi-matrix game must have a Nash equilibrium [38]. According to Lemma 1, there are at least four Pareto–Nash equilibria for any bi-matrix game with interval payoffs.

3. Multi-Objective Bi-Matrix Games with Fuzzy Payoffs

In this section, we discuss the fuzzy Pareto–Nash equilibrium (FPNE) and fuzzy weighted Pareto–Nash equilibrium (FWPNE) of MBGFP. Moreover, we obtain the relationship between FPNE and FWPNE, and provide the necessary and sufficient conditions for the existence of these two equilibria and their calculation methods, respectively.

We focus on the multi-objective bi-matrix game with fuzzy payoffs (MBGFP) $\Gamma = \{(\mathcal{A}_k, \mathcal{B}_k)\}_{k\in\{1,\dots,d\}}$ given by

$$(\mathcal{A}_k, \mathcal{B}_k) = \begin{pmatrix} (\widetilde{a}^k_{11}, \widetilde{b}^k_{11}) & (\widetilde{a}^k_{12}, \widetilde{b}^k_{12}) & \cdots & (\widetilde{a}^k_{1n}, \widetilde{b}^k_{1n}) \\ (\widetilde{a}^k_{21}, \widetilde{b}^k_{21}) & (\widetilde{a}^k_{22}, \widetilde{b}^k_{22}) & \cdots & (\widetilde{a}^k_{2n}, \widetilde{b}^k_{2n}) \\ \cdots & \cdots & \cdots & \cdots \\ (\widetilde{a}^k_{m1}, \widetilde{b}^k_{m1}) & (\widetilde{a}^k_{m2}, \widetilde{b}^k_{m2}) & \cdots & (\widetilde{a}^k_{mn}, \widetilde{b}^k_{mn}) \end{pmatrix} \text{ for } k = 1, \dots, d,$$

where all the components $(\widetilde{a}^k_{ij}, \widetilde{b}^k_{ij}) \in \mathcal{F}(\mathbb{R}) \times \mathcal{F}(\mathbb{R})$. The sets of all mixed strategies for Players I and II are, respectively,

$$\mathcal{S}^m = \{(x_1, \dots, x_m) \in \mathbb{R}^m \mid \sum_{i=1}^{m} x_i = 1, x_i \geq 0, i = 1, \dots, m\}$$

and

$$\mathcal{S}^n = \{(y_1, \dots, y_n) \in \mathbb{R}^n \mid \sum_{i=1}^{n} y_i = 1, y_i \geq 0, i = 1, \dots, n\}.$$

Moreover, the fuzzy-vector-valued expected payoffs for Players I and II are given, respectively, by

$$\widetilde{F}_1(x, y) = (x^T \mathcal{A}_1 y, \dots, x^T \mathcal{A}_d y)$$

and

$$\widetilde{F}_2(x, y) = (x^T \mathcal{B}_1 y, \dots, x^T \mathcal{B}_d y),$$

for $(x, y) \in \mathcal{S}^m \times \mathcal{S}^n$. Clearly, $(\mathcal{S}^m, \mathcal{S}^n, \widetilde{F}_1, \widetilde{F}_2)$ is a two-player multi-objective game with fuzzy payoffs.

For $\alpha_1, \alpha_2 \in [0, 1]$, let $\mathcal{A}^{\alpha_1}_k = ((\widetilde{a}^k_{ij})^{\alpha_1})_{m\times n}$ and $\mathcal{B}^{\alpha_2}_k = ((\widetilde{b}^k_{ij})^{\alpha_2})_{m\times n}$. Then $\Gamma(\alpha_1, \alpha_2) = \{(\mathcal{A}^{\alpha_1}_k, \mathcal{B}^{\alpha_2}_k)\}_{k\in\{1,\dots,d\}}$ constitutes a multi-objective bi-matrix game with interval payoffs (MBGIP). The interval-vector-valued expected payoffs for Players I and II are

$$\widetilde{F}^{\alpha_1}_1(x, y) = ([x^T \underline{\mathcal{A}}^{\alpha_1}_1 y, x^T \overline{\mathcal{A}}^{\alpha_1}_1 y], \dots, [x^T \underline{\mathcal{A}}^{\alpha_1}_d y, x^T \overline{\mathcal{A}}^{\alpha_1}_d y])$$

and

$$\widetilde{F}^{\alpha_2}_2(x, y) = ([x^T \underline{\mathcal{B}}^{\alpha_2}_1 y, x^T \overline{\mathcal{B}}^{\alpha_2}_1 y], \dots, [x^T \underline{\mathcal{B}}^{\alpha_2}_d y, x^T \overline{\mathcal{B}}^{\alpha_2}_d y])$$

respectively, where for $k = 1, \dots, d$,

$$x^T \underline{\mathcal{A}}^{\alpha_1}_k y = \sum_{i=1}^{m} \sum_{j=1}^{n} (\widetilde{a}^k_{ij})_L(\alpha_1) x_i y_j, \; x^T \overline{\mathcal{A}}^{\alpha_1}_k y = \sum_{i=1}^{m} \sum_{j=1}^{n} (\widetilde{a}^k_{ij})_R(\alpha_1) x_i y_j$$

and

$$x^T \underline{\mathcal{B}}^{\alpha_2}_k y = \sum_{i=1}^{m} \sum_{j=1}^{n} (\widetilde{b}^k_{ij})_L(\alpha_2) x_i y_j, \; x^T \overline{\mathcal{B}}^{\alpha_2}_k y = \sum_{i=1}^{m} \sum_{j=1}^{n} (\widetilde{b}^k_{ij})_R(\alpha_2) x_i y_j.$$

3.1. Fuzzy Pareto–Nash Equilibria

We require the following definition of the fuzzy Pareto–Nash equilibrium of MBGFP.

Definition 4. *A strategy profile $(x^*, y^*) \in \mathcal{S}^m \times \mathcal{S}^n$ is called a FPNE of Γ, if for all $(x, y) \in \mathcal{S}^m \times \mathcal{S}^n$, such that*

$$\widetilde{F}_1(x^*, y^*) \not\prec \widetilde{F}_1(x, y^*) \text{ and } \widetilde{F}_2(x^*, y^*) \not\prec \widetilde{F}_2(x^*, y).$$

Theorem 1. *A strategy profile $(x^*, y^*) \in \mathcal{S}^m \times \mathcal{S}^n$ is a FPNE of $\Gamma = \{(\mathcal{A}_k, \mathcal{B}_k)\}_{k\in\{1,\dots,d\}}$ if and only if there exist $\alpha_1, \alpha_2 \in [0, 1]$ and $s, t \in \{1, \dots, d\}$ such that (x^*, y^*) is a Pareto–Nash equilibrium of a BGIP $(\mathcal{A}^{\alpha_1}_s, \mathcal{B}^{\alpha_2}_t)$.*

Proof. Assume that (x^*, y^*) is a FPNE of Γ. By Definition 4, for each $(x, y) \in \mathcal{S}^m \times \mathcal{S}^n$, we have

$$\widetilde{F}_1(x^*, y^*) \not\prec \widetilde{F}_1(x, y^*) \text{ and } \widetilde{F}_2(x^*, y^*) \not\prec \widetilde{F}_2(x^*, y).$$

Added Definition 2, there exist $\alpha_1, \alpha_2 \in [0, 1]$ such that

$$(\widetilde{F}_1^{\alpha_1})_L(x^*, y^*) \not< (\widetilde{F}_1^{\alpha_1})_L(x, y^*) \text{ (or } (\widetilde{F}_1^{\alpha_1})_R(x^*, y^*) \not< (\widetilde{F}_1^{\alpha_1})_R(x, y^*))$$

and

$$(\widetilde{F}_2^{\alpha_2})_L(x^*, y^*) \not< (\widetilde{F}_2^{\alpha_2})_L(x^*, y) \text{ (or } (\widetilde{F}_2^{\alpha_2})_R(x^*, y^*) \not< (\widetilde{F}_2^{\alpha_2})_R(x^*, y)).$$

Furthermore, there exist $s, t \in \{1, \ldots, d\}$ such that

$$x^{*T} \underline{\mathcal{A}}_s^{\alpha_1} y^* \geq x^T \underline{\mathcal{A}}_s^{\alpha_1} y^* \text{ (or } x^{*T} \overline{\mathcal{A}}_s^{\alpha_1} y^* \geq x^T \overline{\mathcal{A}}_s^{\alpha_1} y^*)$$

and

$$x^{*T} \underline{\mathcal{B}}_t^{\alpha_2} y^* \geq x^{*T} \underline{\mathcal{B}}_t^{\alpha_2} y \text{ (or } x^{*T} \overline{\mathcal{B}}_t^{\alpha_2} y^* \geq x^{*T} \overline{\mathcal{B}}_t^{\alpha_2} y),$$

for all $(x, y) \in \mathcal{S}^m \times \mathcal{S}^n$. So (x^*, y^*) is a Pareto–Nash equilibrium of $(\mathcal{A}_s^{\alpha_1}, \mathcal{B}_t^{\alpha_2})$.

Finally, the sufficient part is obvious, according to Definitions 3 and 4. □

Fei and Li [33] proposed a bilinear programming method to solve the Nash equilibrium of the crisp bi-matrix game. With the help of Theorem 1 and Lemma 1, we obtain the following calculation method for FPNE.

Theorem 2. *The strategy profile $(x^*, y^*) \in \mathcal{S}^m \times \mathcal{S}^n$ is a FPNE of $\Gamma = \{(\mathcal{A}_k, \mathcal{B}_k)\}_{k \in \{1,\ldots,d\}}$, if and only if there exist $\alpha_1, \alpha_2 \in [0, 1]$ and $s, t \in \{1, \ldots, d\}$, such that it is an optimal solution to one of the following four bilinear programming models with two parameters:*

$$\max\{x^T \underline{\mathcal{A}}_s^{\alpha_1} y + x^T \underline{\mathcal{B}}_t^{\alpha_2} y - \underline{\mu}(\alpha_1) - \underline{\nu}(\alpha_2)\}$$
$$\text{s.t.} \begin{cases} \underline{\mathcal{A}}_s^{\alpha_1} y \leq \underline{\mu}(\alpha_1) e^m, \\ \underline{\mathcal{B}}_t^{\alpha_2 T} x \leq \underline{\nu}(\alpha_2) e^n, \\ x^T e^m = 1, \\ y^T e^n = 1, \\ x, y \geq 0, \\ \alpha_1, \alpha_2 \in [0, 1]; \end{cases} \tag{1}$$

$$\max\{x^T \underline{\mathcal{A}}_s^{\alpha_1} y + x^T \overline{\mathcal{B}}_t^{\alpha_2} y - \underline{\mu}(\alpha_1) - \overline{\nu}(\alpha_2)\}$$
$$\text{s.t.} \begin{cases} \underline{\mathcal{A}}_s^{\alpha_1} y \leq \underline{\mu}(\alpha_1) e^m, \\ \overline{\mathcal{B}}_t^{\alpha_2 T} x \leq \overline{\nu}(\alpha_2) e^n, \\ x^T e^m = 1, \\ y^T e^n = 1, \\ x, y \geq 0, \\ \alpha_1, \alpha_2 \in [0, 1]; \end{cases} \tag{2}$$

$$\max\{x^T \overline{\mathcal{A}}_s^{\alpha_1} y + x^T \underline{\mathcal{B}}_t^{\alpha_2} y - \overline{\mu}(\alpha_1) - \underline{\nu}(\alpha_2)\}$$
$$\text{s.t.} \begin{cases} \overline{\mathcal{A}}_s^{\alpha_1} y \leq \overline{\mu}(\alpha_1) e^m, \\ \underline{\mathcal{B}}_t^{\alpha_2 T} x \leq \underline{\nu}(\alpha_2) e^n, \\ x^T e^m = 1, \\ y^T e^n = 1, \\ x, y \geq 0, \\ \alpha_1, \alpha_2 \in [0, 1]; \end{cases} \tag{3}$$

$$\max\{x^T\overline{\mathcal{A}}_s^{\alpha_1}y + x^T\overline{\mathcal{B}}_t^{\alpha_2}y - \overline{\mu}(\alpha_1) - \overline{\nu}(\alpha_2)\}$$
$$s.t.\begin{cases} \overline{\mathcal{A}}_s^{\alpha_1}y \le \overline{\mu}(\alpha_1)e^m, \\ \overline{\mathcal{B}}_t^{\alpha_2 T}x \le \overline{\nu}(\alpha_2)e^n, \\ x^Te^m = 1, \\ y^Te^n = 1, \\ x, y \ge 0, \\ \alpha_1, \alpha_2 \in [0,1]; \end{cases} \tag{4}$$

where $e^m = (\underbrace{1, 1, \cdots, 1}_{m})^T$ *and* $e^n = (\underbrace{1, 1, \cdots, 1}_{n})^T$.

Proof. Assume that $(x^*, y^*) \in \mathcal{S}^m \times \mathcal{S}^n$ is a FPNE of Γ. By Theorem 1, there are $\alpha_1, \alpha_2 \in [0,1]$ and $s, t \in \{1, \ldots, d\}$, such that (x^*, y^*) is a Pareto–Nash equilibrium of BGIP $(\mathcal{A}_s^{\alpha_1}, \mathcal{B}_t^{\alpha_2})$. That is, for each $(x, y) \in \mathcal{S}^m \times \mathcal{S}^n$, at least one of the following four conditions are satisfied:

(1) $x^{*T}\underline{\mathcal{A}}_s^{\alpha_1}y^* \ge x^T\underline{\mathcal{A}}_s^{\alpha_1}y^*$ and $x^{*T}\underline{\mathcal{B}}_t^{\alpha_2}y^* \ge x^{*T}\underline{\mathcal{B}}_t^{\alpha_2}y$;
(2) $x^{*T}\underline{\mathcal{A}}_s^{\alpha_1}y^* \ge x^T\underline{\mathcal{A}}_s^{\alpha_1}y^*$ and $x^{*T}\overline{\mathcal{B}}_t^{\alpha_2}y^* \ge x^{*T}\overline{\mathcal{B}}_t^{\alpha_2}y$;
(3) $x^{*T}\overline{\mathcal{A}}_s^{\alpha_1}y^* \ge x^T\overline{\mathcal{A}}_s^{\alpha_1}y^*$ and $x^{*T}\underline{\mathcal{B}}_t^{\alpha_2}y^* \ge x^{*T}\underline{\mathcal{B}}_t^{\alpha_2}y$;
(4) $x^{*T}\overline{\mathcal{A}}_s^{\alpha_1}y^* \ge x^T\overline{\mathcal{A}}_s^{\alpha_1}y^*$ and $x^{*T}\overline{\mathcal{B}}_t^{\alpha_2}y^* \ge x^{*T}\overline{\mathcal{B}}_t^{\alpha_2}y$.

Without loss of generality, we assume that (1) holds. Let

$$\underline{\mu}^*(\alpha_1) = x^{*T}\underline{\mathcal{A}}_s^{\alpha_1}y^*,\ \overline{\mu}^*(\alpha_1) = x^{*T}\overline{\mathcal{A}}_s^{\alpha_1}y^*$$

and

$$\underline{\nu}^*(\alpha_2) = x^{*T}\underline{\mathcal{B}}_t^{\alpha_2}y^*,\ \overline{\nu}^*(\alpha_2) = x^{*T}\overline{\mathcal{B}}_t^{\alpha_2}y^*.$$

Therefore, $(x^*, y^*, \underline{\mu}^*(\alpha_1), \underline{\nu}^*(\alpha_2))$ is an optimal solution of (1).

Conversely, assume that (x^*, y^*) associated with $(\underline{\mu}^*(\alpha_1), \underline{\nu}^*(\alpha_2))$ is an optimal solution of (1). For each $(x, y) \in \mathcal{S}^m \times \mathcal{S}^n$, we have

$$x^T\underline{\mathcal{A}}_s^{\alpha_1}y^* \le \underline{\mu}^*(\alpha_1) \text{ and } x^{*T}\underline{\mathcal{B}}_t^{\alpha_2}y \le \underline{\nu}^*(\alpha_2). \tag{5}$$

Furthermore,

$$x^{*T}\underline{\mathcal{A}}_s^{\alpha_1}y^* + x^{*T}\underline{\mathcal{B}}_t^{\alpha_2}y^* - \underline{\mu}^*(\alpha_1) - \underline{\nu}^*(\alpha_2) \le 0.$$

If the objective function value of (1) is 0 at $(x^*, y^*, \underline{\mu}^*(\alpha_1), \underline{\nu}^*(\alpha_2))$, then

$$\underline{\mu}^*(\alpha_1) = x^{*T}\underline{\mathcal{A}}_s^{\alpha_1}y^* \text{ and } \underline{\nu}^*(\alpha_2) = x^{*T}\underline{\mathcal{B}}_t^{\alpha_2}y^*. \tag{6}$$

For each $(x, y) \in \mathcal{S}^m \times \mathcal{S}^n$, from (5) and (6) it follows that

$$x^{*T}\underline{\mathcal{A}}_s^{\alpha_1}y^* \ge x^T\underline{\mathcal{A}}_s^{\alpha_1}y^* \text{ and } x^{*T}\underline{\mathcal{B}}_t^{\alpha_2}y^* \ge x^{*T}\underline{\mathcal{B}}_t^{\alpha_2}y.$$

Therefore, (x^*, y^*) is a Pareto–Nash equilibrium of BGIP $(\mathcal{A}_s^{\alpha_1}, \mathcal{B}_t^{\alpha_2})$. According to Theorem 1, (x^*, y^*) is a FPNE of Γ.

The other three cases can be verified in the same way. □

Remark 1. *According to the proof of the sufficiency of Theorem 2, we can obtain that the optimal solution that makes the objective function value of (1) equal to 0 is the FPNE of* Γ. *The other three bilinear programming models have similar conclusions.*

3.2. Fuzzy Weighted Pareto–Nash Equilibria

Next, we consider the fuzzy weighted Pareto–Nash equilibrium in MBGFP. Suppose that

$$\Lambda = \left\{(\lambda_1, \ldots, \lambda_d) \in \mathbb{R}^d \mid \sum_{k=1}^{d} \lambda_k = 1, \lambda_k \geq 0, k = 1, \ldots, d\right\}.$$

For $\lambda, \eta \in \Lambda$, the fuzzy weighted expected payoff for Player I is defined by

$$\widetilde{F}_{1\lambda}(x, y) = \sum_{k=1}^{d} x^T(\lambda_k \mathcal{A}_k)y,$$

where each component λ_k of the vector λ can be interpreted as the relative importance of the k-th objective to Player I.

Similarly, the fuzzy weighted expected payoff for Player II is defined by

$$\widetilde{F}_{2\eta}(x, y) = \sum_{k=1}^{d} x^T(\eta_k \mathcal{B}_k)y,$$

where the component η_k of the vector η can be interpreted as the relative importance of the k-th objective to Player II.

Let $\mathcal{A} = \sum_{k=1}^{d} \lambda_k \mathcal{A}_k$ and $\mathcal{B} = \sum_{k=1}^{d} \eta_k \mathcal{B}_k$. Then, $\Gamma_{\lambda\eta} = (\mathcal{A}, \mathcal{B})$ is a bi-matrix game with fuzzy payoffs (BGFP). A fuzzy Pareto–Nash equilibrium of $(\mathcal{A}, \mathcal{B})$ is called a fuzzy weighted Pareto–Nash equilibrium of Γ with weights λ and η. In particular, if $\lambda = e_s$, $\eta = e_t$ $(s, t \in \{1, \ldots, d\})$, then $\mathcal{A} = \mathcal{A}_s$ and $\mathcal{B} = \mathcal{B}_s$, where e_s and e_t are the standard unit vectors in $\mathbb{R}^d$. For this case, the fuzzy weighted Pareto–Nash equilibrium of Γ is its fuzzy Pareto–Nash equilibrium. Thus, a fuzzy Pareto–Nash equilibrium of Γ is a special fuzzy weighted Pareto–Nash- equilibrium of Γ with weights e_s and e_t.

Theorem 3. *A strategy profile $(x^*, y^*) \in \mathcal{S}^m \times \mathcal{S}^n$ is a FWPNE of Γ with weights λ and η if and only if there exist $\alpha_1, \alpha_2 \in [0, 1]$ such that (x^*, y^*) is a Pareto–Nash equilibrium of $(\mathcal{A}^{\alpha_1}, \mathcal{B}^{\alpha_2})$.*

Proof. Assume that $(x^*, y^*) \in \mathcal{S}^m \times \mathcal{S}^n$ is a FWPNE of Γ with weights λ and η, that is, (x^*, y^*) is a FPNE of $(\mathcal{A}, \mathcal{B})$. By Definition 4, we have

$$(x^*)^T \mathcal{A}y^* \not\prec x^T \mathcal{A}y^* \text{ and } (x^*)^T \mathcal{B}y^* \not\prec (x^*)^T \mathcal{B}y,$$

for all $(x, y) \in \mathcal{S}^m \times \mathcal{S}^n$. From Definition 2, there is $\alpha_1, \alpha_2 \in [0, 1]$ such that

$$x^{*T} \underline{\mathcal{A}}^{\alpha_1} y^* \geq x^T \underline{\mathcal{A}}^{\alpha_1} y^* \ (\text{or } x^{*T} \overline{\mathcal{A}}^{\alpha_1} y^* \geq x^T \overline{\mathcal{A}}^{\alpha_1} y^*)$$

and

$$x^{*T} \underline{\mathcal{B}}^{\alpha_2} y^* \geq x^{*T} \underline{\mathcal{B}}^{\alpha_2} y \ (\text{or } x^{*T} \overline{\mathcal{B}}^{\alpha_2} y^* \geq x^{*T} \overline{\mathcal{B}}^{\alpha_2} y).$$

Then, (x^*, y^*) is a Pareto–Nash equilibrium of $(\mathcal{A}^{\alpha_1}, \mathcal{B}^{\alpha_2})$. Finally, the sufficient part is obvious, according to Definition 4. □

Theorem 4. *A strategy profile $(x^*, y^*) \in \mathcal{S}^m \times \mathcal{S}^n$ is a FWPNE of Γ with weights λ and η if and only if there exist $\alpha_1, \alpha_2 \in [0, 1]$ such that it is an optimal solution to one of the following four bilinear programming models with two parameters:*

$$\max\{x^T \underline{\mathcal{A}}^{\alpha_1} y + x^T \underline{\mathcal{B}}^{\alpha_2} y - \underline{\mu}(\alpha_1) - \underline{\nu}(\alpha_2)\}$$
$$s.t. \begin{cases} \underline{\mathcal{A}}^{\alpha_1} y \leq \underline{\mu}(\alpha_1) e^m, \\ \underline{\mathcal{B}}^{\alpha_2 T} x \leq \underline{\nu}(\alpha_2) e^n, \\ x^T e^m = 1, \\ y^T e^n = 1, \\ x, y \geq 0, \\ \alpha_1, \alpha_2 \in [0,1]; \end{cases} \tag{7}$$

$$\max\{x^T \underline{\mathcal{A}}^{\alpha_1} y + x^T \overline{\mathcal{B}}^{\alpha_2} y - \underline{\mu}(\alpha_1) - \overline{\nu}(\alpha_2)\}$$
$$s.t. \begin{cases} \underline{\mathcal{A}}^{\alpha_1} y \leq \underline{\mu}(\alpha_1) e^m, \\ \overline{\mathcal{B}}^{\alpha_2 T} x \leq \overline{\nu}(\alpha_2) e^n, \\ x^T e^m = 1, \\ y^T e^n = 1, \\ x, y \geq 0, \\ \alpha_1, \alpha_2 \in [0,1]; \end{cases} \tag{8}$$

$$\max\{x^T \overline{\mathcal{A}}^{\alpha_1} y + x^T \underline{\mathcal{B}}^{\alpha_2} y - \overline{\mu}(\alpha_1) - \underline{\nu}(\alpha_2)\}$$
$$s.t. \begin{cases} \overline{\mathcal{A}}^{\alpha_1} y \leq \overline{\mu}(\alpha_1) e^m, \\ \underline{\mathcal{B}}^{\alpha_2 T} x \leq \underline{\nu}(\alpha_2) e^n, \\ x^T e^m = 1, \\ y^T e^n = 1, \\ x, y \geq 0, \\ \alpha_1, \alpha_2 \in [0,1]; \end{cases} \tag{9}$$

$$\max\{x^T \overline{\mathcal{A}}^{\alpha_1} y + x^T \overline{\mathcal{B}}^{\alpha_2} y - \overline{\mu}(\alpha_1) - \overline{\nu}(\alpha_2)\}$$
$$s.t. \begin{cases} \overline{\mathcal{A}}^{\alpha_1} y \leq \overline{\mu}(\alpha_1) e^m, \\ \overline{\mathcal{B}}^{\alpha_2 T} x \leq \overline{\nu}(\alpha_2) e^n, \\ x^T e^m = 1, \\ y^T e^n = 1, \\ x, y \geq 0, \\ \alpha_1, \alpha_2 \in [0,1]; \end{cases} \tag{10}$$

where $e^m = \underbrace{(1,1,\ldots,1)}_{m}^T$ *and* $e^n = \underbrace{(1,1,\ldots,1)}_{n}^T$.

Proof. The argument is similar to that of Theorem 2. □

Theorems 1 and 3 illustrate that both the FWPNE and FPNE of MBGFP can be converted into the Pareto–Nash equilibrium of a certain BGIP. From Lemma 1, the Pareto–Nash equilibrium of BGIP can be transformed into four Nash equilibria of crisp bi-matrix games. Through the bilinear programming method, Theorems 2 and 4 further provide the calculation methods for FWPNE and FPNE, which aim to solve bilinear programming problems with parameters (BLPP). The relationship between FWPNE and FPNE, and their calculation methods are presented in Figure 1.

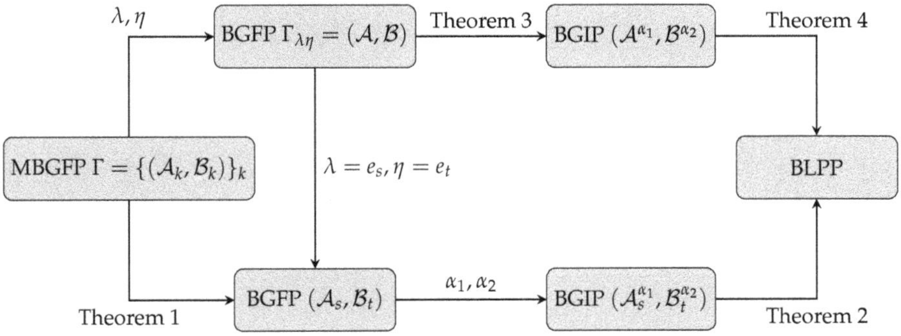

Figure 1. The calculation methods of FWPNE and FPNE for MBGFP.

Remark 2. *Multi-objective matrix games with fuzzy payoffs have been discussed in the literature [24,25], but the idea of obtaining fuzzy (weighted) Pareto–Nash equilibrium based on fuzzy maximum order has never appeared.*

1. *Bigdeli and Hassanpour [25] researched the multi-objective zero-sum matrix games with triangular fuzzy numbers. They first transformed the multi-objective zero-sum matrix game into a multi-objective interval mathematical programming problem. By considering the weight of each objective to the player and the satisfactory crisp equivalent form of interval inequality constraints, the multi-objective interval mathematical programming problem is converted into two linear programming problem. Finally, an interactive algorithm is presented to obtain the satisfactory strategy of the player. However, the interactive algorithm is only applicable when the payoffs are triangular fuzzy numbers. The multi-objective zero-sum matrix game is a special case of multi-objective bi-matrix game. Using the method given in Section 3, we reanalyze and calculate numerical example in [25], see Appendix A.*
2. *Nishizaki and Sakawa [24] mainly discussed multi-objective bi-matrix games with triangular fuzzy numbers (symmetric fuzzy numbers). By introducing the fuzzy goal (linear fuzzy number) of the player, the attainment functions of two players are constructed. Furthermore, the game is transformed into a two-player crisp game. Since the attainment function is nonlinear, the mathematical programming model is relatively complex, and the equilibrium solution of the game is difficult to calculate in practical applications.*

4. Application

In this section, we consider how two companies (Company I and Company II) maximize their profits. They produce two identical clothing types: Type I (casual and natural) and Type II (intellectual elegance). In order to increase profits, they have two strategies: Strategy I (to reduce the price) and Strategy II (advertisement).

Because of the lack of information, managers cannot accurately assess the profits obtained by various clothing types. To handle an uncertain situation, triangular fuzzy numbers are used to represent the profit obtained by each clothing type. Therefore, the problem can be regarded as a two-objective bi-matrix game with fuzzy payoffs. That is, Companies I and II are considered as Players I and II, and two clothing types are considered as the two objectives of Players I and II, respectively.

Example 1. *The market research departments of Companies I and II have respectively established the following fuzzy payoff matrices:*

$$\mathcal{A}_1 = \begin{array}{c} \\ \beta_1 \\ \beta_2 \end{array} \begin{array}{c} \begin{array}{cc} \gamma_1 & \gamma_2 \end{array} \\ \begin{pmatrix} (13.25, 14.00, 14.80) & (14.25, 15.00, 15.50) \\ (15.75, 16.00, 16.25) & (13.50, 14.00, 14.70) \end{pmatrix} \end{array}$$

$$\mathcal{A}_2 = \begin{matrix} & \gamma_1 & \gamma_2 \\ \beta_1 & (12.80, 13.00, 13.50) & (14.45, 15.00, 15.55) \\ \beta_2 & (13.45, 14.00, 14.50) & (12.60, 13.00, 13.80) \end{matrix}$$

$$\mathcal{B}_1 = \begin{matrix} & \gamma_1 & \gamma_2 \\ \beta_1 & (12.60, 13.00, 13.75) & (13.50, 14.00, 14.55) \\ \beta_2 & (13.55, 14.00, 14.25) & (12.30, 13.00, 13.60) \end{matrix}$$

$$\mathcal{B}_2 = \begin{matrix} & \gamma_1 & \gamma_2 \\ \beta_1 & (15.25, 16.00, 16.80) & (17.35, 18.00, 18.20) \\ \beta_2 & (17.20, 18.00, 18.50) & (14.50, 15.00, 15.45) \end{matrix},$$

where β_1, β_2 are two strategies for Company I, and γ_1, γ_2 are two strategies for Company II, respectively.

The MBGFP $\Gamma = \{(\mathcal{A}_k, \mathcal{B}_k)\}_{k\in\{1,2\}}$ with weights λ and η can be transformed into the following BGFP $\Gamma_{\lambda\eta} = (\mathcal{A}, \mathcal{B})$, where λ and η respectively represent the relative importance of two clothing types. Let $\lambda = (0.7, 0.3)$ and $\eta = (0.4, 0.6)$, then

$$\mathcal{A} = 0.7\mathcal{A}_1 + 0.3\mathcal{A}_2 = \begin{pmatrix} (13.115, 13.700, 14.410) & (14.310, 15.000, 15.515) \\ (15.060, 15.400, 15.725) & (13.230, 13.700, 14.430) \end{pmatrix}$$

and

$$\mathcal{B} = 0.4\mathcal{B}_1 + 0.6\mathcal{B}_2 = \begin{pmatrix} (14.190, 14.800, 15.580) & (15.810, 16.400, 16.740) \\ (15.740, 16.400, 16.800) & (13.620, 14.200, 14.710) \end{pmatrix}.$$

The fuzzy-valued expected payoffs for Companies I and II are, respectively

$$\widetilde{F}_{1\lambda}(x, y) = x^T\mathcal{A}y = (x^TA_1y, x^TA_2y, x^TA_3y)$$

and

$$\widetilde{F}_{2\eta}(x, y) = x^T\mathcal{B}y = (x^TB_1y, x^TB_2y, x^TB_3y),$$

for $(x, y) \in \mathcal{S}^2 \times \mathcal{S}^2$, where

$$\begin{aligned} x^TA_1y =& 13.115x_1y_1 + 14.310x_1y_2 + 15.060x_2y_1 + 13.230x_2y_2, \\ x^TA_2y =& 13.700x_1y_1 + 15.000x_1y_2 + 15.400x_2y_1 + 13.700x_2y_2, \\ x^TA_3y =& 14.410x_1y_1 + 15.515x_1y_2 + 15.725x_2y_1 + 14.430x_2y_2, \\ x^TB_1y =& 14.190x_1y_1 + 15.810x_1y_2 + 15.740x_2y_1 + 13.620x_2y_2, \\ x^TB_2y =& 14.800x_1y_1 + 16.400x_1y_2 + 16.400x_2y_1 + 14.200x_2y_2, \\ x^TB_3y =& 15.580x_1y_1 + 16.740x_1y_2 + 16.800x_2y_1 + 14.710x_2y_2. \end{aligned}$$

For $\alpha_1, \alpha_2 \in [0, 1]$, the BGFP $\Gamma_{\lambda\eta}$ can be transformed into the BGIP $\Gamma_{\lambda\eta}(\alpha_1, \alpha_2) = (\mathcal{A}_1^\alpha, \mathcal{B}_2^\alpha)$, where

$$\mathcal{A}^{\alpha_1} = \begin{pmatrix} [13.115 + 0.585\alpha_1, 14.410 - 0.710\alpha_1] & [14.310 + 0.690\alpha_1, 15.515 - 0.515\alpha_1] \\ [15.060 + 0.340\alpha_1, 15.725 - 0.325\alpha_1] & [13.230 + 0.470\alpha_1, 14.430 - 0.730\alpha_1] \end{pmatrix},$$

and

$$\mathcal{B}^{\alpha_2} = \begin{pmatrix} [14.190 + 0.610\alpha_2, 15.580 - 0.780\alpha_2] & [15.810 + 0.590\alpha_2, 16.740 - 0.340\alpha_2] \\ [15.740 + 0.660\alpha_2, 16.800 - 0.400\alpha_2] & [13.620 + 0.580\alpha_2, 14.710 - 0.510\alpha_2] \end{pmatrix}.$$

From Theorem 3 and Lemma 1, for each $\alpha_1, \alpha_2 \in [0, 1]$, MBGFP Γ has four FWPNEs. Due to Theorem 4, we only need to calculate FWPNEs of Γ. From (7)–(10), four bilinear programming models with two parameters are constructed as follows:

$$\max\{(27.305+0.585\alpha_1+0.61\alpha_2)x_1y_1+(30.12+0.69\alpha_1+0.59\alpha_2)x_1y_2+ \\ (30.8+0.34\alpha_1+0.66\alpha_2)x_2y_1+(26.85+0.47\alpha_1+0.58\alpha_2)x_2y_2-\underline{\mu}(\alpha_1)-\underline{\nu}(\alpha_2)\}$$
$$s.t.\begin{cases}(13.115+0.585\alpha_1)y_1+(14.31+0.69\alpha_1)y_2\le\underline{\mu}(\alpha_1),\\(15.06+0.34\alpha_1)y_1+(13.23+0.47\alpha_1)y_2\le\underline{\mu}(\alpha_1),\\(14.19+0.61\alpha_2)x_1+(15.74+0.66\alpha_2)x_2\le\underline{\nu}(\alpha_2),\\(15.81+0.59\alpha_2)x_1+(13.62+0.58\alpha_2)x_2\le\underline{\nu}(\alpha_2),\\x_1+x_2=1,\ y_1+y_2=1,\\x_1,x_2,y_1,y_2\ge0,\ \alpha_1,\alpha_2\in[0,1];\end{cases} \tag{11}$$

$$\max\{(28.695+0.585\alpha_1-0.78\alpha_2)x_1y_1+(31.05+0.69\alpha_1-0.34\alpha_2)x_1y_2+ \\ (31.86+0.34\alpha_1-0.4\alpha_2)x_2y_1+(27.94+0.47\alpha_1-0.51\alpha_2)x_2y_2-\underline{\mu}(\alpha_1)-\overline{\nu}(\alpha_2)\}$$
$$s.t.\begin{cases}(13.115+0.585\alpha_1)y_1+(14.31+0.69\alpha_1)y_2\le\underline{\mu}(\alpha_1),\\(15.06+0.34\alpha_1)y_1+(13.23+0.47\alpha_1)y_2\le\underline{\mu}(\alpha_1),\\(15.58-0.78\alpha_2)x_1+(16.8-0.4\alpha_2)x_2\le\overline{\nu}(\alpha_2),\\(16.74-0.34\alpha_2)x_1+(14.71-0.51\alpha_2)x_2\le\overline{\nu}(\alpha_2),\\x_1+x_2=1,\ y_1+y_2=1,\\x_1,x_2,y_1,y_2\ge0,\ \alpha_1,\alpha_2\in[0,1];\end{cases} \tag{12}$$

$$\max\{(28.6-0.71\alpha_1+0.61\alpha_2)x_1y_1+(31.325-0.515\alpha_1+0.59\alpha_2)x_1y_2+ \\ (31.465-0.325\alpha_1+0.66\alpha_2)x_2y_1+(28.05-0.73\alpha_1+0.58\alpha_2)x_2y_2-\overline{\mu}(\alpha_1)-\underline{\nu}(\alpha_2)\}$$
$$s.t.\begin{cases}(14.41-0.71\alpha_1)y_1+(15.515-0.515\alpha_1)y_2\le\overline{\mu}(\alpha_1),\\(15.725-0.325\alpha_1)y_1+(14.43-0.73\alpha_1)y_2\le\overline{\mu}(\alpha_1),\\(14.19+0.61\alpha_2)x_1+(15.74+0.66\alpha_2)x_2\le\underline{\nu}(\alpha_2),\\(15.81+0.59\alpha_2)x_1+(13.62+0.58\alpha_2)x_2\le\underline{\nu}(\alpha_2),\\x_1+x_2=1,\ y_1+y_2=1,\\x_1,x_2,y_1,y_2\ge0,\ \alpha_1,\alpha_2\in[0,1].\end{cases} \tag{13}$$

$$\max\{(29.99-0.71\alpha_1-0.78\alpha_2)x_1y_1+(32.255-0.515\alpha_1-0.34\alpha_2)x_1y_2+ \\ (32.525-0.325\alpha_1-0.4\alpha_2)x_2y_1+(29.14-0.73\alpha_1-0.51\alpha_2)x_2y_2-\overline{\mu}(\alpha_1)-\overline{\nu}(\alpha_2)\}$$
$$s.t.\begin{cases}(14.41-0.71\alpha_1)y_1+(15.515-0.515\alpha_1)y_2\le\overline{\mu}(\alpha_1),\\(15.725-0.325\alpha_1)y_1+(14.43-0.73\alpha_1)y_2\le\overline{\mu}(\alpha_1),\\(15.58-0.78\alpha_2)x_1+(16.8-0.4\alpha_2)x_2\le\overline{\nu}(\alpha_2),\\(16.74-0.34\alpha_2)x_1+(14.71-0.51\alpha_2)x_2\le\overline{\nu}(\alpha_2),\\x_1+x_2=1,\ y_1+y_2=1,\\x_1,x_2,y_1,y_2\ge0,\ \alpha_1,\alpha_2\in[0,1];\end{cases} \tag{14}$$

The parameters α_1 and α_2 within Formulas (11)–(14) can be controlled by decision makers. Let $\alpha_1=0.2,\alpha_2=0.5$, we can obtain an optimal solution $(x^*,y^*,\underline{\mu}(0.2),\underline{\nu}(0.5))$ of (11) using the Lingo software, where $x^*=(0.5729,0.4271)$, $y^*=(0.3722,0.6278)$, $\underline{\mu}(0.2)=13.9954$, $\underline{\nu}(0.5)=15.1676$. In addition, we obtain an optimal solution of (12), where $x^*=(0.6085,0.3915)$, $y^*=(0.3722,0.6278)$, $\underline{\mu}(0.2)=13.9954$, $\overline{\nu}(0.5)=15.7420$; an optimal solution of (13), where $x^*=(0.5729,0.4271)$, $y^*=(0.4476,0.5524)$, $\overline{\mu}(0.2)=14.8999$, $\underline{\nu}(0.5)=15.1676$; and an optimal solution of (14), where $x^*=(0.6085,0.3915)$, $y^*=(0.4476,0.5524)$, $\overline{\mu}(0.2)=14.8999$, $\overline{\nu}(0.5)=15.7420$. Four fuzzy weighted Pareto–Nash equilibria of Γ and the corresponding fuzzy-valued expected payoffs of Companies I and II are collected in Table 1.

Table 1. The FWPNEs and fuzzy expected payoffs for Companies I and II ($\lambda = (0.7, 0.3)$, $\eta = (0.4, 0.6)$).

x_1^*	x_2^*	y_1^*	y_2^*	$\widetilde{F}_{1(0.7,0.3)}(x^*, y^*)$	$\widetilde{F}_{2(0.4,0.6)}(x^*, y^*)$
0.5729	0.4271	0.3722	0.6278	(13.8848, 14.4378, 15.0218)	(14.8662, 15.4689, 15.9579)
0.6085	0.3915	0.3722	0.6278	(13.8832, 14.4443, 15.0287)	(14.8946, 15.4969, 15.9871)
0.5729	0.4271	0.4476	0.5524	(13.8921, 14.4364, 15.0158)	(14.8645, 15.4707, 15.9751)
0.6085	0.3915	0.4476	0.5524	(13.8824, 14.4349, 15.0162)	(14.8829, 15.4884, 15.9956)

Similar to the calculation method of fuzzy weighted Pareto–Nash equilibria, letting $\lambda = (1, 0)$, $\eta = (1, 0)$, $\alpha_1 = 0.2$, $\alpha_2 = 0.5$, we can obtain fuzzy Pareto–Nash equilibria of $(\mathcal{A}_1, \mathcal{B}_1)$ and the corresponding fuzzy-valued expected payoffs of Companies I and II, see Table 2. Meanwhile, the fuzzy Pareto–Nash equilibria of $(\mathcal{A}_2, \mathcal{B}_2)$, $(\mathcal{A}_1, \mathcal{B}_2)$, and $(\mathcal{A}_2, \mathcal{B}_1)$ are shown in Tables 3–5, respectively.

Table 2. The FWPNEs and fuzzy expected payoffs for Companies I and II ($\lambda = (1, 0)$, $\eta = (1, 0)$).

x_1^*	x_2^*	y_1^*	y_2^*	$\widetilde{F}_{1(1,0)}(x^*, y^*)$	$\widetilde{F}_{2(1,0)}(x^*, y^*)$
0.5422	0.4578	0.2500	0.7500	(14.0286, 14.6356, 15.2163)	(12.9717, 13.5211, 14.0810)
0.4783	0.5217	0.2500	0.7500	(14.0326, 14.6196, 15.2011)	(12.9294, 13.4892, 14.0435)
0.5422	0.4578	0.3500	0.6500	(14.0774, 14.6729, 15.2493)	(12.9801, 13.5127, 14.0674)
0.4783	0.5217	0.3500	0.6500	(14.1022, 14.6761, 15.2485)	(12.9515, 13.4935, 14.0391)

Table 3. The FWPNEs and fuzzy expected payoffs for Companies I and II ($\lambda = (0, 1)$, $\eta = (0, 1)$).

x_1^*	x_2^*	y_1^*	y_2^*	$\widetilde{F}_{1(0,1)}(x^*, y^*)$	$\widetilde{F}_{2(0,1)}(x^*, y^*)$
0.5816	0.4184	0.7231	0.2769	(13.2070, 13.6246, 14.1674)	(16.0913, 16.8113, 17.3834)
0.6402	0.3598	0.7231	0.2769	(13.2062, 13.6147, 14.1535)	(16.0549, 16.7753, 17.3560)
0.5816	0.4184	0.6429	0.3571	(13.2461, 13.6844, 14.2396)	(16.0986, 16.8039, 17.3463)
0.6402	0.3598	0.6429	0.3571	(13.2562, 13.6885, 14.2385)	(16.0848, 16.7914, 17.3398)

Table 4. The FWPNEs and fuzzy expected payoffs for Companies I and II ($\lambda = (1, 0)$, $\eta = (0, 1)$).

x_1^*	x_2^*	y_1^*	y_2^*	$\widetilde{F}_{1(1,0)}(x^*, y^*)$	$\widetilde{F}_{2(0,1)}(x^*, y^*)$
0.5816	0.4184	0.2500	0.7500	(14.0262, 14.6454, 15.2256)	(16.1346, 16.7678, 17.1649)
0.6402	0.3598	0.2500	0.7500	(14.0225, 14.6601, 15.2395)	(16.2313, 16.8704, 17.2608)
0.5816	0.4184	0.3500	0.6500	(14.0621, 14.6709, 15.2498)	(16.1255, 16.7770, 17.2111)
0.6402	0.3598	0.3500	0.6500	(14.0394, 14.6680, 15.2505)	(16.1940, 16.8503, 17.2809)

Table 5. The FWPNEs and fuzzy expected payoffs for Companies I and II ($\lambda = (0, 1)$, $\eta = (1, 0)$).

x_1^*	x_2^*	y_1^*	y_2^*	$\widetilde{F}_{1(0,1)}(x^*, y^*)$	$\widetilde{F}_{2(1,0)}(x^*, y^*)$
0.5422	0.4578	0.7231	0.2769	(13.2075, 13.6313, 14.1768)	(13.0116, 13.4812, 14.0166)
0.4783	0.5217	0.7231	0.2769	(13.2084, 13.6421, 14.1921)	(13.0342, 13.5097, 14.0229)
0.5422	0.4578	0.6429	0.3571	(13.2394, 13.6816, 14.2403)	(13.0048, 13.4879, 14.0275)
0.4783	0.5217	0.6429	0.3571	(13.2284, 13.6770, 14.2414)	(13.0165, 13.5062, 14.0264)

Through the analysis of Tables 1–5, we see that the fuzzy expected payoffs of Companies I and II in Table 4 are better than other situations. Therefore, Company I give priority to the first objective (Type I: casual and natural), and Company II give priority to the second objective (Type II: intellectual elegance).

Taking Table 4 as an example, we illustrate that four FPNEs cannot be replaced by one of them. For example, Company II chooses $y^* = (0.2500, 0.7500)$ and Company I changes its strategy x^* from $(0.6402, 0.3598)$ to $(0.5816, 0.4184)$, the corresponding fuzzy expected payoff of Company I does not become better due to $(14.0225, 14.6601, 15.2395) \not\prec (14.0262, 14.6454, 15.2256)$.

Similarly, Company I chooses $x^* = (0.5816, 0.4184)$ and Company II changes its strategy y^* from $(0.2500, 0.7500)$ to $(0.3500, 0.6500)$. The corresponding fuzzy expected payoff of Company II does not become better due to $(16.1346, 16.7678, 17.1649) \not\prec (16.1255, 16.7770, 17.2111)$. However, simultaneous changes in the strategies of two companies are not within the scope of FPNE's definition. In Table 4, since $(14.0262, 14.6454, 15.2256) \prec (14.0394, 14.6680, 15.2505)$ and $(16.1346, 16.7678, 17.1649) \prec (16.1940, 16.8503, 17.2809)$, Company I can choose to strategy $(0.6402, 0.3598)$ instead of $(0.5816, 0.4184)$, and Company II can also choose to strategy $(0.3500, 0.6500)$ instead of $(0.2500, 0.7500)$. However, this does not affect the fact that both strategy profiles are FPNE. A company's own expected payoffs will not be better if it changes its strategy while another company's strategy remains unchanged. At this time, the strategy combination of the two companies is the FPNE of the competition problem.

5. Conclusions

This paper mainly studies multi-objective bi-matrix games with fuzzy payoffs. This study differs from previous literature in the following two aspects. First, the fuzzy payoff in our game model is a general fuzzy number proposed by Dubois and Prade [37]. Secondly, under the partial order of fuzzy vector values [26], we give the concept of fuzzy Pareto–Nash equilibrium of MBGFP, which is an equilibrium that makes the fuzzy-vector-valued expected payoffs of the players reach Pareto optimality. Furthermore, by considering the weight of each objective in the MBGFP, the weighted MBGFP is transformed into a BGFP, and the concept of fuzzy weighted Pareto–Nash equilibrium is obtained. Finally, the necessary and sufficient conditions for the existence of FPNE and FWPNE of MBGFP and their calculation method are established. This calculation method is feasible for general fuzzy payoffs.

The concepts of FPNE and FWPNE are given based on the fuzzy maximum order. The advantage of this partial order is that it compares two fuzzy numbers while retaining all their characteristics. However, it has relatively high requirements for comparable fuzzy numbers. Some fuzzy numbers that we can intuitively judge as well or bad may not be comparable under this partial order. For example, $\widetilde{a} = (1, 3, 5)$ and $\widetilde{b} = (2, 3, 4)$. We aim to establish a partial order on $\mathcal{F}(\mathbb{R})$ that allows for comparisons between more fuzzy numbers while preserving essential characteristics. Furthermore, under the new partial order, we further refine the equilibrium obtained in this article, which is the focus of our future research.

Author Contributions: Conceptualization, Y.F.; methodology, W.L.; formal analysis, W.L. and D.L.; investigation, Y.F.; data curation, D.Z.; writing—original draft, W.L.; writing—review & editing, W.L.; supervision, D.L.; funding acquisition, D.Z. All authors have read and agreed to the published version of the manuscript.

Funding: This work was supported by the National Natural Science Foundation of China (Nos. [72031009], [12171378]) and the Excellent Youth Foundation of Hubei Scientific Committee (No. 2020CFA079).

Data Availability Statement: Not applicable.

Acknowledgments: The authors are indebted to the four reviewers for their thoughtful comments and suggestions that have helped to significantly improve the quality and presentation of our research.

Conflicts of Interest: The authors declared that they have no conflict of interest to this work.

Appendix A. The Fuzzy Saddle Points of Zero-Sum Matrix Game with Fuzzy Payoffs

According to Definition 4, we can directly obtain the following definition of the saddle point of zero-sum matrix game with fuzzy payoffs (ZMGFP).

Definition A1. *Let $\mathcal{A} = (\widetilde{a}_{ij})_{m\times n}$, where all the components $\widetilde{a}_{ij} \in \mathcal{F}(\mathbb{R})$. A strategy profile $(x^*, y^*) \in \mathcal{S}^m \times \mathcal{S}^n$ is a fuzzy saddle point of ZMGFP $\mathcal{A}$, if for all $(x, y) \in \mathcal{S}^m \times \mathcal{S}^n$, such that*

$$(x^*)^T \mathcal{A} y \not\succ (x^*)^T \mathcal{A} y^* \not\succ x^T \mathcal{A} y^*.$$

Theorem A1. *A strategy profile $(x^*, y^*) \in \mathcal{S}^m \times \mathcal{S}^n$ is a fuzzy saddle point of ZMGFP $\mathcal{A}$ if and only if there exists $\alpha \in [0,1]$ such that it is an optimal solution to one of the following two bilinear programming models with a parameter:*

$$\max\{x^T \underline{\mathcal{A}}^\alpha y - \underline{\mu}(\alpha)\}$$
$$\text{s.t.}\begin{cases} \underline{\mathcal{A}}^\alpha y \leq \underline{\mu}(\alpha) e^m, \\ \underline{\mathcal{A}}^{\alpha T} x \geq \underline{\mu}(\alpha) e^n, \\ x^T e^m = 1, \\ y^T e^n = 1, \\ x, y \geq 0, \\ \alpha \in [0,1]; \end{cases} \tag{A1}$$

$$\max\{x^T \overline{\mathcal{A}}^\alpha y - \overline{\mu}(\alpha)\}$$
$$\text{s.t.}\begin{cases} \overline{\mathcal{A}}^\alpha y \leq \overline{\mu}(\alpha) e^m, \\ \overline{\mathcal{A}}^{\alpha T} x \geq \overline{\mu}(\alpha) e^n, \\ x^T e^m = 1, \\ y^T e^n = 1, \\ x, y \geq 0, \\ \alpha \in [0,1]; \end{cases} \tag{A2}$$

where $e^m = (\underbrace{1, 1, \ldots, 1}_{m})^T$ and $e^n = (\underbrace{1, 1, \ldots, 1}_{n})^T$.

Example A1. *Below we use the method in Theorem A1 to calculate the fuzzy saddle points of the two-objective zero-sum matrix game with fuzzy payoffs in [25].*

$$\mathcal{A}_1 = \begin{pmatrix} (175, 180, 190) & (150, 156, 158) \\ (80, 90, 100) & (175, 180, 190) \end{pmatrix}$$

and

$$\mathcal{A}_2 = \begin{pmatrix} (125, 130, 135) & (120, 130, 135) \\ (120, 130, 135) & (150, 160, 170) \end{pmatrix}.$$

Let $\lambda = \eta = (0.5, 0.5)$, then

$$\mathcal{A} = 0.5\mathcal{A}_1 + 0.5\mathcal{A}_2 = \begin{pmatrix} (150.0, 155.0, 162.5) & (135.0, 143.0, 146.5) \\ (100.0, 110.0, 117.5) & (162.5, 170.0, 180.0) \end{pmatrix}.$$

The fuzzy-valued expected payoff for the game is expressed as follows:

$$\widetilde{F}_\lambda(x, y) = x^T \mathcal{A} y = (x^T A_1 y, x^T A_2 y, x^T A_3 y)$$

for $(x, y) \in \mathcal{S}^2 \times \mathcal{S}^2$, where

$$\begin{aligned} x^T A_1 y =& 150.0 x_1 y_1 + 135.0 x_1 y_2 + 100.0 x_2 y_1 + 162.5 x_2 y_2, \\ x^T A_2 y =& 155.0 x_1 y_1 + 143.0 x_1 y_2 + 110.0 x_2 y_1 + 170.0 x_2 y_2, \\ x^T A_3 y =& 162.5 x_1 y_1 + 146.5 x_1 y_2 + 117.5 x_2 y_1 + 180.0 x_2 y_2. \end{aligned}$$

For $\alpha \in [0,1]$, the ZMGFP $\mathcal{A}$ can be transformed into the zero-sum matrix game with interval payoffs $\mathcal{A}^{\alpha}$, where

$$\mathcal{A}^{\alpha} = \begin{pmatrix} [150.0+5.0\alpha, 162.5-7.5\alpha] & [135.0+8.0\alpha, 146.5-3.5\alpha] \\ [100.0+10.0\alpha, 117.5-7.5\alpha] & [162.5+7.5\alpha, 180.0-10.0\alpha] \end{pmatrix}.$$

From Theorem A1, we can specifically calculate fuzzy saddle points of $\mathcal{A}$. From (A1) and (A2), two bilinear programming models with a parameter are constructed as follows:

$$\max\{(150.0+5.0\alpha)x_1y_1+(135.0+8.0\alpha)x_1y_2+(100.0+10.0\alpha)x_2y_1+ (162.5+7.5\alpha)x_2y_2-\underline{\mu}(\alpha)\}$$

$$\text{s.t.}\begin{cases} (150.0+5.0\alpha)y_1+(135.0+8.0\alpha)y_2 \le \underline{\mu}(\alpha), \\ (100.0+10.0\alpha)y_1+(162.5+7.5\alpha)y_2 \le \underline{\mu}(\alpha), \\ (150.0+5.0\alpha)x_1+(100.0+10.0\alpha)x_2 \ge \underline{\mu}(\alpha), \\ (135.0+8.0\alpha)x_1+(162.5+7.5\alpha)x_2 \ge \underline{\mu}(\alpha), \\ x_1+x_2=1,\ y_1+y_2=1, \\ x_1,x_2,y_1,y_2 \ge 0,\ \alpha \in [0,1]; \end{cases} \tag{A3}$$

$$\max\{(162.5-7.5\alpha)x_1y_1+(146.5-3.5\alpha)x_1y_2+(117.5-7.5\alpha)x_2y_1+ (180.0-10.0\alpha)x_2y_2-\overline{\mu}(\alpha)\}$$

$$\text{s.t.}\begin{cases} (162.5-7.5\alpha)y_1+(146.5-3.5\alpha)y_2 \le \overline{\mu}(\alpha), \\ (117.5-7.5\alpha)y_1+(180.0-10.0\alpha)y_2 \le \overline{\mu}(\alpha), \\ (162.5-7.5\alpha)x_1+(117.5-7.5\alpha)x_2 \ge \overline{\mu}(\alpha), \\ (146.5-3.5\alpha)x_1+(180.0-10.0\alpha)x_2 \ge \overline{\mu}(\alpha), \\ x_1+x_2=1,\ y_1+y_2=1, \\ x_1,x_2,y_1,y_2 \ge 0,\ \alpha \in [0,1]; \end{cases} \tag{A4}$$

Let $\alpha = 1, 0.5, 0$, we can obtain five fuzzy saddle points of $\mathcal{A}$. The five fuzzy saddle points of $\mathcal{A}$ and the corresponding fuzzy expected payoffs are collected in Table A1.

Table A1. The fuzzy saddle points and the corresponding fuzzy expected payoffs ($\lambda = (0.5, 0.5)$).

α	x_1^*	x_2^*	y_1^*	y_2^*	$\widetilde{F}_{(0.5,0.5)}(x^*, y^*)$
1	0.8333	0.1667	0.3750	0.6250	(140.3645, 147.5000, 153.1772)
0.5	0.8140	0.1860	0.4020	0.5980	(140.3502, 147.4624, 153.2934)
0.5	0.8194	0.1806	0.3645	0.6355	(140.3323, 147.5105, 153.2145)
0	0.8065	0.1935	0.3548	0.6452	(140.3226, 147.5390, 153.2697)
0	0.7962	0.2038	0.4268	0.5732	(140.2654, 147.3615, 153.3280)

References

1. Zadeh, L.A. Fuzzy sets. *Inf. Control* **1965**, *8*, 338–353. [CrossRef]
2. Butnariu, D. Fuzzy games: A description of the concept. *Fuzzy Sets Syst.* **1978**, *1*, 181–192. [CrossRef]
3. Dubois, D.; Prade, H. *Fuzzy Sets and Systems: Theory and Applications*; Mathematics in Sciences and Engineering Series; Academic Press: New York, NY, USA, 1980; Volume 144.
4. Campos, L. Fuzzy linear programming models to solve fuzzy matrix games. *Fuzzy Sets Syst.* **1989**, *32*, 275–289. [CrossRef]
5. Maeda, T. Characterization of equilibirum strategy of the bimatrix game with fuzzy payoff. *J. Math. Anal. Appl.* **2000**, *251*, 885–896. [CrossRef]
6. González, A.; Vila, M.A. A discrete method for studying indifference and order relations between fuzzy numbers. *Inf. Sci.* **1991**, *56*, 245–258. [CrossRef]
7. Ramík, J.; Římánek, J. Inequality relation between fuzzy numbers and its use in fuzzy optimization. *Fuzzy Sets Syst.* **1985**, *16*, 123–150. [CrossRef]
8. Clemente, M.; Fernández, F.R.; Puerto, J. Pareto-optimal security strategies in matrix games with fuzzy payoffs. *Fuzzy Sets Syst.* **2011**, *176*, 36–45. [CrossRef]
9. Li, D.F. A fast approach to compute fuzzy values of matrix games with payoffs of triangular fuzzy numbers. *Eur. J. Oper. Res.* **2012**, *223*, 421–429. [CrossRef]

10. Chandra, S.; Aggarwal, A. On solving matrix games with payoffs of triangular fuzzy numbers: Certain observations and generalizations. *Eur. J. Oper. Res.* **2015**, *246*, 575–581. [CrossRef]
11. An, J.J.; Li, D.F. A linear programming approach to solve constrained bi-matrix games with intuitionistic fuzzy payoffs. *Int. J. Fuzzy Syst.* **2019**, *21*, 908–915. [CrossRef]
12. Bhaumik, A.; Roy, S.K.; Li, D.F. (α, β, γ)-cut set based ranking approach to solving bi-matrix games in neutrosophic environment. *Soft Comput.* **2021**, *25*, 2729–2739. [CrossRef]
13. Feng, Z.; Tan, C. Credibilistic bimatrix games with loss aversion and triangular fuzzy payoffs. *Int. J. Fuzzy Syst.* **2020**, *22*, 1635–1652. [CrossRef]
14. Kacher, F.; Larbani, M. Existence of equilibrium solution for a non-cooperative game with fuzzy goals and parameters. *Fuzzy Sets Syst.* **2008**, *159*, 164–176. [CrossRef]
15. Larbani, M. Solving bi-matrix games with fuzzy payoffs by introducing nature as a third player. *Fuzzy Sets Syst.* **2009**, *160*, 657–666. [CrossRef]
16. Larbani, M. Non cooperative fuzzy games in normal form: A survey. *Fuzzy Sets Syst.* **2009**, *160*, 3184–3210. [CrossRef]
17. Maeda, T. On characterization of equilibrium strategy of two-person zero-sum games with fuzzy payoffs. *Fuzzy Sets Syst.* **2003**, *139*, 283–296. [CrossRef]
18. Li, S.Y.; Tu, G.P. Bi-Matrix Games with General Intuitionistic Fuzzy Payoffs and Application in Corporate Environmental Behavior. *Symmetry* **2022**, *14*, 671. [CrossRef]
19. Zeleny, M. Game with multiple payoffs. *Int. J. Game Theory* **1976**, *4*, 179–191. [CrossRef]
20. Buckley, J.J. Multiple goals non cooperative conflict under uncertainty: A fuzzy set approach. *Fuzzy Sets Syst.* **1984**, *13*, 107–124. [CrossRef]
21. Bellman, R.E.; Zadeh, L.A. Decision making in a fuzzy environment. *Manag. Sci.* **1970**, *17*, 141–164. [CrossRef]
22. Sakawa, M.; Nishizaki, I. Max-min solutions for fuzzy multiobjective matrix games. *Fuzzy Sets Syst.* **1994**, *67*, 53–69. [CrossRef]
23. Fernandez, F.R.; Puerto, J. Vector linear programming in zero-sum multicriteria matrix games. *J. Optim. Theory Appl.* **1996**, *89*, 115–127. [CrossRef]
24. Nishizaki, I.; Sakawa, M. Equilibrium solutions in multiobjective bimatrix games with fuzzy payoffs and fuzzy goals. *Fuzzy Sets Syst.* **2000**, *111*, 99–116. [CrossRef]
25. Bigdeli, H.; Hassanpour, H. A satisfactory strategy of multiobjective two person matrix games with fuzzy payoffs. *Iran. J. Fuzzy Syst.* **2016**, *13*, 17–33.
26. Li, W.; Li, D.Y.; Feng, Y.Q.; Zou, D. Existence and stability of fuzzy Pareto-Nash equilibria for fuzzy constrained multi-objective games with fuzzy payoffs. *AIMS Math.* **2023**, *8*, 15907–15931. [CrossRef]
27. Moore, R.E. *Interval Analysis*; Prentice-Hall: Englewood Cliffs, NJ, USA, 1966.
28. Alefeld, G.; Mayer, G. Interval analysis: Theory and applications. *J. Comput. Appl. Math.* **2000**, *121*, 421–464. [CrossRef]
29. Chalco-Cano, Y.; Rufián-Lizana, A.; Román-Flores, H.; Jiménez-Gamero, M.D. Calculus for interval-valued functions using generalized Hukuhara derivative and applications. *Fuzzy Sets Syst.* **2013**, *219*, 49–67. [CrossRef]
30. Moore, R.E. *Method and Applications of Interval Analysis*; Society for Industrial and Applied Mathematics: Philadelphia, PA, USA, 1987.
31. Oppenheimer, E.P.; Michel, A.N. Application of interval analysis techniques to linear systems. Part I: Fundamental results. *IEEE Trans. Circuits Syst.* **1988**, *35*, 1129–1138. [CrossRef]
32. Stefanini, L.; Bede, B. Generalized Hukuhara differentiability of interval-valued functions and interval differential equations. *Nonlinear Anal.* **2009**, *71*, 1311–1328. [CrossRef]
33. Fei, W.; Li, D.F. Bilinear Programming Approach to Solve Interval Bimatrix Games in Tourism Planning Management. *Int. J. Fuzzy Syst.* **2016**, *18*, 504–510. [CrossRef]
34. Collins, W.D.; Hu, C. Studying interval valued matrix games with fuzzy logic. *Soft Comput.* **2008**, *12*, 147–155. [CrossRef]
35. Li, D.F. Linear programming approach to solve interval-valued matrix games. *Omega* **2011**, *39*, 655–666. [CrossRef]
36. Liu, S.T.; Kao, C. Matrix games with interval data. *Comput. Ind. Eng.* **2009**, *56*, 1697–1700. [CrossRef]
37. Dubois, D.; Prade, H. Operations on fuzzy numbers. *Int. J. Syst. Sci.* **1978**, *9*, 613–626. [CrossRef]
38. Nash, J. Non-cooperative games. *Ann. Math.* **1951**, *54*, 286–295. [CrossRef]

mathematics MDPI

Article

New Hybrid EC-Promethee Method with Multiple Iterations of Random Weight Ranges: Applied to the Choice of Policing Strategies

Marcio Pereira Basilio [1,2,*], Valdecy Pereira [2,*] and Fatih Yigit [3,*]

1 Controladoria-Geral do Estado do Rio de Janeiro (CGE), Avenida Erasmo Braga, 118, Centro, Rio de Janeiro 20020-000, Brazil
2 Department of Production Engineering, Fluminense Federal University (UFF), Niteroi 24210-240, Brazil
3 Department of Industrial Engineering, Altinbas University, 34218 Istanbul, Turkey
* Correspondence: mbasilio@cge.rj.gov.br (M.P.B.); valdecypereira@id.uff.br (V.P.); fatih.yigit@altinbas.com.tr (F.Y.); Tel.: +55-(21)99637-9803 (M.P.B.)

Abstract: The decision-making process is part of everyday life for people and organizations. When modeling the solutions to problems, just as important as the choice of criteria and alternatives is the definition of the weights of the criteria. This study will present a new hybrid method for weighting criteria. The technique combines the ENTROPY and CRITIC methods with the PROMETHE method to create EC-PROMETHEE. The innovation consists of using a weight range per criterion. The construction of a weight range per criterion preserves the characteristics of each technique. Each weight range includes lower and upper limits, which combine to generate random numbers, producing "t" sets of weights per criterion, allowing "t" final rankings to be obtained. The alternatives receive a value corresponding to their position with each ranking generated. At the end of the process, they are ranked in descending order, thus obtaining the final ranking. The method was applied to the decision support problem of choosing policing strategies to reduce crime. The model used a decision matrix with twenty criteria and fourteen alternatives evaluated in seven different scenarios. The results obtained after 10,000 iterations proved consistent, allowing the decision maker to see how each alternative behaved according to the weights used. The practical implication observed concerning traditional models, where a single final ranking is generated for a single set of weights, is the reversal of positions after "t" iterations compared to a single iteration. The method allows managers to make decisions with reduced uncertainty, improving the quality of their decisions. In future research, we propose creating a web tool to make this method easier to use, and propose other tools are produced in Python and R.

Keywords: ENTROPY; CRITIC; PROMETHEE; policing strategy; decision maker; MCDA; operations research

MSC: 90B50; 91B06

1. Introduction

Making decisions is an action that permeates human life. Some decisions are simple, like choosing which tie to wear. Others are complex and impact the lives of people, organizations, economies, and countries, like selecting a policing strategy to reduce the crime rate. Deciding implies making choices that are not always easy to make. The decision maker is not immune to macro-environment variables and can be influenced by organizational and personal objectives. Over the last four decades, researchers have developed and applied decision support methods that allow large volumes of information to be systematized, presenting the decision maker with the alternatives that, when compared pair-by-pair and criterion-by-criterion under the influence of weights, are best classified.

Basilio et al. [1] affirm that MCDA methods solve decision-making problems in various areas, including information and communication technology, business intelligence, environmental risk analysis, water resources management, remote sensing, flood risk management, health technology assessment, climate change, energy, international law, human resources policy, financial management, supplier selection, e-commerce and mobile commerce, agriculture and horticulture, chemical and biochemical engineering, software evaluation, flood risk management, health, transportation research, nanotechnology research, climate change, energy, human resources, financial management, performance and benchmarking, supplier selection, chemical and biochemical engineering, education and social policy, and public safety.

In their research, Basilio et al. [1] report that AHP, TOPSIS, VIKOR, PROMETHEE, and ANP are the methods most frequently used by authors in their respective studies. An essential issue in the decision-making process that profoundly impacts the evaluation of alternatives is the weights to be assigned to the criteria. Experts classify weighting methods as objective, subjective, and hybrid [2]. The AHP [1,3] is the method most researchers use when integrating methods for measuring weights with methods for ordering alternatives. This is followed by DEMATEL [4], SWARA [5–7], ANP [4], ENTROPY [8], CRITIC [9], BWM [10], CILOS [11], IDOCRIW [11], FUCOM [12,13], LBWA [14], SAPEVO-M [15], and MEREC [16,17]. From the taxonomy described by Ayan [2], we can infer that hybrid weight measurement methods are used to find a resulting position between the techniques used. However, generating a weight for each criterion reduces a certain degree of uncertainty, which, when inserted into the ordering method, will produce a ranking of the alternatives.

This study aims to combine objective and subjective methods, not to produce a single weight per criterion. Instead, this study aims to build a weight range for each criterion, preserving the characteristics of each technique. Each weight range comprises lower and upper limits, which can be combined to generate random numbers, producing "t" sets of weights per criterion, and making it possible to obtain "t" final rankings. The alternatives are given a value corresponding to their position in each ranking generated. At the end of the process, they will be ranked in descending order, thus obtaining the final definitive ranking. In this way, managers can analyze the behavior of each alternative throughout the process, and the final ranking will be more consistent due to the incorporation of the variations observed due to the influence of the weight of the criteria on the alternatives. In this study, we chose the ENTROPY-CRITIC methods and the weights generated by the decision makers to deal with the problem of selecting a policing strategy to reduce crime rates.

The CRITIC method aims to define weights by using the contract intensity and the conflicting character of the evaluation criteria. The CRITIC method is proposed by Diakoulaki et al. [18]. CRITIC is one of the most frequently used objective methods for criterion weight determination [9]. Since its first introduction, research has focused mainly on two topics. The first area aims to improve the CRITIC model, and the improvements focus on the normalization procedure. The studies focus on using vague information by employing fuzzy logic and alternative similarity and distance measures. By utilizing different approaches, new studies are performed. Normalization procedures are performed using various methods; to name a few, employing fuzzy logic [19], logarithmic normalization [20], and alternative rankings [21] are used. Another point for improvement is the weighting technique. The model is limited to deficiency in capturing the correlation between criteria [22]. A recent study employed a new D-CRITIC approach to overcome this limitation [9]. The proposed research aims to integrate different strategies to overcome such constraints using a hybrid system.

Another approach used for weight determination is the entropy approach. Entropy is based on a different discipline. The technique has its origins in the field of Thermodynamics [23]. The entropy approach was proposed first by Clausius [24]. Shannon and Weaver [25] proposed the entropy concept. The method employs a measure of uncertainty in information formulated regarding probability theory. The entropy method evaluates the relative contrast intensities of the criteria [23]. The approach does not consider the decision makers but the value of each alternative per criterion.

Since its introduction, the entropy model has been applied in different areas. To name a few, cryptocurrency evaluation [26], supplier selection [23], study of poverty alleviation [27], and industrial arc robot selection [28]. Other studies have focused on improving the entropy method. Szmidt and Kacprzyk [29] proposed an entropy measure for intuitionistic fuzzy sets (IFS) that was extended. The difference between normalized Euclidean distance and normalized Hamming distance is investigated. A new entropy method was proposed by Liu and Ren [30], which considered both the uncertainty and hesitancy degree. Thakur et al. [31] proposed a new approach using the COPRAS Model under IFS. As the literature shows, entropy is used in calculating weights [32].

The second stage of the proposed model uses the PROMETHEE approach to classify the alternatives. This model was proposed by Brans et al. [33]. A few years later, several versions of the PROMETHEE methods were developed such as PROMETHEE III, PROMETHEE IV, PROMETHEE V [34], PROMETHEE VI [35], PROMETHEE GDSS [36], and the GAIA interactive visual module for graphical representation [37]. These versions were developed to help with more complicated decision-making situations [38]. Like other methods, applications in new areas are carried out simultaneously, including cryptocurrency portfolio allocation [39], a barrier assessment framework for carbon sink project implementation [40], and an application of hybrid composites [41].

The motivation for developing the proposed model is based on the need to reduce uncertainty in the decision-making process without dehumanizing the process. The proposed method combines objective and subjective methods to strengthen the results presented to the decision maker. The methods chosen are widely disseminated among the scientific community and are easy to understand and implement. The concept used allows for expansion and integration with other methods. By using hybrid approaches, the results are supposed to be more efficient and balance the subjectivity of the decision makers. EC-PROMETHEE does not use combined weights between the three methods. However, it will operate with a range of weights based on the upper and lower limits of the values obtained in the three methods. The final ranking will not be accepted by applying a single set of weights, but with "m" iterations using a set of random weights produced within the respective weight ranges, criterion by criterion.

In this article, we will revisit the research developed by Basilio et al. [42–45], which dealt with identifying and choosing policing strategies customized to local criminal demands. The research was conducted in Rio de Janeiro, Brazil, and analyzed the criminal demand from 2016 to 2019. The authors used the PROMETHEE method, Electre IV, and Electre I to identify the most appropriate policing strategies for the observed criminal demands. At the time, the researchers used equal weights for each criterion. In the present research, we seek to answer the question: how can using objective weighting methods influence the ranking of policing strategies in the case studied? In response, the authors developed the EC-PROMETHEE method, which combines objective and subjective methods of weighting criteria, implementing a range of weights for criteria, and defining the final ranking from a certain number of iterations.

This article is divided into five parts. The first part is described above, where we contextualize concepts about the multi-criteria methods used and the importance of decision-making in the decision-making process, and present the problem that will be studied. Then, in the second section, we will describe the methods and algorithms we will use to solve the problem. In the third section, we describe the results found. In the fourth section, we present the discussions about the nuances of the new method concerning the traditional models. Finally, in the fifth section, we will conclude the research report and indicate possibilities for future research.

2. Materials and Methods

This section presents the concepts for formulating the hybrid EC-PROMETHEE method. Figure 1 illustrates the description of the proposed method by subdividing it into eight steps.

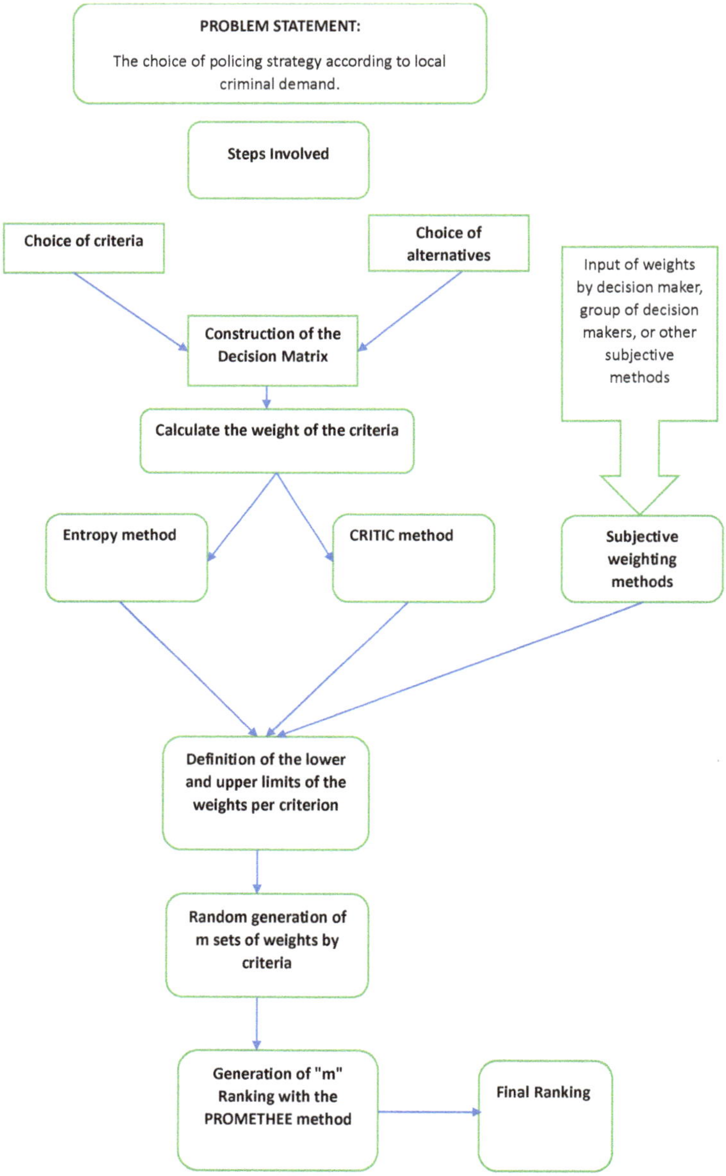

Figure 1. Methodological scheme.

Step 1—Identification of criteria

In the first stage, we identified twenty criteria. The specified criteria are taken from the studies of Basilio and Pereira [42–48]. The criteria show the most recurrent types of crime, misdemeanors, and urban disorder. The Public Security Institute (ISP) performs statistical analysis and monitoring. Table 1 shows the list of crime types used in the proposed modeling.

Table 1. List of criminal lawsuits.

Criteria Caption	Code
Murder	C1
Robbery	C2
Vehicle theft	C3
Theft residence	C4
Street robbery	C5
Cargo theft	C6
Bank robbery	C7
Theft to a commercial establishment	C8
Theft	C9
Kidnapping	C10
Drug seizure	C11
Seizure of weapons	C12
Threat	C13
Use of narcotic	C14
Drug traffic	C15
Disruption to quietness	C16
Traffic accident	C17
Illegal weapon	C18
Domestic violence	C19
Bank alarm trip	C20

Source: Adapted from Basilio et al. [42].

Step 2—Identification of alternatives

Table 2 shows fourteen policing strategies taken from the study carried out by Basilio et al. [42–46]. The data presented in Table 2 originates from the literature review produced by Basilio et al. [48].

Table 2. Types of policing strategies.

Types of Strategies	Random	Oriented
Foot patrol	Strategy_1	Strategy_5
Radio patrol	Strategy_2	Strategy_6
Motorcycle patrol	Strategy_3	Strategy_7
Horse patrol	Strategy_4	Strategy_8
Preventive action operation	Not applied	Strategy_9
Operation of repressive action (scouring)	Not applied	Strategy_10
Operation of repressive action (search and capture)	Not applied	Strategy_11
Operation of repressive action (to search)	Not applied	Strategy_12
Operation of repressive action (siege)	Not applied	Strategy_13
Transit operations	Not applied	Strategy_14

Source: Adapted from Basilio et al. [42,49].

Step 3—Construction of the decision matrix

In this step, we will use the data from the research reported by Basilio et al. [42–47], which were obtained by applying 430 questionnaires to decision makers distributed at the strategic, tactical, and operational levels of the Military Police of the State of Rio de Janeiro. The reported research covered thirty-nine operational units in the State of Rio de Janeiro/Brazil territory. The questionnaire obtained the decision makers' perception of the effectiveness of policing strategies (Table 2) in impacting criminal demands (Table 1).

The researchers used a five-point Likert scale to systematize the collection of respondents' perceptions. The scale was established as follows: (5) contribute an extreme amount; (4) contribute very much; (3) contribute moderately; (2) contribute little; (1) contribute very little. The data were subjected to descriptive statistical treatment, and the statistical measure of the central tendency "mode" was used to identify the predominant perception of the respondents regarding the set of evaluations performed [42].

In the current research, in addition to the "mode", we will use other measures, such as the average, median, consensus_mode, consensus_average, consensus_median, and the Likert scale, to increase the information power of each alternative and verify how they influence the final ordering of policing strategies. Tables 3–9 show the data used in the decision matrices used in the proposed model.

Step 4—Calculation of the weights of the criteria

Weight denotes the importance of each criterion in the decision-making process. Changes in criteria weight may lead to different results. Thus, selecting a suitable method for assigning accurate weights to different criteria is crucial [3]. Subjective, Objective, and Integrated weighting methods are some of the different methods used for assigning weights [50]. In subjective weighting methods, experts' opinions are used. The main disadvantages are that it is time-consuming and may offer conflicting opinions, according to Mahajan et al. [51]. The analytic Hierarchy Process (AHP) is a widely used method for subjective weighting. It uses pairwise comparison questions to elicit a matrix of relative preference judgments between each pair of alternatives with respect to each criterion, and a matrix of relative importance of each criterion. The judgements are derived from nominal group discussions or the Delphi technique, which may result in bias [51]. With the increase in the number of criteria, pairwise comparisons increase, resulting in hefty computation. Due to these limitations, the current study proposes the use of objective weighting methods. Weights are derived using mathematical computation without the intercession of a decision maker when objective weighting methods are employed. Entropy method, Criteria Importance Through Intercriteria Correlation (CRITIC method), and FANMA method are among the most commonly used methods for objective weighting methods [13,50,51]. In this article, we have considered the Entropy and CRITIC methods to assess the criteria weights.

Step 4.1 The ENTROPY method

The criteria weights are based on the predefined decision matrix that includes the information regarding the set of alternatives. Entropy in information theory is a model for the uncertainty volume served by a discrete probability distribution [51,52]. Salwa et al. [53] used the entropy method to calculate criterion weight to select optimal starch as the matrix in green composites for single-use food packaging applications [53]. The Entropy of the normalized decision matrix (NDM) criterion is given in Equation (1):

$$E_j = -\frac{\left[\sum_{i=1}^{m} P_{ij}\ln(P_{ij})\right]}{\ln(m)}; j = 1,2,\ldots,n \; and \; i = 1,2,\ldots,m \tag{1}$$

where P_{ij} is NDM, which is given by Equation (2):

$$P_{ij} = \frac{x_{ij}}{\sum_{i=1}^{m} x_{ij}}; j = 1,2,\ldots,n \; and \; i = 1,2,\ldots,m \tag{2}$$

where x_{ij} corresponds to the criteria value for each alternative in DM. The criteria weight, W_j^E can be calculated using Equation (3):

$$W_j^E = \frac{1 - E_j}{\sum_{j=1}^{n} 1 - E_j}; j = 1,2,\ldots,n \tag{3}$$

where $(1 - E_j)$ denotes the degree of diversity of the information in the jth criterion outcome.

Table 3. Decision matrix of the impact of policing strategies versus criminal demand ("Mode").

Alternatives\Criteria	C1	C2	C3	C4	C5	C6	C7	C8	C9	C10	C11	C12	C13	C14	C15	C16	C17	C18	C19	C20
Strategy_1	1	1	2	2	5	1	2	5	5	1	2	2	1	3	3	2	1	2	1	1
Strategy_2	2	3	3	3	3	3	3	3	2	1	3	3	2	3	3	3	2	3	1	1
Strategy_3	2	3	3	3	4	3	3	3	3	1	3	3	2	3	3	3	1	3	1	1
Strategy_4	1	1	1	1	3	1	1	1	1	1	1	1	1	1	1	1	1	1	1	1
Strategy_5	1	3	3	3	5	1	4	5	5	1	3	3	2	4	3	3	1	3	1	1
Strategy_6	4	4	5	4	5	5	5	5	3	3	4	4	3	4	4	4	4	4	1	1
Strategy_7	3	4	5	4	5	5	4	5	5	3	4	4	3	4	4	3	4	4	1	1
Strategy_8	1	1	1	1	5	1	1	3	4	1	1	1	1	3	1	1	1	3	1	1
Strategy_9	3	4	4	4	4	4	4	4	3	1	3	3	1	4	3	3	3	3	1	1
Strategy_10	3	3	4	3	3	4	3	3	3	1	5	5	1	4	5	1	1	5	1	1
Strategy_11	3	2	4	3	3	4	3	3	1	1	5	5	1	4	5	1	1	5	1	1
Strategy_12	3	4	4	3	4	4	3	4	4	1	5	5	1	5	5	1	1	5	1	1
Strategy_13	3	3	5	3	3	5	4	4	3	1	4	4	1	4	4	1	1	5	1	1
Strategy_14	1	2	5	1	3	5	3	3	1	1	3	3	1	3	3	1	5	4	1	1

Source: Adapted from Basilio et al. [42].

Table 4. Decision matrix of the impact of policing strategies versus criminal demand ("Average").

Alternatives\Criteria	C1	C2	C3	C4	C5	C6	C7	C8	C9	C10	C11	C12	C13	C14	C15	C16	C17	C18	C19	C20
Strategy_1	1.94	2.58	2.35	2.53	3.93	1.70	2.99	3.50	3.31	1.74	2.44	2.44	2.24	3.14	2.58	2.60	1.72	2.48	1.68	1.86
Strategy_2	2.52	3.09	3.46	2.91	3.27	3.18	3.13	3.23	2.62	2.16	2.81	2.88	2.23	3.03	2.89	2.73	2.21	2.80	1.86	2.15
Strategy_3	2.47	2.98	3.48	2.84	3.49	2.94	3.04	3.28	2.78	2.14	2.79	2.88	2.23	3.01	2.81	2.62	2.34	2.77	1.78	2.12
Strategy_4	1.52	1.79	1.77	1.89	2.81	1.59	1.93	2.28	2.21	1.50	1.87	1.90	1.80	2.35	2.05	2.07	1.49	2.03	1.45	1.56
Strategy_5	2.39	3.01	2.81	3.27	4.39	2.31	3.70	4.10	3.72	2.09	3.05	3.03	2.51	3.56	3.16	3.02	2.13	3.00	1.93	2.31
Strategy_6	3.20	3.73	4.22	3.79	4.11	4.14	4.08	4.16	3.40	2.73	3.76	3.78	2.85	3.81	3.86	3.44	2.85	3.58	2.39	2.77
Strategy_7	3.03	3.67	4.21	3.73	4.30	3.81	4.01	4.20	3.54	2.67	3.62	3.59	2.75	3.77	3.63	3.33	2.95	3.51	2.25	2.65
Strategy_8	1.95	2.34	2.38	2.62	3.48	2.05	2.67	3.05	2.91	1.89	2.33	2.35	2.06	2.85	2.47	2.49	1.83	2.43	1.76	1.92
Strategy_9	2.84	3.34	3.90	3.37	3.88	3.82	3.69	3.80	3.30	2.63	3.09	3.16	2.50	3.44	3.36	3.05	2.88	3.16	2.24	2.38
Strategy_10	2.97	3.00	3.53	2.82	2.97	3.63	2.80	2.83	2.55	2.51	4.29	4.33	2.36	3.78	4.28	2.53	2.05	4.07	1.98	1.85
Strategy_11	3.20	3.15	3.62	2.88	3.02	3.52	2.88	2.94	2.59	2.62	4.27	4.31	2.46	3.74	4.26	2.38	1.98	4.06	1.95	1.84
Strategy_12	3.08	3.40	4.08	3.01	3.50	3.91	3.14	3.27	2.85	2.70	4.28	4.33	2.28	3.84	4.08	2.24	2.18	4.14	1.84	1.79
Strategy_13	2.86	3.05	4.01	2.99	3.05	3.84	3.15	3.06	2.53	2.79	3.65	3.70	2.14	3.16	3.57	2.10	2.16	3.57	1.81	1.82
Strategy_14	2.15	2.57	4.09	2.30	2.92	3.83	2.74	2.73	2.30	2.47	3.09	3.24	1.90	2.72	2.95	1.90	3.66	3.28	1.61	1.68

Table 5. Decision matrix of the impact of policing strategies versus criminal demand ("Median").

Alternatives\Criteria	C1	C2	C3	C4	C5	C6	C7	C8	C9	C10	C11	C12	C13	C14	C15	C16	C17	C18	C19	C20
Strategy_1	2	3	2	2	4	1	3	4	3	1	2	2	2	3	3	2	1	2	1	1
Strategy_2	2	3	3	3	3	3	3	3	3	2	3	3	2	3	3	3	2	3	2	2
Strategy_3	2	3	4	3	4	3	3	3	3	2	3	3	2	3	3	3	2	3	1	2
Strategy_4	1	1	1	2	3	1	2	2	2	1	2	2	1.5	2	2	2	1	2	1	1
Strategy_5	2	3	3	3	5	2	4	4	4	2	3	3	2	4	3	3	2	3	2	2
Strategy_6	3	4	4	4	4	4	4	4	3	3	4	4	3	4	4	4	3	4	2	3
Strategy_7	3	4	4	4	5	4	4	4	4	3	4	4	3	4	4	3	3	4	2	3
Strategy_8	2	2	2	3	4	2	3	3	3	1	2	2	2	3	2	2	1.5	2	1	1
Strategy_9	3	3	4	3.5	4	4	4	4	3	3	3	3	2	4	3	3	3	3	2	2
Strategy_10	3	3	4	3	3	4	3	3	2.5	2	4	4.5	2	4	5	2	2	4	2	1
Strategy_11	3	3	4	3	3	4	3	3	3	2	4	4	2	4	4.5	2	2	4	2	1
Strategy_12	3	4	4	3	4	4	3	3	3	3	4	5	2	4	4	2	2	4	1	1
Strategy_13	3	3	4	3	3	4	3	3	2	3	4	4	2	3	4	2	2	4	1	1
Strategy_14	2	2	4	2	3	4	3	3	2	2	3	3	2	3	3	2	4	3	1	1

Table 6. Decision matrix of the impact of policing strategies versus criminal demand ("Likert Scale").

Alternatives\Criteria	C1	C2	C3	C4	C5	C6	C7	C8	C9	C10	C11	C12	C13	C14	C15	C16	C17	C18	C19	C20
Strategy_1	556	739	672	723	1123	487	854	1001	947	499	698	697	641	898	738	743	493	710	480	533
Strategy_2	722	885	989	831	935	909	891	925	750	617	805	825	638	868	826	782	633	800	533	615
Strategy_3	705	852	996	813	998	841	866	937	796	612	799	824	638	862	804	749	669	792	510	606
Strategy_4	435	512	507	541	803	454	550	653	632	428	534	543	516	672	587	592	427	580	414	445
Strategy_5	683	862	804	936	1255	660	1054	1174	1064	598	871	868	717	1019	905	865	609	859	552	662
Strategy_6	915	1066	1207	1083	1175	1185	1162	1191	972	780	1076	1081	814	1091	1103	985	816	1023	683	791
Strategy_7	868	1051	1205	1066	1229	1091	1142	1202	1013	765	1034	1028	786	1078	1039	953	843	1005	644	758
Strategy_8	558	670	681	748	995	586	761	873	833	540	666	673	590	814	707	712	524	694	503	549
Strategy_9	813	954	1114	963	1110	1093	1053	1088	943	751	884	905	716	984	962	873	823	903	641	682
Strategy_10	849	858	1011	806	849	1037	799	808	730	718	1227	1238	675	1080	1224	723	586	1164	566	528
Strategy_11	916	900	1036	824	863	1007	820	841	741	748	1221	1233	703	1071	1219	681	565	1162	558	527
Strategy_12	880	971	1167	861	1001	1118	895	934	814	772	1223	1239	652	1097	1166	642	624	1184	527	511
Strategy_13	817	871	1147	855	873	1097	897	874	724	797	1044	1057	613	903	1020	601	618	1022	519	521
Strategy_14	614	735	1170	657	836	1095	780	780	659	706	884	927	542	779	844	544	1048	938	460	480

Table 7. Decision matrix of the impact of policing strategies versus criminal demand ("Consensus_mode").

Alternatives\Criteria	C1	C2	C3	C4	C5	C6	C7	C8	C9	C10	C11	C12	C13	C14	C15	C16	C17	C18	C19	C20
Strategy_1	0.724	0.674	1.404	1.355	3.427	0.756	1.340	3.341	3.173	0.748	1.392	1.458	0.702	2.033	2.032	1.322	0.769	1.428	0.761	0.732
Strategy_2	1.427	2.220	2.159	2.211	2.218	2.111	2.205	2.257	1.425	0.718	2.260	2.251	1.511	2.245	2.168	2.144	1.441	2.238	0.755	0.703
Strategy_3	1.405	2.184	2.075	2.162	2.860	2.111	2.180	2.218	2.054	0.713	2.222	2.222	1.494	2.227	2.109	2.102	0.685	2.167	0.761	0.692
Strategy_4	0.816	0.771	0.764	0.757	1.996	0.790	0.735	0.700	0.687	0.810	0.778	0.769	0.762	0.700	0.732	0.712	0.822	0.750	0.823	0.795
Strategy_5	0.685	2.182	2.103	2.129	3.903	0.678	2.834	3.842	3.509	0.704	2.161	2.176	1.388	2.914	2.085	2.073	0.716	2.210	0.731	0.667
Strategy_6	2.686	3.045	4.013	3.008	3.920	3.882	3.913	3.908	2.032	1.919	3.084	3.114	2.064	3.107	3.031	2.775	2.723	2.977	0.670	0.640
Strategy_7	1.969	2.898	3.929	2.915	3.958	3.566	3.110	3.889	3.382	1.928	3.035	3.017	2.044	3.005	2.831	2.058	2.704	2.954	0.686	0.634
Strategy_8	0.748	0.681	0.672	0.659	3.208	0.715	0.642	1.952	2.539	0.708	0.702	0.687	0.737	1.979	0.664	0.661	0.754	2.104	0.747	0.714
Strategy_9	2.030	2.799	3.175	2.826	3.108	2.994	2.916	3.026	2.010	0.637	2.100	2.097	0.677	2.800	2.099	2.046	2.010	2.107	0.665	0.638
Strategy_10	2.043	2.047	2.861	2.077	2.093	2.871	2.041	2.074	2.027	0.638	4.001	4.045	0.682	2.898	3.920	0.671	0.709	3.816	0.717	0.713
Strategy_11	1.921	1.299	2.855	2.062	2.040	2.757	1.998	2.031	0.658	0.629	4.010	4.032	0.660	2.812	3.911	0.690	0.729	3.784	0.730	0.705
Strategy_12	2.040	2.817	3.199	2.180	2.727	3.017	2.062	2.702	2.611	0.626	4.018	4.067	0.678	3.538	3.829	0.702	0.682	3.888	0.735	0.731
Strategy_13	2.026	2.065	3.800	2.094	2.033	3.542	2.681	2.690	2.031	0.606	2.880	2.931	0.696	2.639	2.754	0.730	0.676	3.416	0.745	0.726
Strategy_14	0.723	1.375	3.822	0.702	2.029	3.501	2.019	2.075	0.692	0.615	2.066	2.113	0.739	2.067	2.086	0.746	3.135	2.799	0.778	0.747

Table 8. Decision matrix of the impact of policing strategies versus criminal demand ("Consensus_average").

Alternatives\Criteria	C1	C2	C3	C4	C5	C6	C7	C8	C9	C10	C11	C12	C13	C14	C15	C16	C17	C18	C19	C20
Strategy_1	1.408	1.741	1.649	1.713	2.692	1.287	2.001	2.339	2.102	1.305	1.699	1.776	1.574	2.128	1.748	1.718	1.325	1.773	1.277	1.364
Strategy_2	1.802	2.290	2.488	2.141	2.417	2.236	2.298	2.433	1.869	1.550	2.120	2.164	1.685	2.271	2.087	1.954	1.594	2.086	1.408	1.512
Strategy_3	1.732	2.169	2.409	2.048	2.495	2.069	2.209	2.422	1.906	1.526	2.069	2.134	1.666	2.237	1.976	1.835	1.601	2.000	1.357	1.467
Strategy_4	1.240	1.380	1.355	1.433	1.868	1.254	1.418	1.598	1.519	1.212	1.453	1.460	1.375	1.644	1.503	1.475	1.227	1.521	1.191	1.238
Strategy_5	1.637	2.192	1.971	2.322	3.426	1.565	2.620	3.154	2.611	1.472	2.194	2.202	1.740	2.595	2.199	2.090	1.524	2.213	1.410	1.544
Strategy_6	2.148	2.838	3.387	2.848	3.221	3.217	3.191	3.255	2.302	1.745	2.901	2.942	1.958	2.963	2.923	2.390	1.942	2.662	1.601	1.769
Strategy_7	1.992	2.662	3.310	2.717	3.401	2.721	3.116	3.269	2.396	1.719	2.743	2.711	1.873	2.831	2.571	2.286	1.993	2.595	1.545	1.679
Strategy_8	1.460	1.595	1.599	1.724	2.232	1.466	1.713	1.986	1.849	1.336	1.634	1.616	1.521	1.877	1.642	1.647	1.381	1.702	1.314	1.370
Strategy_9	1.924	2.334	3.091	2.379	3.016	2.860	2.693	2.878	2.209	1.672	2.164	2.212	1.694	2.408	2.353	2.082	1.928	2.217	1.491	1.521
Strategy_10	2.022	2.047	2.529	1.951	2.071	2.602	1.908	1.953	1.725	1.601	3.433	3.502	1.609	2.736	3.356	1.696	1.452	3.106	1.418	1.316
Strategy_11	2.051	2.044	2.585	1.980	2.052	2.427	1.916	1.991	1.704	1.646	3.424	3.476	1.621	2.633	3.334	1.642	1.440	3.075	1.424	1.300
Strategy_12	2.092	2.391	3.263	2.188	2.387	2.948	2.158	2.206	1.858	1.691	3.436	3.524	1.545	2.714	3.122	1.577	1.488	3.219	1.355	1.307
Strategy_13	1.929	2.097	3.048	2.087	2.068	2.717	2.110	2.055	1.714	1.690	2.629	2.708	1.492	2.083	2.455	1.533	1.460	2.442	1.353	1.322
Strategy_14	1.552	1.767	3.127	1.614	1.977	2.681	1.842	1.887	1.594	1.519	2.129	2.283	1.400	1.877	2.052	1.418	2.297	2.295	1.251	1.254

Table 9. Decision matrix of the impact of policing strategies versus criminal demand ("Consensus_Median").

Alternatives\Criteria	C1	C2	C3	C4	C5	C6	C7	C8	C9	C10	C11	C12	C13	C14	C15	C16	C17	C18	C19	C20
Strategy_1	1.448	2.021	1.404	1.355	2.742	0.756	2.010	2.673	1.904	0.748	1.392	1.458	1.405	2.033	2.032	1.322	0.769	1.428	0.761	0.732
Strategy_2	1.427	2.220	2.159	2.211	2.218	2.111	2.205	2.257	2.138	1.437	2.260	2.251	1.511	2.245	2.168	2.144	1.441	2.238	1.511	1.406
Strategy_3	1.405	2.184	2.767	2.162	2.860	2.111	2.180	2.218	2.054	1.426	2.222	2.222	1.494	2.227	2.109	2.102	1.369	2.167	0.761	1.385
Strategy_4	0.816	0.771	0.764	1.515	1.996	0.790	1.470	1.400	1.375	0.810	1.556	1.538	1.143	1.399	1.464	1.425	0.822	1.500	0.823	0.795
Strategy_5	1.371	2.182	2.103	2.129	3.903	1.356	2.834	3.074	2.807	1.408	2.161	2.176	1.388	2.914	2.085	2.073	1.432	2.210	1.461	1.334
Strategy_6	2.014	3.045	3.210	3.008	3.136	3.106	3.130	3.126	2.032	1.919	3.084	3.114	2.064	3.107	3.031	2.775	2.042	2.977	1.341	1.919
Strategy_7	1.969	2.898	3.143	2.915	3.958	2.853	3.110	3.111	2.706	1.928	3.035	3.017	2.044	3.005	2.831	2.058	2.028	2.954	1.372	1.901
Strategy_8	1.497	1.362	1.343	1.977	2.567	1.431	1.925	1.952	1.904	0.708	1.403	1.374	1.475	1.979	1.328	1.323	1.131	1.402	0.747	0.714
Strategy_9	2.030	2.099	3.175	2.473	3.108	2.994	2.916	3.026	2.010	1.910	2.100	2.097	1.353	2.800	2.099	2.046	2.010	2.107	1.330	1.275
Strategy_10	2.043	2.047	2.861	2.077	2.093	2.871	2.041	2.074	1.689	1.276	3.201	3.641	1.363	2.898	3.920	1.342	1.418	3.053	1.433	0.713
Strategy_11	1.921	1.948	2.855	2.062	2.040	2.757	1.998	2.031	1.974	1.258	3.208	3.225	1.319	2.812	3.519	1.380	1.458	3.027	1.460	0.705
Strategy_12	2.040	2.817	3.199	2.180	2.727	3.017	2.062	2.026	1.958	1.879	3.214	4.067	1.356	2.831	3.063	1.405	1.364	3.111	0.735	0.731
Strategy_13	2.026	2.065	3.040	2.094	2.033	2.834	2.011	2.018	1.354	1.819	2.880	2.931	1.393	1.979	2.754	1.459	1.351	2.733	0.745	0.726
Strategy_14	1.445	1.375	3.058	1.405	2.029	2.801	2.019	2.075	1.384	1.231	2.066	2.113	1.477	2.067	2.086	1.491	2.508	2.099	0.778	0.747

Step 4.2 The CRITIC method

In this section, the researchers briefly describe the CRITIC method. The CRITIC method proposed by [52] aims to determine the criteria weights. The main stages of this technique are described below:

Step 4.2.1. A decision matrix, Z, with m rows as the number of alternatives and n column as the number of criteria, is defined by Equation (4):

$$Z = (r_{ij})_{mxn}; i = 1, \ldots, m; j = 1, \ldots, n \tag{4}$$

where r_{ij} is the correlation of the ith alternative and of the jth criterion.

Step 4.2.2. Each criterion can be considered beneficial or non-beneficial [54–56]. A criterion takes value in some bounded range. Sharkasi and Rezakhah [22] assert that for a beneficial, $j \in F^+$, the criterion is normalized by dividing its distance from the minimum value by the length of the range. In contrast, a non-beneficial one, $j \in F^-$, is normalized by dividing its distance from the maximum value by the length of the range. The elements of the decision matrix are normalized as given in Equations (5) and (6) for the positive or beneficial criteria and the negative or non-beneficial ones.

$$x_{ij}^+ = \frac{r_{ij} - r_j^-}{r_j^+ - r_j^-}; i = 1, \ldots, m; j = 1, \ldots, n \; if \; j \in F^+ \tag{5}$$

$$x_{ij}^- = \frac{r_j^+ - r_{ij}}{r_j^+ - r_j^-}; i = 1, \ldots, m; j = 1, \ldots, n \; if \; j \in F^- \tag{6}$$

where $r_j^+ = \max(r_{1j}, r_{2j}, \ldots, r_{mj})$ and $r_j^- = \min(r_{1j}, r_{2j}, \ldots, r_{mj})$, and x_{ij} which is either x_j^+ or x_j^- represents the normalized value of the ij element of the decision matrix.

Step 4.2.3. The Pearson correlation coefficient between two criteria, j and k, is computed as Equation (7)

$$\rho_{jk} = \frac{\sum_{i=1}^{m} \left(x_{ij} - \underline{x_j}\right)\left(x_{ik} - \underline{x_k}\right)}{\sqrt{\sum_{i=1}^{m} \left(x_{ij} - \underline{x_j}\right)^2 \sum_{i=1}^{m} \left(x_{ik} - \underline{x_k}\right)^2}} \tag{7}$$

where $\underline{x_j}$ and $\underline{x_k}$ represent the mean of jth and kth criteria Equation (8):

$$\underline{x_k} = \frac{1}{n} \sum_{i=1}^{m} x_{ik}; k = 1, \ldots, n. \tag{8}$$

The Pearson correlation coefficient captures linear correlations.

Step 4.2.4. The standard deviation of each criterion is estimated by Equation (9):

$$\sigma_j = \sqrt{\frac{1}{n-1} \sum_{i=1}^{m} \left(x_{ij} - \underline{x_j}\right)^2}; \; j = 1, \ldots, n \tag{9}$$

Step 4.2.5. The index of the jth criteria, E_j, is evaluated by Equation (10)

$$E_j = \sigma_j \sum_{k=1}^{n} \left(1 - \rho_{jk}\right); j = 1, \ldots, n. \tag{10}$$

Step 4.2.6. The weights of the criteria are determined by Equation (11)

$$W_j^C = \frac{E_j}{\sum_{j=1}^{n} E_j}; j = 1, \ldots, n. \tag{11}$$

Finally, the ranking of the weights of the criteria is obtained. The ranking identifies the importance given to each criterion.

Step 5—Definition of the lower and upper limits of the weights per criterion

After generating the weights of each criterion using the Entropy and CRITIC methods, which constitute the objective methods, the model opens the door to input weights from subjective methods, which can be obtained by a single decision maker or a group of decision makers, with or without the use of subjective methods [2] such as AHP; SAPEVO-M; FUCOM; and MEREC among others.

In this step, we define the lower-limit vector. Ll_j where criterion j will store the smallest weight value obtained from the set of values formed by $\left\{W_j^E, W_j^C, W_j^{DM}\right\}$, as shown in Equation (12)

$$Ll_j = Min\left\{W_j^E, W_j^C, W_j^{DM}\right\} \tag{12}$$

Next, we will define the upper limit vector. Ul_j, which for each criterion j will store the highest weight value obtained from the set of values formed by $\left\{W_j^E, W_j^C, W_j^{DM}\right\}$, as shown in Equation (13)

$$Ul_j = Max\left\{W_j^E, W_j^C, W_j^{DM}\right\} \tag{13}$$

Step 6—Random generation of "t" sets of weights by criteria

The Randomised Weight Matrix RWm of dimension t × n will be generated in this phase where t is the total number of rows, corresponding to the total number of iterations inserted in the model by the decision maker, and where n is the total number of columns of the matrix. The RWm matrix is obtained by generating different random numbers limited for each criterion by the limits Ll_j and Ul_j, as shown in Equation (14):

$$RWm_{ij} = \left(\left(Ul_j - Ll_j\right) * Rnd\right) + Ll_j\right) \tag{14}$$

Next, the matrix RWm_{ij} is normalized by Equation (15):

$$RWm_{ij}^n = \frac{x_{ij}}{\sum_{j=1}^n x_{ij}} \tag{15}$$

Step 7—Generation of "t" ranking with the PROMETHEE method

The literature identifies seven types of methods that integrate the PROMETHEE family [33,57], as recorded in recent research: PROMETHEE I [58]; PROMETHEE II [59,60]; PROMETHEE III; PROMETHEE IV; PROMETHEE V [61]; PROMETHEE VI [62]; and PROMETHEE GAIA [63].

The PROMETHEE II method consists of constructing an outranking relation of values. As Fontana and Cavalcante [64] state, the main advantage of PROMETHEE II is that it is a relatively simple ranking method in design and application compared to other multi-criteria analysis methods. It is well suited to issues where a finite number of alternatives should be ranked considering criteria. This method stands out since it seeks to involve concepts and parameters with some physical or economic interpretation, easily understood by the decision maker.

In their research, 217 papers are analyzed by Behzadian et al. [65] identifying studies that applied the PROMETHEE method. The following areas are given as fields of study: Environment Management, Manufacturing and Assembly, Hydrology and Water Management, Chemistry Logistics and Transportation, Business and Financial Management, Energy Management, and Social and Public Security.

The method is implemented in five steps. In the first step, there is a function showing the decision makers' preference concerning share "a" compared with share "b". The second step compares the suggested alternatives to the pairs for the preference function. The PROMETHEE proposes the six following types (shapes) of preference functions, as shown in Table 10:

Table 10. Types of preference function.

Type	Generalized Criterion	Condition	Quantification of Preference	Parameter to Fix
Type I—Usual preference function	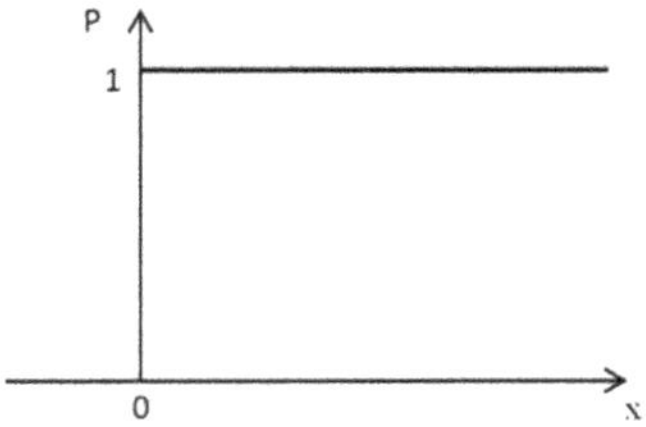	$g(a)-g(b)>0$ $g(a)-g(b)\leq 0$	$P_j(a,b)=1$ $P_j(a,b)=0$	-
Type II—U-shape preference function	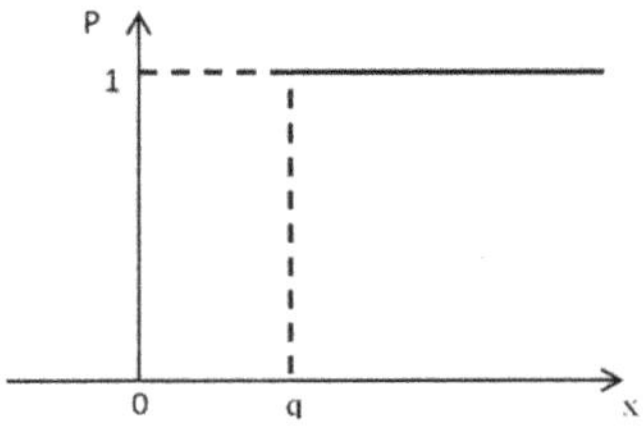	$g(a)-g(b)>q$ $g(a)-g(b)\leq q$	$P_j(a,b)=1$ $P_j(a,b)=0$	q
Type III—V-shape preference function	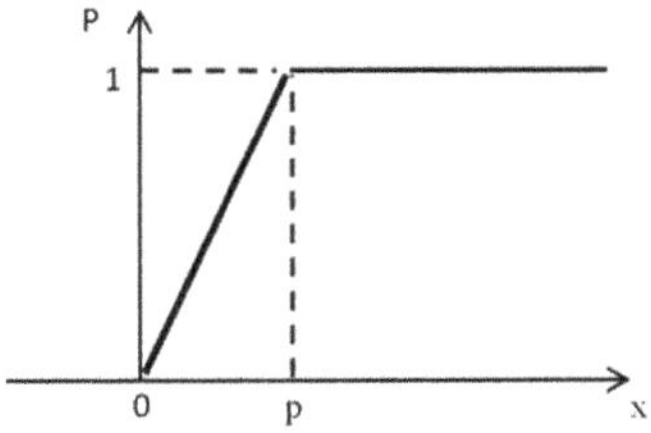	$g(a)-g(b)>p$ $g(a)-g(b)\leq p$ $g(a)-g(b)\leq 0$	$P_j(a,b)=1$ $P_j(a,b)=\frac{\lvert g(a)-g(b)\rvert}{p}$ $P_j(a,b)=0$	p
Type IV—Level preference function	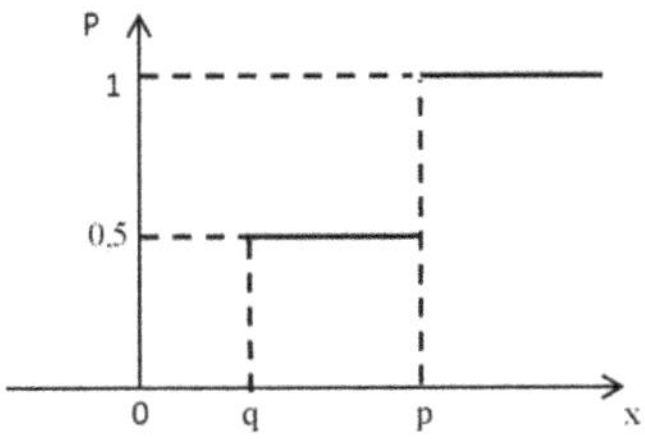	$\lvert g(a)-g(b)\rvert>p$ $q<\lvert g(a)-g(b)\rvert\leq p$ $\lvert g(a)-g(b)\rvert\leq q$	$P_j(a,b)=1$ $P_j(a,b)=\frac{1}{2}$ $P_j(a,b)=0$	p, q
Type V—Linear preference function	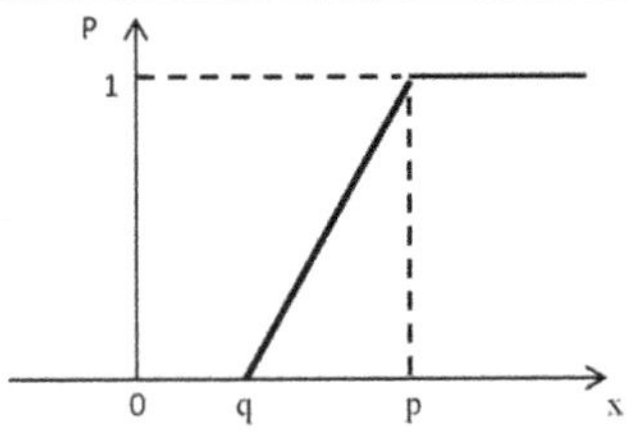	$\lvert g(a)-g(b)\rvert>p$ $q<\lvert g(a)-g(b)\rvert\leq p$ $\lvert g(a)-g(b)\rvert\leq q$	$P_j(a,b)=1$ $P_j(a,b)=\frac{[\lvert g(a)-g(b)\rvert-q]}{(p-q)}$ $P_j(a,b)=0$	p, q
Type VI—Gaussian preference function	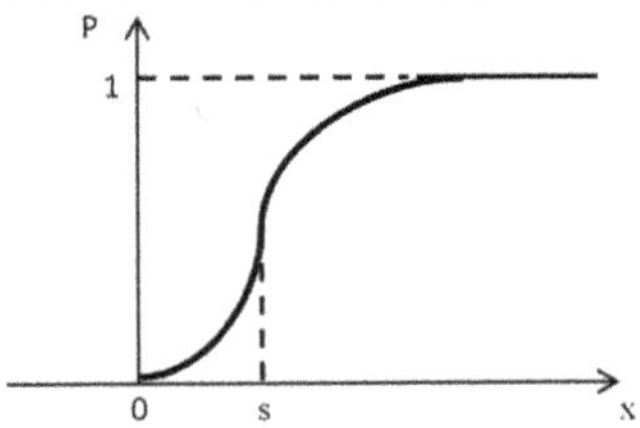	$g(a)-g(b)>0$ $g(a)-g(b)\leq 0$	$P_j(a,b)=$ $1-e^{\{\frac{-(g(a)-g(b))^2}{2s^2}\}}$ $P_j(a,b)=0$	s

Source: Adapted from Basilio et al. [42].

As a third step, the results of this comparison are presented in an evaluation matrix as the estimated values of each criterion for each alternative. The classification is performed in two final steps: a partial ranking in the fourth step and then a total ranking of alternatives in the fifth step, as follows:

Step 7.1. Determination of deviations based on pairwise comparations

$$d_j(a,b) = g_j(a) - g_j(b) \tag{16}$$

where $d_j(a,b)$ denotes the difference between the evaluations of a and b on each criterion.

Step 7.2. Application of the preference function

$$P_j(a,b) = F_j\left[d_j(a,b)\right] \quad j = 1,\ldots,k \tag{17}$$

where $P_j(a,b)$ denotes the preference of alternative a with regard to alternative b on each criterion as a function of $d_j(a,b)$.

Step 7.3. Calculation of an overall or global preference index

$$\forall a, b \epsilon A, \;\; \pi(a,b) = \sum_{j=1}^{k} P_j(a,b) w_j \tag{18}$$

where $\pi(a,b)$ of a over b is defined as the weighted sum $P_j(a,b)$ of for each criterion and w_j is the weight associated with the jth criterion.

Step 7.4. Calculation of outranking flows/The PROMETHEE II partial ranking

$$\varphi^{+}(a) = \sum_{x \in A} \pi(a,b) \tag{19}$$

And

$$\varphi^{-}(a) = \sum_{x \in A} \pi(b,a) \tag{20}$$

where $\varphi^{+}(a)$ and $\varphi^{-}(a)$ denotes the positive outranking and negative outranking flow for each alternative, respectively.

Step 7.5. Calculation of net outranking flow/The PROMETHEE II complete ranking

$$\varphi(a) = \varphi^{+}(a) - \varphi^{-}(a) \tag{21}$$

where $\varphi(a)$ denotes the net outranking flow for each alternative.

Step 8—Definition of final ranking

In this step, we present the second novelty of this new method. In step 6, we present the matrix. The matrix RWm_{ij}^{n} contains t sets of weights per criterion. The innovative point of this method is to generate t sets of rankings as different sets of weights are used, varying within the range of weights for each criterion, as dealt with in Step 5. In this sense, $\varphi(a)$ is transformed into an ordinal value. The $\varphi(a)$ is sorted in descending order, assigning 1st place to the alternative (a) that has the highest $\varphi(a)$, and so on until the last alternative m. The final ranking matrix FRm is of dimension t x m, where m is the number of columns composed of each alternative (a). Where "t" is the number of rows representing the ranking generated by the PROMETHEE method for each iteration, a_{ij} is the ordinal value of the ranking that alternative "j" obtained in iteration "i", as shown in Equation (22):

$$FRm_{ij} = a_{ij}, \forall i = 1,2,\ldots,t \; and \; j = 1,2,\ldots,m \tag{22}$$

Then, the value of each rank-ordering a_{ij} will be replaced by a score, as follows: 1st = m, 2nd = (m – 1), . . ., nth = $-m - -m - 1$). Thus, the final ranking vector FRv of dimension j is obtained. The final position of each alternative will be obtained by summing the scores of the t iterations of each alternative, as shown in Equation (23):

$$FRv_j = \sum_{i=1}^{t} FRm_{ij}, \; j = 1,2,\ldots,m \tag{23}$$

The final ranking will be obtained in descending order among the total scores of each alternative j of the vector FRv_j.

3. Results

In this section, we report the results found by applying the EC-PROMETHEE model to the problem of policing strategies and compare them with the results obtained in previous research [42]. In addition to the comparison, we address another latent issue: constructing the decision matrix. In this case, in addition to the statistical measure "Mode", we used the mean, median, consensus concept, and the Likert scale, as shown in Tables 3–9.

Initially, we emulated the model with the parameters common to those used in Basilio et al. [42] to maintain the comparison conditions, as described in Table 11.

Table 11. Common parameters inserted in the EC-PROMETHEE model.

Criterion	Code	Objective	Unit	Scale	Preference	Thresholds	Weight (W_j^{DM})
Murder	C1	Max	Scalar	R	Usual	Absolute	0.05
Robbery	C2	Max	Scalar	R	Usual	Absolute	0.05
Vehicle theft	C3	Max	Scalar	R	Usual	Absolute	0.05
Theft residence	C4	Max	Scalar	R	Usual	Absolute	0.05
Street robbery	C5	Max	Scalar	R	Usual	Absolute	0.05
Cargo theft	C6	Max	Scalar	R	Usual	Absolute	0.05
Bank robbery	C7	Max	Scalar	R	Usual	Absolute	0.05
Theft to a commercial establishment	C8	Max	Scalar	R	Usual	Absolute	0.05
Theft	C9	Max	Scalar	R	Usual	Absolute	0.05
Kidnapping	C10	Max	Scalar	R	Usual	Absolute	0.05
Drug seizure	C11	Max	Scalar	R	Usual	Absolute	0.05
Seizure of weapons	C12	Max	Scalar	R	Usual	Absolute	0.05
Threat	C13	Max	Scalar	R	Usual	Absolute	0.05
Use of narcotic	C14	Max	Scalar	R	Usual	Absolute	0.05
Drug traffic	C15	Max	Scalar	R	Usual	Absolute	0.05
Disruption to quietness	C16	Max	Scalar	R	Usual	Absolute	0.05
Traffic accident	C17	Max	Scalar	R	Usual	Absolute	0.05
Illegal weapon	C18	Max	Scalar	R	Usual	Absolute	0.05
Domestic violence	C19	Max	Scalar	R	Usual	Absolute	0.05
Bank alarm trip	C20	Max	Scalar	R	Usual	Absolute	0.05

Source: Adapted from Basilio et al. [42].

Following the model illustrated in Figure 1, we obtained the criteria and alternatives described in Tables 1 and 2, corresponding to steps 1–2. In step 3, we used Tables 3–9 as decision matrices to evaluate and compare different biases. In the fourth step, we introduced the values of the decision matrices from Tables 3–9 and the parameters in Table 11, and employed Algorithm (Appendix A), which executed Equations (1)–(11) and obtained the weights of the Entropy and CRITIC method, which are the input variables for steps 5 and 6 of the model, as recorded in Table 12.

In step 5, after obtaining the weights using the Entropy and CRITIC methods and with the external input of the weights of the decision makers (Table 11), we applied Equations (12) and (13) and the definition of the lower and upper limits of the weights per criterion, as shown in Table 13.

In step 6, with the data input from step 5, Equations (14) and (15) were applied to generate t iterations of weights for each criterion. In the proposed solution to the problem, we used t = 10,000 iterations. In this sense, a matrix of weights was generated with a total of 10,000, which will be applied using the PROMETHEE method to generate the final ranking. Figures 2–5 show the set of weights and how they vary between the scenarios used in this study.

Table 12. Table of weights per criterion generated by the EC-PROMETHEE hybrid method.

Scenario	Method	C1	C2	C3	C4	C5	C6	C7	C8	C9	C10	C11	C12	C13	C14	C15	C16	C17	C18	C19	C20
Mode	W_j^E	0.045	0.049	0.050	0.051	0.062	0.040	0.052	0.057	0.045	0.044	0.050	0.050	0.046	0.059	0.052	0.039	0.022	0.055	0.067	0.067
	W_j^C	0.049	0.048	0.046	0.045	0.104	0.061	0.035	0.044	0.078	0.059	0.048	0.048	0.065	0.032	0.047	0.069	0.072	0.050	0.001	0.001
	W_j^{DM}	0.050	0.050	0.050	0.050	0.050	0.050	0.050	0.050	0.050	0.050	0.050	0.050	0.050	0.050	0.050	0.050	0.050	0.050	0.050	0.050
Average	W_j^E	0.050	0.050	0.049	0.051	0.051	0.047	0.050	0.051	0.051	0.050	0.049	0.049	0.052	0.051	0.050	0.051	0.048	0.050	0.051	0.051
	W_j^C	0.040	0.027	0.051	0.030	0.079	0.059	0.034	0.051	0.071	0.047	0.057	0.058	0.034	0.041	0.051	0.059	0.068	0.058	0.032	0.051
	W_j^{DM}	0.050	0.050	0.050	0.050	0.050	0.050	0.050	0.050	0.050	0.050	0.050	0.050	0.050	0.050	0.050	0.050	0.050	0.050	0.050	0.050
Median	W_j^E	0.051	0.049	0.048	0.053	0.054	0.044	0.054	0.054	0.053	0.046	0.051	0.050	0.054	0.054	0.051	0.052	0.046	0.051	0.047	0.040
	W_j^C	0.040	0.031	0.044	0.037	0.078	0.056	0.036	0.050	0.060	0.047	0.052	0.047	0.035	0.037	0.047	0.057	0.049	0.052	0.084	0.059
	W_j^{DM}	0.050	0.050	0.050	0.050	0.050	0.050	0.050	0.050	0.050	0.050	0.050	0.050	0.050	0.050	0.050	0.050	0.050	0.050	0.050	0.050
Consensus_Mode	W_j^E	0.048	0.048	0.048	0.050	0.061	0.040	0.051	0.055	0.046	0.047	0.048	0.048	0.046	0.059	0.049	0.039	0.028	0.055	0.067	0.067
	W_j^C	0.042	0.041	0.041	0.039	0.067	0.048	0.031	0.038	0.061	0.050	0.044	0.043	0.056	0.029	0.042	0.057	0.062	0.043	0.081	0.083
	W_j^{DM}	0.050	0.050	0.050	0.050	0.050	0.050	0.050	0.050	0.050	0.050	0.050	0.050	0.050	0.050	0.050	0.050	0.050	0.050	0.050	0.050
Consensus_Average	W_j^E	0.051	0.050	0.047	0.050	0.050	0.047	0.049	0.049	0.051	0.052	0.047	0.047	0.052	0.051	0.048	0.051	0.051	0.049	0.053	0.052
	W_j^C	0.041	0.027	0.054	0.030	0.067	0.055	0.036	0.052	0.062	0.040	0.068	0.069	0.037	0.036	0.063	0.051	0.070	0.062	0.029	0.051
	W_j^{DM}	0.050	0.050	0.050	0.050	0.050	0.050	0.050	0.050	0.050	0.050	0.050	0.050	0.050	0.050	0.050	0.050	0.050	0.050	0.050	0.050
Consensus_Median	W_j^E	0.052	0.049	0.047	0.052	0.052	0.044	0.053	0.052	0.053	0.048	0.050	0.048	0.054	0.053	0.049	0.051	0.049	0.051	0.049	0.044
	W_j^C	0.047	0.032	0.045	0.034	0.066	0.055	0.038	0.050	0.058	0.044	0.055	0.052	0.045	0.034	0.052	0.053	0.050	0.051	0.080	0.059
	W_j^{DM}	0.050	0.050	0.050	0.050	0.050	0.050	0.050	0.050	0.050	0.050	0.050	0.050	0.050	0.050	0.050	0.050	0.050	0.050	0.050	0.050
Likert Scale	W_j^E	0.050	0.050	0.049	0.051	0.051	0.047	0.050	0.051	0.051	0.050	0.049	0.049	0.052	0.051	0.050	0.051	0.048	0.050	0.051	0.051
	W_j^C	0.040	0.027	0.051	0.030	0.079	0.059	0.034	0.051	0.071	0.047	0.057	0.058	0.034	0.041	0.051	0.059	0.068	0.058	0.032	0.051
	W_j^{DM}	0.050	0.050	0.050	0.050	0.050	0.050	0.050	0.050	0.050	0.050	0.050	0.050	0.050	0.050	0.050	0.050	0.050	0.050	0.050	0.050

Table 13. Table of the lower and upper limits of the weight system of the EC-PROMETHEE method.

Criteria	Scenario													
	Mode		Average		Median		Consensus Mode		Consensus Average		Consensus Median		Likert Scale	
	Ll_j	Ul_j	Ll_j	Ul_j	Ll_j	Ul_j	Ll_j	Ul_j	Ll_j	Ul_j	Ll_j	Ul_j	Ll_j	Ul_j
C1	0.045	0.050	0.040	0.050	0.040	0.051	0.042	0.050	0.041	0.051	0.047	0.052	0.040	0.050
C2	0.048	0.050	0.027	0.050	0.031	0.050	0.041	0.050	0.027	0.050	0.032	0.050	0.027	0.050
C3	0.046	0.050	0.049	0.051	0.044	0.050	0.041	0.050	0.047	0.054	0.045	0.050	0.049	0.051
C4	0.045	0.051	0.030	0.051	0.037	0.053	0.039	0.050	0.030	0.050	0.034	0.052	0.030	0.051
C5	0.050	0.104	0.050	0.079	0.050	0.078	0.050	0.067	0.050	0.067	0.050	0.066	0.050	0.079
C6	0.040	0.061	0.047	0.059	0.044	0.056	0.040	0.050	0.047	0.055	0.044	0.055	0.047	0.059
C7	0.035	0.052	0.034	0.050	0.036	0.054	0.031	0.051	0.036	0.050	0.038	0.053	0.034	0.050
C8	0.044	0.057	0.050	0.051	0.050	0.054	0.038	0.055	0.049	0.052	0.050	0.052	0.050	0.051
C9	0.045	0.078	0.050	0.071	0.050	0.060	0.046	0.061	0.050	0.062	0.050	0.058	0.050	0.071
C10	0.044	0.059	0.047	0.050	0.046	0.050	0.047	0.050	0.040	0.052	0.044	0.050	0.047	0.050
C11	0.048	0.050	0.049	0.057	0.050	0.052	0.044	0.050	0.047	0.068	0.050	0.055	0.049	0.057
C12	0.048	0.050	0.049	0.058	0.047	0.050	0.043	0.050	0.047	0.069	0.048	0.052	0.049	0.058
C13	0.046	0.065	0.034	0.052	0.035	0.054	0.046	0.056	0.037	0.052	0.045	0.054	0.034	0.052
C14	0.032	0.059	0.041	0.051	0.037	0.054	0.029	0.059	0.036	0.051	0.034	0.053	0.041	0.051
C15	0.047	0.052	0.050	0.051	0.047	0.051	0.042	0.050	0.048	0.063	0.049	0.052	0.050	0.051
C16	0.039	0.069	0.050	0.059	0.050	0.057	0.039	0.057	0.050	0.051	0.050	0.053	0.050	0.059
C17	0.022	0.072	0.048	0.068	0.046	0.050	0.028	0.062	0.050	0.070	0.049	0.050	0.048	0.068
C18	0.050	0.055	0.050	0.058	0.050	0.052	0.043	0.055	0.049	0.062	0.050	0.051	0.050	0.058
C19	0.001	0.067	0.032	0.051	0.047	0.084	0.050	0.081	0.029	0.053	0.049	0.080	0.032	0.051
C20	0.001	0.067	0.050	0.051	0.040	0.059	0.050	0.083	0.050	0.052	0.044	0.059	0.050	0.051

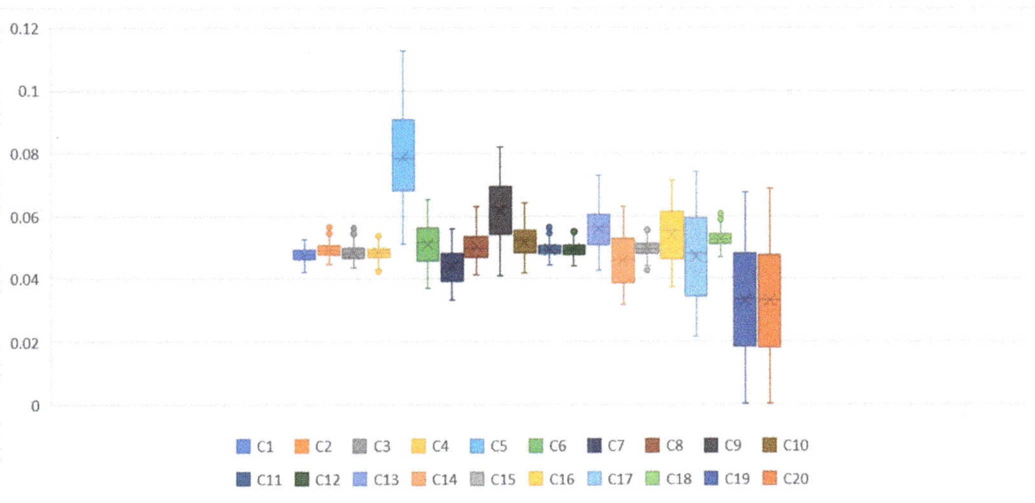

Figure 2. Graphical representation of the set of 10,000 iterations for the "Mode" Scenario.

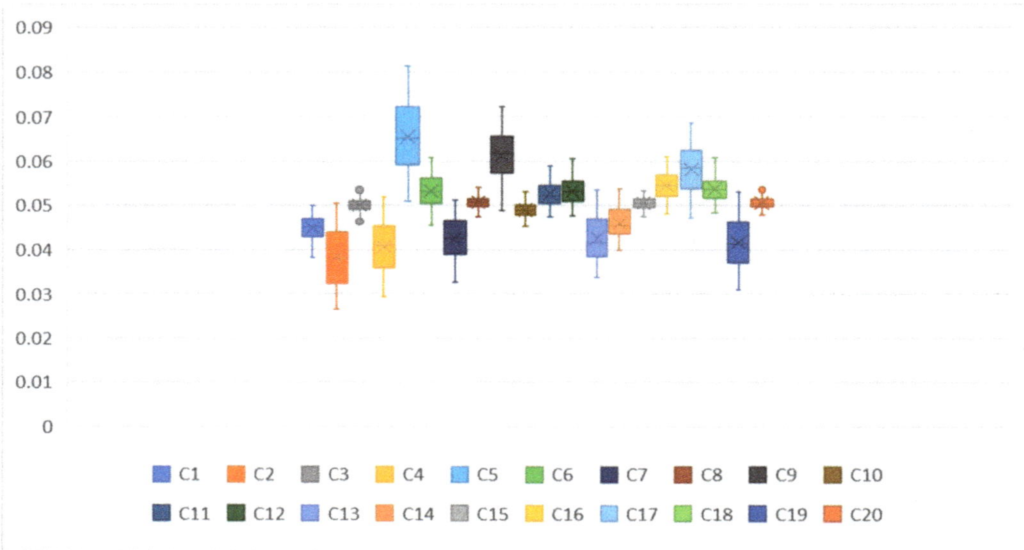

Figure 3. Graphical representation of the set of 10,000 iterations for the "Average" Scenario.

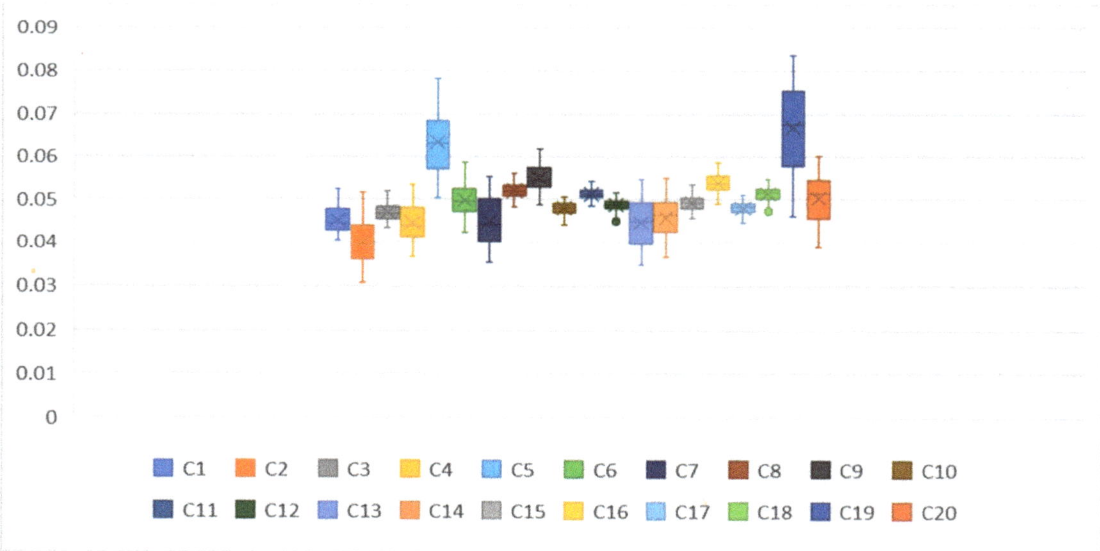

Figure 4. Graphical representation of the set of 10,000 iterations for the "Median" Scenario.

In step 7, we introduced the criteria, alternatives, decision matrix, weight matrices, and parameters into the PROMETHEE method. We ran Equations (16)–(21) in t = 10,000 iterations and obtained the t ranking for each scenario proposed in this study. We then applied Equations (22) and (23) and the rules prescribed in step 8 and obtained the final ranking for each scenario, as shown in Table 14. We then calculated the standard deviations of the t iterations of each criterion in all the proposed scenarios, as shown in Figure 6. Finally, we calculated the Spearman correlation between the final rankings for each scenario, as shown in Table 15.

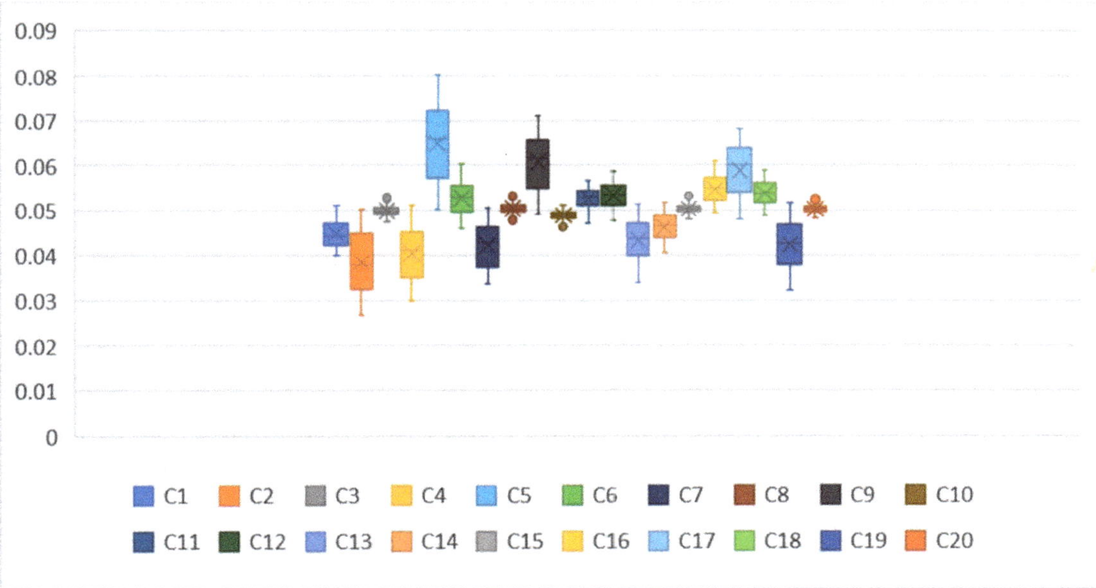

Figure 5. Graphical representation of the set of 10,000 iterations for the "Likert Scale" Scenario.

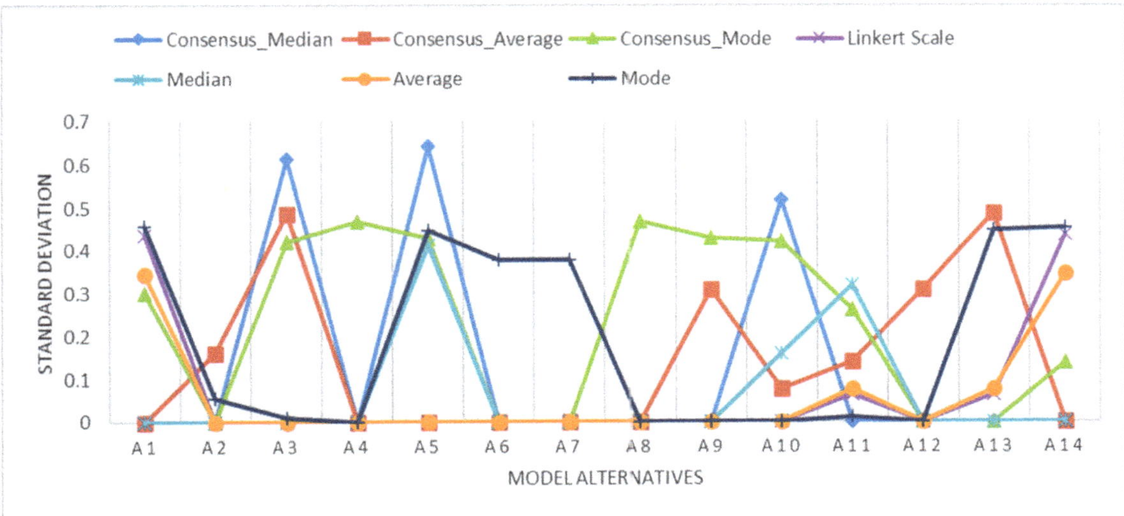

Figure 6. The standard deviation of the t iterations of the criteria in different scenarios.

Regardless of its complexity, the decision-making process involves identifying criteria and alternatives and obtaining the weights for each criterion. The definition of criteria weights is a critical stage in decision-making, as they can influence the final result. The literature presents readers with three methods for defining criteria weights: objective, subjective, and hybrid. Around this discussion is a current of thought that proposes reducing the discretionary power of the decision maker, assigning this task to mathematical methods, such as the CRITIC, ENTROPY, and SWARA methods. On the other hand, some experts claim that the subjectivity of the decision maker's discretion is fundamental, as it comes with the added layer of experience, culture, information, maturity, and underlying knowledge of the business that mathematical methods cannot measure, such as AHP SAPEVO. However, a third stream of researchers has combined the concepts of objective and subjective methods to form a third stream: the hybrids. These use mathematical methods associated with the weights assigned by the decision makers or group of decision makers. EC-PROMETHEE is a flexible method, as it can take on the role of an objective method and use only the combination of the ENTROPY (E) and CRITIC (C) methods to obtain the range of weights. However, it can also add weights generated by subjective methods or even weights assigned directly by the decision makers to be classified as a hybrid method. We believe that EC-PROMETHEE is inaugurating a fourth class of methods, which we call flexible.

Table 14. Final ranking of the EC-PROMETHE model for different decision matrix schemes.

Scenarios	Ranking													
	1st	2nd	3rd	4th	5th	6th	7th	8th	9th	10th	11th	12th	13th	14th
Consensus_Median	a6	a7	a12	a9	a2	a10	a5	a3	a11	a13	a14	a1	a8	a4
Consensus_Average	a6	a7	a9	a12	a5	a10	a2	a11	a3	a13	a14	a1	a8	a4
Consensus_Mode	a6	a7	a12	a2	a9	a5	a3	a10	a13	a11	a1	a14	a4	a8
Likert_Scale	a6	a7	a12	a9	a5	a11	a13	a10	a2	a3	a14	a1	a8	a4
Median	a7	a6	a9	a12	a5	a11	a10	a13	a3	a2	a14	a1	a8	a4
Average	a6	a7	a12	a9	a5	a11	a13	a10	a2	a3	a14	a1	a8	a4
Mode	a6	a7	a12	a9	a5	a13	a10	a11	a3	a2	a1	a14	a8	a4
Model *	a6	a7	a12	a9	a11	a5	a10	a13	a2	a3	a14	a1	a8	a4

Note: * This is the final modeling ranking from the report by Basilio et al. [42], which is used as a comparison model for EC-PROMETHEE.

Table 15. Spearman correlation of the final ranking of the proposed scenarios.

	Consensus_Median	Consensus_Average	Consensus_Mode	Likert Scale	Median	Average	Mode
Consensus_Median	1	0.973626374	0.969230769	0.8989011	0.894505	0.898901	0.89011
Consensus_Average		1	0.92967033	0.94725275	0.956044	0.947253	0.934066
Consensus_Mode			1	0.86813187	0.846154	0.868132	0.872527
Likert Scale				1	0.982418	1	0.978022
Median					1	0.982418	0.969231
Average						1	0.978022
Mode							1

Sensitivity Analysis of EC-PROMETHEE

In this section, the sensitivity analysis of the model proposed in the research was based on the methodology applied by Basilio et al. [42]. The sensitivity analysis was carried out using the script described in Appendix A. In each scenario, an alternative was removed, and the behavior of the others was verified. Then, the dropped alternative was reintroduced into the model, another alternative was dropped, and the process restarted until the last alternative was tested. After that, the results were analyzed and checked for order reversal and the model's sensitivity to changes. Seven scenarios were created from the data in Tables 3–9. The process was carried out sequentially from alternative "a1" to "a14". The expected result was that when an alternative is removed, the subsequent alternatives in the ranking improve one position, and the previous alternatives do not change their positions. Table 16 and Figure 7 illustrate the sensitivity analysis considering the seven scenarios used to emulate EC-PROMETHEE. The authors decided to conduct this analysis based on the percentage changes expected throughout the process. The percentage was defined as follows: Considering the "n" position in the ranking of the alternative obtained in each scenario, we inferred that the number of changes predicted would be equal to (n − 1) throughout the process in each scenario. The percentage is obtained by dividing (n − 1) by the total number of alternatives in the model minus the subtracted alternative. Table 16 shows the values found. The values highlighted in yellow represent that the alternatives do not align with the expected values. We can infer that the total changes correspond to twenty-seven percent of the process. In particular, we can say that the changes observed do not invalidate the final rankings of each scenario, as the main positions have remained the same in the case of the first and second positions (a6 and a7). Nine of the fourteen alternatives only displayed a change from the expected value in the proposed scenarios, which are as follows: "a1"; "a4"; "a6"; "a7"; "a8"; "a11"; "a12"; "a13"; and "a14". As with the first positions, the last ones were also preserved from the eleventh to the fourteenth, as we can see by looking at the alternatives "a14" > "a1" > "a8" > "a4". The greatest instability occurred in the middle positions of the ranking. Concerning the proposed decision matrix construction scenarios, we can say that the "Likert-Scale" and "Average" scenarios only saw one change in position. However, the "Consensus_Mode" and "Mode" scenarios had the biggest changes—seven and six positions, respectively.

Table 16. Expected percentages of variation in the EC-PROMETHEE sensitivity analysis.

Scenarios/Alternatives	a1	a2	a3	a4	a5	a6	a7	a8	a9	a10	a11	a12	a13	a14
Consensus_Median	85%	31%	23%	100%	23%	0%	8%	92%	23%	31%	62%	15%	69%	77%
Consensus_Average	85%	31%	31%	100%	31%	0%	8%	92%	46%	31%	54%	46%	69%	77%
Consensus_Mode	77%	69%	31%	77%	62%	0%	8%	85%	38%	69%	69%	15%	62%	85%
Likert_Scale	85%	62%	69%	100%	38%	0%	8%	92%	23%	54%	38%	15%	46%	77%
Median	85%	69%	62%	100%	69%	0%	8%	92%	15%	77%	100%	23%	54%	77%
Average	85%	62%	69%	100%	38%	0%	8%	92%	23%	54%	38%	15%	46%	77%
Mode	31%	69%	62%	100%	46%	8%	15%	92%	23%	46%	54%	15%	85%	100%

Note: Areas highlighted in yellow indicate changes in position that are not in line with the expected value.

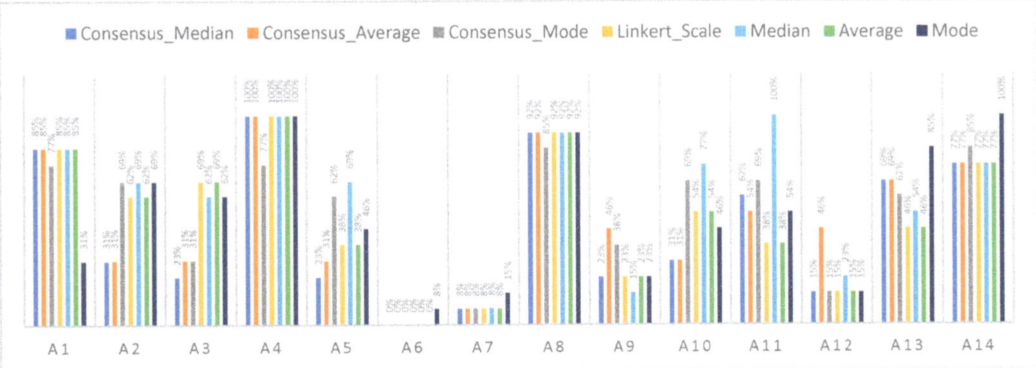

Figure 7. Graphical representation of the EC-PROMETHEE sensitivity analysis.

4. Discussion

In this paper, we revisited the problem of planning policing strategies to reduce crime in a given locality. The relevance of this topic is based on the impact of crime reduction on the social and economic life of cities. Our research used the data shared by Basilio et al. [42], in which they applied the PROMETHEE II method and obtained the following final ranking: a6 > a7 > a12 > a9 > a11 > a5 > a10 > a13 > a2 > a3 > a14 > a1 > a8 > a4. A detail that needs to be noted is that Basilio et al. [42] used the statistical measure "Mode" in constructing their decision matrix. In the current proposal, the researchers used six other measures (Tables 4–9) based on the questionnaire data used by Basilio et al. [42]. The initial motivation for using these measures was to check the stability of the final composition of the ranking. The stability can be seen in Table 15, which shows the result of the Spearman correlation between the proposed models, which we will discuss later.

The primary difference between the model applied in Basilio et al. [42] and the model proposed in this paper is that equal weights were applied to each criterion, as indicated in the eighth column of Table 11. Figure 1 illustrates that in the EC-PROMETHEE model, we combined two objective methods to obtain the weights for each criterion, which can be associated with the results of subjective methods or direct data input from the decision makers' evaluation. In the current case, to maintain parity in the analysis between the two results, we inserted the weights used by Basilio et al. [42], which consisted of equal weights for each criterion, as shown in Table 11. The weights generated provided the random weight generation model with inner and upper limits, which allowed us to obtain the weight ranges. Table 12 shows the weights obtained in the model, and Table 13 shows the lower and upper limits. Table 13 shows that the limits comprise the values corresponding to the methods used. The limits integrate the information in the objective methods and the intrinsic knowledge of each decision maker. This combination is a new development compared to the other models, which combine objective methods only to obtain weights. In the proposed method, working with a range of weights, the model can observe the consistency of the final rankings through t iterations. In Figures 2–5, the reader can see how the weights were generated and behaved for each criterion in the boxplot graphs. We can see that because the decision matrices differ, the weights generated behave differently.

Table 14 shows an overview of the final ranking in the seven proposed scenarios. Graphically, we can see how each alternative behaved over the 10,000 iterations with the set of weights. We can see that there were changes in the ranking of the alternatives in at least one of the scenarios. Based on the information recorded in the tenth line of Table 14, which corresponds to the final ranking of the evaluation of policing strategies according to criminal demand in the city of Rio de Janeiro, Brazil [43], the authors considered equal weight for the criteria and the decision matrix was built based on the Mode statistical measure. We then compared it with the result of the ninth row, in which we kept the same decision matrix but applied the random weight range with 10,000 iterations, which was the innovation proposed in the EC-PROMETHEE model, and found that there were changes in position between the fifth and sixth positions and changes in position between the ninth and twelfth positions.

In contrast, the first four positions of the original ranking remained unchanged, as did the last and penultimate positions. This result demonstrates that when the decision maker integrates objective and subjective methods for obtaining criteria weights and starts working with a set of weights obtained randomly within the upper and lower limits established by integrating methods for obtaining weights, combined with a strategy of emulating "t" iterations, the distortions in the ranking can be observed with just one weight and one iteration of the original method. In this way, EC-PROMETHEE generates a more consistent ranking, which makes it easier to choose the best alternatives for solving a problem. A second point we would like to highlight in this research concerns the choice of statistical measure for processing data from questionnaires designed to build a decision matrix. The statistical measure used to represent the perception of a group of experts on a given topic, in this case, the policing strategies that have the greatest impact on reducing a given criminal demand, directly affects the ranking of the alternatives evaluated. This is shown in Table 14. In general, there will be changes in the rankings of all the measures chosen. Analyzing the first four positions in the ranking, we can say that compared with the model presented by Basilio et al. [42], there was no reversal of position using the following scales: Consensus_Median, Likert_Scale, Average, and Mode. In the Consensus_Average measure, we observed a reversal between the third and fourth positions in the ranking. Concerning the median positions in the ranking, all the scales showed changes. Regarding the final section of the ranking, the scales produced with the following statistical measures remained unchanged between the 11th and 14th positions: Consensus_Median, Consensus_Average, Likert_Scale, Median, and Average.

Figure 6 shows a graphical representation of each scenario's standard deviations from the 10,000 iterations. From this data, we can see that there are primarily changes between positions in the ranking. This information corroborates the proposal of the new EC-PROMETHEE method, which reinforces the consistency of the final ranking, offering decision makers greater certainty in the decision-making process and reducing the uncertainty of the decision-making process. Table 15 shows the Spearman correlation between the final rankings for each scenario. The final ranking of the results of the research reported by Basilio et al. [42] compared to EC-PROMETHEE's "Mode" scenario shows a Spearman correlation of 0.96044. This is a high correlation, but with a change in ranking. Sperman's correlation reveals to the decision maker that the choice of statistical method to systematize the information from the questionnaires influences the decision-making process in the final definition of the ranking of alternatives. Among the proposed scenarios, we can say that only the Likert Scale and Average had the same ranking. In the other scenarios, we had high correlations ranging from 0.84–0.98, which reaffirms that the choice of measurement for constructing the decision matrix in the case of obtaining data through questionnaires, combined with the choice of methods for obtaining weights, can influence the final ranking of the alternatives.

5. Conclusions

Over the last four decades, studies and applications of multi-criteria methods to support decision-making in various branches of science and organizations have multiplied. If one word can define the current state-of-the-art in operations research about decision support methods, it would be integration.

Specifically, concerning the central theme of this article, which is the integration of objective and subjective techniques for assigning weights to criteria, we can say that in recent years, we have seen the integration of weighting methods with classical methods. Along these lines, we have also seen the combination of weighting methods to reduce existing uncertainty. In this case, we used the objective methods ENTROPY and CRITIC to obtain their respective weights associated with the subjective weights derived from the decision makers' evaluation. The difference is that we didn't reduce the information from these three inputs into a single weight value for each criterion. Instead, we created a weight range for each criterion.

These weight ranges have lower and upper limits which were defined based on the values obtained in each method. The lower limit is the lowest value obtained for the criterion. Likewise, the upper limit is the highest value among the values generated for a specific criterion. With this, we adapted the characteristics of each method to the model. Our intention was always to reduce uncertainty in the decision-making process. With the help of random generation, the proposed method can produce "t" possible iterations defined by the decision maker. The innovation is that we did not have just one final ranking but "t" sets of rankings. With this measure, the manager will be able to observe the behavior of the alternatives as a function of the various sets of weights, respecting the limits defined in the method. Step 8 of the methodology, using Equations (22) and (23), describes and defines the final ranking.

The methodology was tested using real data from the article "Ranking policing strategies as a function of criminal complaints: application of the PROMETHEE II method in the Brazilian context", published in 2021. In the chosen article, the researchers used equal weight for each criterion, as it makes no sense to assign zero weight, which would invalidate the criterion, so using equal weight does not interfere with the relationship between the criteria. For comparison purposes, we preserved the data and information used in 2021. The classic method used was PROMETHEE. So, we applied EC-PROMETHEE to maintain the same conditions, setting t = 10,000 iterations. In this research, we developed the script presented in Appendix A in Visual Basic for Excel (VBA). Seven scenarios were emulated, and the results were compared, which allowed us to affirm that the production of "t" final rankings with variations in the set of weights for each criterion revealed that there was a reversal of positions in the ranking compared to just a single iteration of the traditional methods with a single set of weights for the criteria. We therefore consider the results produced by EC-PROMETHEE to be consistent, presenting the decision maker with a tool that reduces the uncertainties of the process and presents a robust ranking for decision-making.

The research does not end with this publication, as there are still gaps to be filled with future research, such as analyzing the choice of preference types in the PROMETHEE method, integrating other objective and subjective methods into the model, comparing it with different sorting methods, building a web platform to disseminate the technique, and compiling the algorithm in other languages such as python and R.

Author Contributions: Conceptualization, M.P.B. and V.P.; methodology, M.P.B.; software, M.P.B.; validation, V.P., F.Y. and M.P.B.; formal analysis, M.P.B.; investigation, V.P.; resources, M.P.B.; data curation, F.Y.; writing—original draft preparation, F.Y.; writing—review and editing, F.Y.; visualization, M.P.B.; supervision, V.P.; project administration, M.P.B.; funding acquisition, M.P.B. All authors have read and agreed to the published version of the manuscript.

Funding: This research received no external funding.

Data Availability Statement: Appendix A describes the EC-PROMETHEE software script in Excel VBA.

Conflicts of Interest: The authors declare no conflict of interest.

Abbreviations

The following abbreviations are used in this manuscript:

AHP	Analytic Hierarchy Process
ANP	Analytical Network Process
BWM	Best-Worst Method
CILOS	Criterion Impact Loss
COPRAS	Complex Proportional Assessment
CRITIC	Criteria Importance Through Intercriteria Correlation
D-CRITIC	Distance Correlation-based CRITIC
DEMATEL	Decision-Making Trial and Evaluation Laboratory
DM	Decision-making
EC-PROMETHEE	Entropy-Critic-PROMETHEE
ELECTRE	ÉLimination et Choix Traduisant la REalité (French)
FUCOM	Full Consistency Method
GAIA	Geometrical Analysis for Interactive Aid
IDOCRIW	Integrated Determination of Objective CRIteria Weights
IFS	Intuitionistic fuzzy sets
LBWA	Level Based Weight Assessment
MCDA	Multi-Criteria Decision Analysis
MCDM	Multi-Criteria Decision-Making
MEREC	Method Based on the Removal Effects of Criteria
PROMETHEE	Preference Ranking Organization Method for Enrichment of Evaluation
SAPEVO-M	Simple Aggregation of Preferences Expressed by Ordinal Vectors—Multi-Decision Makers
SWARA	Step-wise Weight Assessment Ratio Analysis
TOPSIS	Technique for Order of Preference by Similarity to Ideal Solution
VIKOR	VlseKriterijumska Optimizacija I Kompromisno Resenje (Serbian)

Appendix A

```
Private Sub CommandButton1_Click()
'Developer: Marcio Pereira Basilio, PhD
                    'Company: Military Police of the State of Rio de Janeiro
                    'Product: EC-PROMETHEE Hybrid Method
'###########################################################################
'Obtaining the initial parameters
n = Sheets("EC").Cells(2, 2).Value 'Parameter of the number of criteria
```

```
t = Sheets("EC").Cells(3, 2).Value 'Parameter of the number of alternatives
m = Sheets("EC").Cells(4, 2).Value 'Parameter of iteration quantity
'##############################################################################
'Variable sizing
'##############################################################################
Dim LI As Double
Dim LS As Double
Dim MD(100, 100) As Variant 'Decision matrix
Dim MN(100, 100) As Variant 'Standardisation Matrix
Dim WM(20,000, 20,000) As Double 'Matrix of random weights obtained between lower
and upper limits
Dim VDjT As Double
Dim Cont As Variant 'Summation X
Dim Cont1 As Double 'Summation Y
Dim Cont2 As Double 'Summation X^2
Dim Cont3 As Double 'Summation Y^2
Dim Cont4 As Double 'Summation X*Y
Dim Cont5 As Double 'Summation Ej
Dim Cont6 As Variant 'Auxiliary summation
ReDim TP(1 To n) As Double 'Vector type of preference
ReDim VN(1 To t) As Double 'Auxiliary vector summation
ReDim VE(1 To n) As Double 'Entropy calculation vector
ReDim VDj(1 To n) As Double 'Vector of the calculation of the parameter Dj
ReDim VWe(1 To n) As Double 'Vector of the calculation of the Weight per criterion in
the entropy method
ReDim VWc(1 To n) As Double 'Vector of the calculation of the Weight per criterion in
the CRITIC method
ReDim VWdm(1 To n) As Double 'Weight vector obtained from decision makers
ReDim Best(1 To n) As Variant
ReDim Worst(1 To n) As Variant
ReDim TCrit(1 To n) As Variant 'Type of criteria "0" Benefit and "1" Cost
ReDim Average(1 To n) As Variant 'Vector storing the average of each criterion
ReDim SD(1 To n) As Variant 'Vector storing the Standard deviation
ReDim MC(n, n) As Double 'Correlation matrix
ReDim Ej(1 To n) As Double 'Ej of the CRITIC method formula
ReDim MCT(n, n) As Double 'Auxiliary correlation matrix
ReDim MP(t, t) As Double 'Promethee preference matrix
ReDim Phi_row(t) As Double 'Phi+
ReDim Phi_column(t) As Double 'Phi-
ReDim Phi_total(t) As Variant 'Phi total auxiliary
ReDim Phi_total_A(t) As Variant 'Phi total
ReDim Phi_total_ord(m, t) As Variant 'Phi total ordinal
ReDim VrankG(t) As Variant 'Total ranking vector based on Likert
ReDim VrankGF(t) As Variant
ReDim VrankA(t) As Variant
'##############################################################################
For k = 1 To n
        TCrit(k) = Sheets("EC").Cells(27, 10 + k).Value
Next
'##############################################################################
'Obtaining the decision matrix
For k = 1 To n
        For p = 1 To t
                MD(p, k) = Sheets("EC").Cells(30 + p, 1 + k).Value
```

```
        Next
Next
'##########################################################################
'Weight calculation by the Entropy method
'##########################################################################
'Step_1 Normalization of the decision matrix
Cont = 0
For k = 1 To n
        For p = 1 To t
                Cont = Cont + MD(p, k)
        Next
        'Standardization
        For j = 1 To t
                MN(j, k) = MD(j, k)/Cont
        Next
        Cont = 0
Next
'Calculation of parameter h
E = 2.718282 'Euler parameter
h = 1/(Log(10)/Log(E))
'Sheets("EC").Cells(1, 10).Value = h
'Step 2 Calculation of entropy (e)
For k = 1 To n
        For p = 1 To t
        VE(k) = VE(k) + (MN(p, k) * (Log(MN(p, k)/Log(E))))
        Next
        VE(k) = VE(k) * (−h)
        VDj(k) = Abs((1 − VE(k)))
        VDjT = VDjT + VDj(k)
Next
'Step 3 calculation of weight per criterion
For k = 1 To n
        VWe(k) = VDj(k)/VDjT
Next
'############################################################
'Weight calculation by the CRITIC method
'############################################################
'Step 1—Identification of the Highest and Lowest criterion value i
For k = 1 To n
        Best(k) = 0
Next
For k = 1 To n
        For p = 1 To (t)
                If MD(p, k) > Best(k) Then
                        Best(k) = MD(p, k)
                End If
        Next
Next
For k = 1 To n
        Worst(k) = 99999
Next
For k = 1 To n
        For p = 1 To (t)
                If MD(p, k) < Worst(k) Then
```

```
                    Worst(k) = MD(p, k)
                End If
        Next
Next
'Step 2—Normalization of the Decision Matrix
For k = 1 To n
        For p = 1 To t
                If TCrit(k) = 0 Then
                MN(p, k) = (MD(p, k) − Worst(k) + 0.0001)/(Best(k) − Worst(k) + 0.01)
                Else
                        MN(p, k) = (Best(k) − MD(p, k))/(Best(k) − Worst(k))
                End If
        Next
Next
'Step 3—Calculation of standard deviation
'Step 3.1—Calculation of the average
For k = 1 To n
                Cont = 0
        For p = 1 To t
                Cont = Cont + MN(p, k)
        Next
                Average(k) = (Cont/t)
Next
'Step 3.2—Calculation of standard deviation
For k = 1 To n
                Cont = 0
        For p = 1 To t
                Cont = Cont + ((MN(p, k) − Average(k)) ^ (2))
        Next
                SD(k) = Sqr(Cont/t)
Next
'Step 4—Calculation of correlation between criteria
For k = 1 To n
        For p = 1 To n
                Cont = 0
                Cont1 = 0
                Cont2 = 0
                Cont3 = 0
                Cont4 = 0
                For Z = 1 To t
                Cont = Cont + MN(Z, k)
                Cont1 = Cont1 + MN(Z, p)
                Cont2 = Cont2 + (MN(Z, k) ^ 2)
                Cont3 = Cont3 + (MN(Z, p) ^ 2)
                Cont4 = Cont4 + (MN(Z, k) * MN(Z, p))
                Next
                MC(k, p) = ((t * Cont4) − (Cont * Cont1))/(Sqr((t * Cont2) − (Cont ^ 2))
* Sqr((t * Cont3) − (Cont1 ^ 2)))
        Next
Next
For p = 1 To n
        For k = 1 To n
                MCT(p, k) = (1 − MC(p, k))
        Next
```

```
Next
Cont5 = 0
For p = 1 To n
        For k = 1 To n
                Ej(p) = Ej(p) + MCT(p, k)
        Next
        Ej(p) = Ej(p) * SD(p)
        Cont5 = Cont5 + Ej(p)
Next
For k = 1 To n
        VWc(k) = Ej(k)/Cont5
Next
'#################################################################
'Obtaining the Decision-Maker weight vector
'#################################################################
Cont6 = 0
For j = 1 To n
        VWdm(j) = Sheets("EC").Cells(29, 1 + j).Value
        Sheets("EC").Cells(18, 2 + j).Value = VWe(j)
        Sheets("EC").Cells(19, 2 + j).Value = VWc(j)
        Cont6 = Cont6 + VWdm(j)
Next
For i = 1 To n
        VWdm(i) = VWdm(i)/Cont6
        Sheets("EC").Cells(20, 2 + i).Value = VWdm(i)
Next
'####################################################################
'By generating the matrix of random weights between lower and upper bounds obtained from the outputs of the Entropy and CRITIC methods, this is a solution set to be utilized by the ranking method.
'####################################################################
For k = 1 To n
        If VWe(k) < VWc(k) Then
                If VWe(k) < VWdm(k) Then
                    If VWc(k) < VWdm(k) Then
                        LI = VWe(k)
                        LS = VWdm(k)
                    Else
                        LI = VWe(k)
                        LS = VWc(k)
                    End If
                Else
                    LI = VWdm(k)
                    LS = VWc(k)
                End If
        Else
                If VWdm(k) < VWc(k) Then
                    LI = VWdm(k)
                    LS = VWe(k)
                Else
                    If VWdm(k) < VWe(k) Then
                        LI = VWc(k)
                        LS = VWe(k)
                    Else
```

```
                        LI = VWc(k)
                        LS = VWdm(k)
                    End If
                End If
            End If
            Sheets("EC").Cells(21, 2 + k).Value = LI
            Sheets("EC").Cells(22, 2 + k).Value = LS
    For p = 1 To m
            Randomize
            WM(p, k) = (((LS − LI) * Rnd) + LI)
    Next
Next
'Normalization of the random weight matrix
For p = 1 To m
            Z = 0
            Cont5 = 0
            For k = 1 To n
                    Z = Z + WM(p, k)
            Next
            For k = 1 To n
                    WM(p, k) = (WM(p, k)/Z)
                    Sheets("Result").Cells(4 + p, k).Value = WM(p, k)
                    Cont5 = Cont5 + WM(p, k)
            Next
            Sheets("Result").Cells(4 + p, n + 1).Value = Cont5
Next
Cont5 = 0
'###################################################################
'                       Method PROMETHEE II
'###################################################################
'Obtaining the types of preferences:
For k = 1 To n
            TP(k) = Sheets("EC").Cells(28, 1 + k).Value
Next
p = 0.5
q = 0.15
s = 0.6
'Step 1—Determination of deviations based on pairwise comparations
'Step 2—Application of the preference function
'Step 3—Calculation of an overall or global preference index
Cont5 = 0
Cont6 = 0
For y = 1 To m 'Loop da iteração
            For k = 1 To t
                    Phi_row(k) = 0
                    Phi_column(k) = 0
                    Phi_total(k) = 0
                    Phi_total_A(k) = 0
            Next
For k = 1 To t
            For j = 1 To t
                    For i = 1 To n
                            Dj = (MN(k, i) - MN(j, i))
                            If TP(i) = 1 Then
```

```
                                        If Dj > 0 Then
                                                Cont6 = Cont6 + (1 * WM(y, i))
                                        End If
                                End If
'________________________________________
                                If TP(i) = 2 Then
                                        If Dj > q Then
                                            Cont6 = Cont6 + (1 * WM(y, i))
                                        End If
                                End If
'________________________________________
                                If TP(i) = 3 Then
                                        If Dj > p Then
                                            Cont6 = Cont6 + (1 * WM(y, i))
                                        Else
                                            If Dj > 0 And Dj ≤ p Then
                                                Cont6 = Cont6 + ((Dj/p) * WM(y, i))
                                        End If
                                End If
                            End If
'________________________________________
                        If TP(i) = 4 Then
                                If Dj > p Then
                                        Cont6 = Cont6 + (1 * WM(y, i))
                                Else
                                        If Dj > q And Dj ≤ p Then
                                            Cont6 = Cont6 + (0.5 * WM(y, i))
                                        End If
                                End If
                        End If
'________________________________________
                        If TP(i) = 5 Then
                                If Dj > p Then
                                        Cont6 = Cont6 + (1 * WM(m, i))
                                Else
                                        If Dj > q And Dj ≤ p Then
                                           Cont6 = Cont6 + (((Dj − q)/(p − q)) * WM(y, i))
                                        End If
                                End If
                        End If
'________________________________________
                        If TP(i) = 6 Then
                                If Dj > 0 Then
                                        Cont5 = ((Dj ^ 2) * (−1))/(2 * (s ^ 2))
                                        Cont6 = Cont6 + ((1 − (E ^ (Cont5))) * WM(y, i))
                                        Cont5 = 0
                                End If
                        End If
'________________________________________
                        Next
                        MP(k, j) = Cont6
                        Cont6 = 0
                Next
        Next
```

```
For k = 1 To t
        For j = 1 To t
                Sheets("EC").Cells(3 + k, 17 + j).Value = MP(k, j)
        Next
Next
'Step 4. Calculation of outranking flows/The PROMETHEE II partial ranking
For k = 1 To t
        For j = 1 To t
                Phi_row(k) = Phi_row(k) + MP(k, j)
                Phi_column(k) = Phi_column(k) + MP(j, k)
        Next
Next
For k = 1 To t
                Sheets("EC").Cells(3 + k, 26).Value = Phi_row(k)
                Sheets("EC").Cells(12, 17 + k).Value = Phi_column(k)
        Next
'Step 5. Calculation of net outranking flow/The PROMETHEE II complete ranking
For k = 1 To t
        Phi_total(k) = Phi_row(k) − Phi_column(k)
        Sheets("EC").Cells(3 + k, 28).Value = Phi_total(k)
Next
'Sorting the final ranking
Cont6 = −9999
For k = 1 To t
        Phi_total_A(k) = Phi_total(k)
Next
For k = 1 To t
        For p = k To t
            If Phi_total_A(k) < Phi_total_A(p) Then
                Cont6 = Phi_total_A(k)
                Phi_total_A(k) = Phi_total_A(p)
                Phi_total_A(p) = Cont6
            End If
        Next
        Sheets("EC").Cells(3 + k, 30).Value = Phi_total_A(k)
Next
For k = 1 To t
        For p = 1 To t
            If Phi_total(k) = Phi_total_A(p) Then
                Phi_total_ord(y, k) = p
            End If
        Next
        Sheets("Ranking").Cells(5 + y, k).Value = Phi_total_ord(y, k)
        VrankG(k) = VrankG(k) + Phi_total_ord(y, k) 'Operation that sums
ordinal values.
Next
Next
'Sorting the final ordinal ranking Likert version
For k = 1 To t
        VrankA(k) = VrankG(k)
Next
For k = 1 To t
        For p = k To t
            If VrankA(k) > VrankA(p) Then ' <
```

```
                Cont6 = VrankA(k)
                VrankA(k) = VrankA(p)
                VrankA(p) = Cont6
            End If
        Next
Next
For k = 1 To t
        For p = 1 To t
            If VrankG(k) = VrankA(p) Then
                VrankGF(k) = p
            End If
        Next
Next
For k = 1 To t
                Sheets("Ranking").Cells(1, k).Value = "a" & k
                Sheets("Ranking").Cells(3, k).Value = VrankGF(k)
                Sheets("Ranking").Cells(4, k).Value = VrankG(k)
Next
End Sub
```

References

1. Basílio, M.P.; Pereira, V.; Costa, H.G.; Santos, M.; Ghosh, A. A Systematic Review of the Applications of Multi-Criteria Decision Aid Methods (1977–2022). *Electronics* **2022**, *11*, 1720. [CrossRef]
2. Ayan, B.; Abacıoğlu, S.; Basilio, M.P. A Comprehensive Review of the Novel Weighting Methods for Multi-Criteria Decision-Making. *Information* **2023**, *14*, 285. [CrossRef]
3. Saaty, R.W. The analytic hierarchy process—What it is and how it is used. *Math. Model.* **1987**, *9*, 161–176. [CrossRef]
4. Singh, M.; Pant, M. A review of selected weighing methods in MCDM with a case study. *Int. J. Syst. Assur. Eng. Manag.* **2021**, *12*, 126–144. [CrossRef]
5. Keršulienė, V.; Zavadskas, E.K.; Turskis, Z. Selection of rational dispute resolution method by applying new step-wise weight assessment ratio analysis (Swara) by applying new step-wise weight assessment ratio. *J. Bus. Econ. Manag.* **2010**, *11*, 243–258. [CrossRef]
6. Mardani, A.; Nilashi, M.; Zakuan, N.; Loganathan, N.; Soheilirad, S.; Saman, M.Z.M.; Ibrahim, O. A systematic review and meta-Analysis of SWARA and WASPAS methods: Theory and applications with recent fuzzy developments. *Appl. Soft Comput.* **2017**, *57*, 265–292. [CrossRef]
7. Fernandes, P.G.; Quelhas, O.L.G.; Gomes, C.F.S.; Júnior, E.L.P.; Bella, R.L.F.; Junior, C.d.S.R.; Pereira, R.C.A.; Basilio, M.P.; dos Santos, M. Product Engineering Assessment of Subsea Intervention Equipment Using SWARA-MOORA-3NAG Method. *Systems* **2023**, *11*, 125. [CrossRef]
8. Mukhametzyanov, I. Specific character of objective methods for determining weights of criteria in MCDM problems: Entropy, CRITIC and SD. *Decis. Making: Appl. Manag. Eng.* **2021**, *4*, 76–105. [CrossRef]
9. Krishnan, A.R.; Kasim, M.M.; Hamid, R.; Ghazali, M.F. A Modified CRITIC Method to Estimate the Objective Weights of Decision Criteria. *Symmetry* **2021**, *13*, 973. [CrossRef]
10. Rezaei, J. Best-worst multi-criteria decision-making method. *Omega* **2015**, *53*, 49–57. [CrossRef]
11. Zavadskas, E.K.; Podvezko, V. Integrated Determination of Objective Criteria Weights in MCDM. *Int. J. Inf. Technol. Decis. Mak.* **2016**, *15*, 267–283. [CrossRef]
12. Pamučar, D.; Stević, Ž.; Sremac, S. A New Model for Determining Weight Coefficients of Criteria in Mcdm Models: Full Consistency Method (Fucom). *Symmetry* **2018**, *10*, 393. [CrossRef]
13. Khan, F.; Ali, Y.; Pamucar, D. A new fuzzy FUCOM-QFD approach for evaluating strategies to enhance the resilience of the healthcare sector to combat the COVID-19 pandemic. *Kybernetes* **2022**, *51*, 1429–1451. [CrossRef]
14. Žižović, M.; Pamucar, D. New Model for Determining Criteria Weights: Level Based Weight Assessment (LBWA). *Model. Decis. Mak. Appl. Manag. Eng.* **2019**, *2*, 126–137.
15. Tenório, F.M.; Moreira, M.L.; Costa, I.P.d.A.; Gomes, C.F.S.; dos Santos, M.; Silva, F.C.A.; da Silva, R.F.; Basilio, M.P. SADEMON: The Computational Web Platform to the SAPEVO-M Method. *Procedia Comput. Sci.* **2022**, *214*, 125–132. [CrossRef]
16. Keshavarz-Ghorabaee, M.; Amiri, M.; Zavadskas, E.K.; Turskis, Z.; Antucheviciene, J. Determination of Objective Weights Using a New Method Based on the Removal Effects of Criteria (MEREC). *Symmetry* **2021**, *13*, 525. [CrossRef]
17. Banik, B.; Alam, S.; Chakraborty, A. Comparative study between GRA and MEREC technique on an agricultural-based MCGDM problem in pentagonal neutrosophic environment. *Int. J. Environ. Sci. Technol.* **2023**, *20*, 1–16. [CrossRef]

18. Diakoulaki, D.; Mavrotas, G.; Papayannakis, L. Determining objective weights in multiple *criteria* problems: The critic method. *Comput. Oper. Res.* **1995**, *22*, 763–770. [CrossRef]
19. Zhang, L.; Gao, L.; Shao, X.; Wen, L.; Zhi, J. A PSO-Fuzzy group decision-making support system in vehicle performance evaluation. *Math. Comput. Model.* **2010**, *52*, 1921–1931. [CrossRef]
20. Jahan, A.; Edwards, K.L. A state-of-the-art survey on the influence of normalization techniques in ranking: Improving the materials selection process in engineering design. *Mater. Des.* **2015**, *65*, 335–342. [CrossRef]
21. Ouenniche, J.; Pérez-Gladish, B.; Bouslah, K. An out-of-sample framework for TOPSIS-based classifiers with application in bankruptcy prediction. *Technol. Forecast. Soc. Chang.* **2018**, *131*, 111–116. [CrossRef]
22. Sharkasi, N.; Rezakhah, S. A modified CRITIC with a reference point based on fuzzy logic and hamming distance. *Knowl.-Based Syst.* **2022**, *255*, 109768. [CrossRef]
23. Shemshadi, A.; Shirazi, H.; Toreihi, M.; Tarokh, M. A fuzzy VIKOR method for supplier selection based on entropy measure for objective weighting. *Expert Syst. Appl.* **2011**, *38*, 12160–12167. [CrossRef]
24. Clausius, R. Ueber verschiedene für die Anwendung bequeme Formen der Hauptgleichungen der mechanischen Wärmetheorie. *Annalen der physic* 1865. [CrossRef]
25. Shannon, C.; Weaver, W. *The Mathematical Theory of Communication*; The University of Illinois Press: Urbana, IL, USA, 1949.
26. Bulduk, S.; Ecer, F. Entropi-ARAS Yaklaşımıyla Kripto Para Yatırım Alternatiflerinin Değerlendirilmesi. *Süleyman Demirel Üniversitesi Vizyoner Derg.* **2023**, *14*, 314–333. [CrossRef]
27. Chen, W.; Feng, D.; Chu, X. Study of Poverty Alleviation Effects for Chinese Fourteen Contiguous Destitute Areas Based on Entropy Method. *Int. J. Econ. Finance* **2015**, *7*, 89–98. [CrossRef]
28. Chodha, V.; Dubey, R.; Kumar, R.; Singh, S.; Kaur, S. Selection of industrial arc welding robot with TOPSIS and Entropy MCDM techniques. *Mater. Today Proc.* **2021**, *50*, 709–715. [CrossRef]
29. Szmidt, E.; Kacprzyk, J. Entropy for intuitionistic fuzzy sets. *Fuzzy Sets Syst.* **2001**, *118*, 467–477. [CrossRef]
30. Liu, M.; Ren, H. A New Intuitionistic Fuzzy Entropy and Application in Multi-Attribute Decision Making. *Information* **2014**, *5*, 587–601. [CrossRef]
31. Thakur, P.; Kizielewicz, B.; Gandotra, N.; Shekhovtsov, A.; Saini, N.; Saeid, A.B.; Sałabun, W. A New Entropy Measurement for the Analysis of Uncertain Data in MCDA Problems Using Intuitionistic Fuzzy Sets and COPRAS Method. *Axioms* **2021**, *10*, 335. [CrossRef]
32. Sampathkumar, S.; Augustin, F.; Kaabar, M.K.; Yue, X.-G. An integrated intuitionistic dense fuzzy Entropy-COPRAS-WASPAS approach for manufacturing robot selection. *Adv. Mech. Eng.* **2023**, *15*, 16878132231160265. [CrossRef]
33. Brans, J.P.; Vincke, P.; Mareschal, B. How to select and how to rank projects: The Promethee method. *Eur. J. Oper. Res.* **1986**, *24*, 228–238. [CrossRef]
34. Brans, J.P.; Mareschal, B. Promethee V: Mcdm Problems With Segmentation Constraints. *INFOR Inf. Syst. Oper. Res.* **1992**, *30*, 85–96. [CrossRef]
35. Brans, J.-P.; Mareschal, B. The PROMETHEE VI procedure: How to differentiate hard from soft multicriteria problems. *J. Decis. Syst.* **1995**, *4*, 213–223. [CrossRef]
36. Macharis, C.; Brans, J.P.; Mareschal, B. The GDSS PROMETHEE procedure: A PROMETHEE-GAIA based procedure for group decision support. *J. Decis. Syst.* **1998**, *7*, 283–307.
37. Mareschal, B.; Brans, J.-P. Geometrical representations for MCDA. *Eur. J. Oper. Res.* **1988**, *34*, 69–77. [CrossRef]
38. Brans, J.-P.; Mareschal, B. The PROMCALC & GAIA decision support system for multicriteria decision aid. *Decis. Support Syst.* **1994**, *12*, 297–310. [CrossRef]
39. Zolfani, S.H.; Taheri, H.M.; Gharehgozlou, M.; Farahani, A. An asymmetric PROMETHEE II for cryptocurrency portfolio allocation based on return prediction. *Appl. Soft Comput.* **2022**, *131*, 109829. [CrossRef]
40. Wei, Q.; Zhou, C.; Liu, Q.; Zhou, W.; Huang, J. A barrier evaluation framework for forest carbon sink project implementation in China using an integrated BWM-IT2F-PROMETHEE II method. *Expert Syst. Appl.* **2023**, *230*, 120612. [CrossRef]
41. Ajin, M.; Moses, J.; Dharshini, M.P. Tribological and machining characteristics of AA7075 hybrid composites and optimizing utilizing modified PROMETHEE approach. *Mater. Res. Express* **2023**, *10*, 046509. [CrossRef]
42. Basilio, M.P.; Pereira, V.; de Oliveira, M.W.C.; Neto, A.F.d.C. Ranking policing strategies as a function of criminal complaints: Application of the PROMETHEE II method in the Brazilian context. *J. Model. Manag.* **2021**, *16*, 1185–1207. [CrossRef]
43. Basilio, M.P.; Pereira, V. Operational research applied in the field of public security: The ordering of policing strategies such as the ELECTRE IV. *J. Model. Manag.* **2020**, *15*, 1227–1276. [CrossRef]
44. Basilio, M.P.; Brum, G.S.; Pereira, V. A model of policing strategy choice: The integration of the Latent Dirichlet Allocation (LDA) method with ELECTRE I. *J. Model. Manag.* **2020**, *15*, 849–891. [CrossRef]
45. Basilio, M.P.; Pereira, V.; Brum, G. Identification of operational demand in law enforcement agencies: An application based on a probabilistic model of topics. *Data Technol. Appl.* **2019**, *53*, 333–372. [CrossRef]
46. Basilio, M.P. *O Modelo Multicritério de Ordenação de Estratégias de Policiamento: Uma Aplicação dos Métodos da Família Electre*; Universidade Federal Fluminense: Niterói, Brazil, 2019.
47. Basilio, M.P.; Pereira, V. Estudo sobre a premiação das áreas de segurança pública no Rio de Janeiro via método multicritério: Uma aplicação do método Electre III. *Exacta* **2019**, *18*, 130–164. [CrossRef]

48. Basilio, M.P.; Pereira, V.; de Oliveira, M.W.C.M. Knowledge discovery in research on policing strategies: An overview of the past fifty years. *J. Model. Manag.* **2021**, *17*, 1372–1409. [CrossRef]
49. Do, E.; do, P.M.; de Janeiro, R. *Diretriz Geral de Operações [General Operational Guideline]*; Polícia Militar Do Estado Do Rio De Janeiro: Rio De Janeiro, Brazil, 1982.
50. Odu, G. Weighting methods for multi-criteria decision making technique. *J. Appl. Sci. Environ. Manag.* **2019**, *23*, 1449. [CrossRef]
51. Mahajan, A.; Binaz, V.; Singh, I.; Arora, N. Selection of Natural Fiber for Sustainable Composites Using Hybrid Multi Criteria Decision Making Techniques. *Compos. Part C Open Access* **2022**, *7*, 100224. [CrossRef]
52. Lau, K.-T.; Hung, P.-Y.; Zhu, M.-H.; Hui, D. Properties of natural fibre composites for structural engineering applications. *Compos. Part B Eng.* **2018**, *136*, 222–233. [CrossRef]
53. Salwa, H.N.; Sapuan, S.M.; Mastura, M.T.; Zuhri, M.Y.M. Application of Shannon's entropy-analytic hierarchy process (AHP) for the selection of the most suitable starch as matrix in green biocomposites for takeout food packaging design. *BioResources* **2020**, *15*, 4065–4088. [CrossRef]
54. Wu, H.-W.; Zhen, J.; Zhang, J. Urban rail transit operation safety evaluation based on an improved CRITIC method and cloud model. *J. Rail Transp. Plan. Manag.* **2020**, *16*, 100206. [CrossRef]
55. Vafaei, N.; Ribeiro, R.A.; Matos, L.M.C. Data normalisation techniques in decision making: Case study with TOPSIS method. *Int. J. Inf. Decis. Sci.* **2018**, *10*, 19. [CrossRef]
56. Chakraborty, S. A Simulation-Based Comparative Study of Normalization Procedures in Multiattribute Decision Making. In Proceedings of the 6th WSEAS International Conference on Artificial Intelligence, Knowledge Engineering and Data Bases, Corfu Island, Greece, 16–19 February 2007.
57. Vincke, J.P.; Brans, P. A preference ranking organization method. The PROMETHEE method for MCDM. *Manag. Sci.* **1985**, *31*, 647–656.
58. Makan, A.; Fadili, A. Sustainability assessment of large-scale composting technologies using PROMETHEE method. *J. Clean. Prod.* **2020**, *261*, 121244. [CrossRef]
59. de Almeida, A.T.; Costa, A.P.C.S. Modelo de decisão multicritério para priorização de sistemas de informação com base no método PROMETHEE. *Gestão Produção* **2002**, *9*, 201–214. [CrossRef]
60. Basilio, M.P.; de Freitas, J.G.; Kämpffe, M.G.F.; Rego, R.B. Investment portfolio formation via multicriteria decision aid: A Brazilian stock market study. *J. Model. Manag.* **2018**, *13*, 394–417. [CrossRef]
61. Ghandi, M.; Roozbahani, A. Risk Management of Drinking Water Supply in Critical Conditions Using Fuzzy PROMETHEE V Technique. *Water Resour. Manag.* **2020**, *34*, 595–615. [CrossRef]
62. Alencar, L.; Almeida, A. Supplier Selection Based on the PROMETHEE VI Multicriteria Method. In *Lecture Notes in Computer Science, Proceedings of the Evolutionary Multi-Criterion Optimization: 6th International Conference, EMO 2011, Ouro Preto, Brazil, 5–8 April 2011*; Takahashi, R.H.C., Deb, K., Wanner, E.F., Greco, S., Eds.; Springer: Berlin/Heidelberg, Germany, 2011; Volume 6576, pp. 608–618. [CrossRef]
63. Anwar, M.; Rasul, M.G.; Ashwath, N. The efficacy of multiple-criteria design matrix for biodiesel feedstock selection. *Energy Convers. Manag.* **2019**, *198*, 111790. [CrossRef]
64. Fontana, M.E.; Cavalcante, C.A.V. Use of Promethee method to determine the best alternative for warehouse storage location assignment. *Int. J. Adv. Manuf. Technol.* **2014**, *70*, 1615–1624. [CrossRef]
65. Behzadian, M.; Kazemzadeh, R.; Albadvi, A.; Aghdasi, M. PROMETHEE: A comprehensive literature review on methodologies and applications. *Eur. J. Oper. Res.* **2010**, *200*, 198–215. [CrossRef]

mathematics

MDPI

Article

An Exploration of Prediction Performance Based on Projection Pursuit Regression in Conjunction with Data Envelopment Analysis: A Comparison with Artificial Neural Networks and Support Vector Regression

Xiaohong Yu [1,2] and Wengao Lou [3,*]

1 College of Humanities and Law, Shanghai Business School, Shanghai 200235, China; yuxiaohong@nuaa.edu.cn
2 College of Economics and Management, Nanjing University of Aeronautics and Astronautics, Nanjing 210016, China
3 School of Information Management, Shanghai Lixin University of Accounting and Finance, Shanghai 201209, China
* Correspondence: louwengao@lixin.edu.cn

Abstract: Data envelopment analysis (DEA) is a leading approach in performance analysis and discovering newer benchmarks, and the traditional DEA models cannot forecast the future efficiency of decision-making units (DMUs). Machine learning, such as the artificial neural networks (ANNs), support vector machine/regression (SVM/SVR), projection pursuit regression (PPR), etc., have been viewed as beneficial for managers in predicting system behaviors. PPR is especially suitable for small and non-normal distribution samples, the usual cases in DEA analysis. This paper integrates DEA and PPR to cover the shortcomings we faced while using DEA and DEA-BPNN, DEA-SVR, etc. This study explores the advantages of combining these complementary methods into an integrated performance measurement and prediction model. Firstly, the DEA approach is used to evaluate and rank the efficiency of DMUs. Secondly, we establish two DEA-PPR combined models to describe the DEA efficiency scores (also called the production function) and the DEA-efficient frontier function. The first combined model's input variables are input–output indicators in the DEA model, and the output variable is the DEA efficiency. In the second model, its input variables are input or output indicators in the DEA model, and the output variable is the optimal input indicator for input-oriented DEA or the output indicator for output-oriented DEA. We conducted positive research on two examples with actual data and virtual small, medium-sized, and large samples. Compared with the DEA-BPNN and DEA-SVR models, the results show that the DEA-PPR combined model has more vital global optimization ability, better convergence, higher accuracy, and a simple topology. The DEA-PPR model can obtain robust results for both small and large cases. The DEA-BPNN and DEA-SVR models cannot obtain robust results for small and medium-sized samples due to overfitting. For large samples, the DEA-PPR model outperforms DEA-BPNN, DEA-SVR, etc. The DEA-PPR combined model possesses better suitability, applicability, and reliability than the DEA-BPNN model, the DEA-SVR model, etc.

Keywords: projection pursuit regression (PPR); data envelopment analysis (DEA); artificial neural networks (ANNs); support vector machine/regression (SVM/SVR); efficiency measure; decision-making units (DMUs); combined model

MSC: 68T07; 90B50; 68V99

1. Introduction

The measurement of the efficiency or performance of homogeneous production or services, namely decision-making units (DMUs), has long been a hotspot or subject of interest for researchers in many fields. Since Charnes et al. [1] proposed the radial-based data envelopment analysis (called DEA, DEA-CCR, or CCR) under input-oriented cases with a constant return-to-scale (CRS), DEA is widely and successfully used to measure the performance or efficiency of DMUs [2,3]. DEA is a nonparametric basis for evaluating the multiple input–output efficiency of different DMUs. To better meet various situations and conditions, a variety of DEA models have been proposed successively [4]. Distinct calculation principles, a simple structure, fewer input–output indicators, low-cost data collection, and easy application characterize them.

Meanwhile, the DEA model also has some shortcomings. For example, DEA is very sensitive to data noise and cannot be used for prediction [5]. When new DMUs are added, they must be remodeled, and the efficiency of the original DMUs is changed. At the same time, the shortcomings of these DEA models happen to be the advantage of another nonparametric modeling approach—the artificial neural network (ANN) [6]. Furthermore, DEA models treat DMUs as a black box; inputs enter and outputs exit without considering the intervening steps [7]. Therefore, Athanassopoulos and Curram [8] proposed to combine the DEA and BPNN models for efficiency analysis and conducted empirical research on the operating efficiency of the Cobb–Douglas production function and bank; the results show that the DEA model has better performance in simulating production functions, and the BPNN model is equivalent to DEA in the efficiency ranking of DMUs; they conducted in-depth research on how to combine DEA and BPNN better to give full play to their respective advantages. Since then, extensive theoretical and applied research has been conducted on the combination of ANN and DEA [4,9–16]. Additionally, other machine learning methods (such as SVR and SVM) are also combined with DEA in applied and comparative studies [4,11,17–20]. Some scholars have optimal experimental research on the integrated model of RSM (response surface methodology) and ANN-DEA. So far, extensive and in-depth research has been conducted on combined models such as ANN-DEA, achieving good results. However, there are still areas to be improved and perfected. First, for small and medium-sized samples in DEA, the input–output data and the evaluation results (efficiency scores) usually do not obey normal distribution, and it is not easy to guarantee the generalization ability and practical value of ANN or SVR combined models. Although some studies divide the sample data into training and test subsets [11,13,21], all studies do not use validation subsets to monitor the training process to avoid over-training. In training the ANN-DEA combined model, we cannot judge whether over-training has occurred, so it is not easy to guarantee its generalization ability and practical value. Second, to establish SVR (SVM) and ANN, the principles of determining input and output variables in different studies are often inconsistent, resulting in unconvincing modeling results. Some scholars [11,13,21–23] took the input–output indicators of the DEA model as the input variables of the ANN model. The DEA model evaluated the efficiency scores as the output variables of the ANN for modeling, but did not discuss how to reasonably determine the number of hidden-layer neurons and avoid over-training. Farahmand et al. [24] and Zhu et al. [11] used the input–output indicators of the DEA model as the input variables of SVR and the efficiency value evaluated via DEA as the output variables of SVR. Through empirical research, Zhu et al. [11] believed that the performance of the ANN-DEA model was better than that of SVR-DEA. Zhong et al. [13] argued that the performance of the ANN-DEA model was better than that of SVR-DEA, random forest (RF-) DEA, gradient boosting regressor (GBR-) DEA, and other models. Bose [15] established an input-oriented CCR-DEA model, took values uniformly within the output-indicator range of DEA-efficient samples, generated a certain number of modeling virtual samples, and used the output indicators of DEA as the input variables of the ANN model and the input indicators of the DEA model as the output variables of the ANN model; without a test subset, the DEA-efficient frontier ANN model was established.

It can be seen from those mentioned above that although the academic community has carried out extensive research on the combined prediction model of ANN-DEA (including SVR-, RF-, etc.), there are still many problems to be further researched, especially on the efficiency of small and medium-sized samples that do not obey normal distribution; a better model should be established with a better generalization ability and robustness. On the other hand, projection pursuit regression (PPR) is particularly applied to modeling with nonlinear, small, and medium-sized samples that do not obey normal distribution [25–28]. Therefore, this paper introduces PPR into efficiency research to establish a DEA-PPR combined model to obtain more reliable and robust results. It expands not only the efficiency evaluation method but also the application field of the PPR model.

This study fulfills a practical need and improves benchmarking and decision-making processes by exploring an innovative performance measurement and prediction framework using DEA-PPR. The proposed combined model utilizes DEA as a preprocessor, and the subsequent PPR model conducts prediction tasks for the best performance output for each DMU. This paper provides insight into two examples to build the DEA-PPR combined models and compare the performance of different combined models such as DEA-BPNN, DEA-SVR, etc.

In summary, the primary purpose and motivations for this research paper are fivefold:

(1) To present a new DEA-PPR combined model for performance measurement and prediction, thus bridging the research gap through methodological advancement;
(2) To provide empirical support for the proposed model using two datasets through streamlining sequential processes of DEA measurement and DEA-PPR prediction;
(3) To provide a way of thinking to improve managerial efficiency and enhance administrative flexibility in selecting actionable options from theoretical and practically feasible alternatives and potential progress monitoring via the DEA-PPR combined model;
(4) To discuss the advantages and disadvantages of machine learning, such as BPNN, SVR, PPR, RF, etc.;
(5) To put forward the basic principles and some matters needing attention for building machine learning, such as BPNN, PPR, SVR, etc.

The rest of this paper is organized as follows: Section 2 is a review of the literature on DEA combined models and PPR; Section 3 is about CCR-DEA and PPR modeling principles; Section 4 is on the empirical research on the production function of combined models of DEA and the DEA-efficient frontier function; Section 5 is the results and discussion; and Section 6 describes the limitations and future research fields.

2. Literature Review

2.1. DEA and Its Combined Models

DEA, proposed in 1978 [1], provides an effective tool for measuring the efficiency of the homogeneous DMUs of non-profit (e.g., schools, hospitals, and local authorities) and for-profit (e.g., banks, public houses, corporates, listed companies) organizations or persons (e.g., scholars, employees). Since then, many new DEA models have been introduced, such as BCC, SBM, super-efficiency (SE-) DEA, SE-SBM, (SE-)SBM with undesirable output, dynamic DEA, and the window analysis technique [2–4]. DEA models with multiple input and output indicators have been widely used to assess DMUs' efficiency.

On the other hand, the DEA models used for performance assessment have inherent disadvantages. First, DEA models are easily affected by data noise. Second, when new DMUs are added, we must re-establish the DEA, and the efficiency of all the original DMUs must be changed. It is not easy to compare the efficiency of different datasets. To overcome the above disadvantages, Athanassopoulos and Curram [8] thought that DEA and artificial neural networks (ANNs) were nonparametric methods, in that no assumption is made, and the inputs and outputs are used to describe an operational process. They started work to combine DEA with BPNN and studied the known Cobb–Douglas production function with two inputs and one output (simulated data). A series of 16 datasets were artificially generated and tested using the ANN-DEA combined models. The number of nodes in the input and output layers of the ANN model was equal to the DEA model. The results showed that the DEA performed more satisfactorily as a tool for estimating empirical production functions. Furthermore, they researched 250 commercial bank branches, using four inputs and three outputs. Take the network (topology) structure of the BPNN model as 4-10-3 (that is, the input layer, hidden layer, and output layer have four, ten, and three neurons, respectively, and the number of network connection weights is $5 \times 10 + 11 = 61$); the input-layer and output-layer neurons of the BPNN model are the same as the input–output indicators of the DEA model; that is, the BPNN model is used to simulate the operation of the bank branch. Randomly select a 20% (50) proportion of test (validation) subset data; it is believed that the more validation data, the better the model's generalization ability is. The error changes in the validation data are used to monitor the training process. When the error reaches the minimum and does not decrease with further training, we take the network connection weights with the minimum error of the validation data (called the early-stopping method), eliminate the influence of over-training as much as possible, and improve the model's generalization ability and practical value. The results show that the ANN is instead a tool for obtaining relative rankings of DMUs based on their predicted outcome. Since then, scholars have extensively researched other DEA combined models. (These previous studies added meaningful value to the existing literature, as summarized in Table 1).

Table 1. The detailed information on the articles establishing a combined DEA model with machine learning such as BPNN, RF, SVR, etc.

References	DEA Model	Topology Employed in a Combined Model	Combined Model	Number of Samples (Validation, Test)	To Obey the Rule of Thumb	Model for Efficiency Score (ES) and Efficient Frontier (EF)
Wu et al. [5]	CCR (2, 3)	5-10-1, 5-4-1	two BPNNs	142 (84)	no	ES
Athanassopoulos et al. [8]	CCR (2, 1) *, DEA (4, 3)	2-3-1 **, 4-10-3	BPNN	250 (50)	yes	ES, DEA > BPNN; efficiency rank, BPNN > DEA
Na et al. [9]	CCR (5, 2)	5-3-2	BPNN	13N, N	no	ES
Ma et al. [10]	CCR (6, 1)	4-4-3	BPNN	38 (5)	no	ES
Zhu et al. [11]	CCR (2, 2)	4-?-1	BPNN, GANN, SVR, ISVR	948 (48)	yes	ES, GANN > BPNN > ISVM > SVM
Olanrewaju [12]	CCR (1, 1)	1-5-1	BPNN	8	no	ES
Zhong et al. [13]	SESBM (3, 1)	3-?-1	15 MLs such as CART, CIT, Bagging, RF, BPNN, etc.	710 (3:1)	no	EF, BPNN > ET > RF > GBR
Anouze et al. [14]	CCR (5, 4), and 15 environmental variables	9-5-1	15 MLs such as CART, CIT, Bagging, RF, BPNN, etc.	151 (2:1)	no	The bagging and RF are better than BPNN, CIT, etc.
Bose et al. [15]	CCR (2, 2)	2-3-2-2, 5-4-3-2	Two BPNNs	12, 99	no	EF
Kwon [16]	two DEAs, CCRs (4, 1)	5-7-1, 5-8-1, 5-8-1, 5-7-1; 4-3-1, 4-2-1, 4-4-1, 4-5-1	eight BPNNs	56 (17)	no	ES, EF
Hong et al. [17]	CCR (4,4)	/	SOM[c]	50	/	/
Yang et al. [20]	SBM (5, 5)	trial-and-error	SVM[c]	creating at most 500 instances, 10-fold CV	/	ES
Saeidi et al. [22]	CCR (4, 2)	6-?-1	BPNN	26	no	ES and EF
Kwon et al. [23]	CCR (3, 3), CCR (3, 1)	15-9-1, 6-30-1, 15-3-1, 15-3-1	four BPNNs	181 (37, 36) #	no	ES, EF
Ren et al. [29]	CCR (5, 5)	10-21-1	BPNN	5N, N	no	ES
Tsolas et al. [30]	eight DEAs, CCR (3, 2)	3-3-1, 6-?-1	two BPNNs	160 (4:1)	no	ES, EF

Table 1. *Cont.*

References	DEA Model	Topology Employed in a Combined Model	Combined Model	Number of Samples (Validation, Test)	To Obey the Rule of Thumb	Model for Efficiency Score (ES) and Efficient Frontier (EF)
Zhang et al. [31]	SESBM (3, 2)	5-10-1; 5-10-20-1; 5-10-10-10-1; 5-10-10-20-1; 5-10-20-30-1	11 MLs such as BPNN, SVR, etc.	420 (30), 5-fold CV	no	ES, BPNN is the optimal
Fallahpour et al. [32]	CCR (3, 3)	trial-and-error	SVM	48 (12)	/	ES
Yazdanparast et al. [33]	Z-DEA (1, 17)	17-?-1	BPNN	150 (45)	no	EF
Sreekumar et al. [34]	CCR (3, 8), BCC (3, 8)	11-?-1	GRNN	49	/	ES
Kao et al. [35]	CCR (10, 2)	12-?-1	two SVMs^{c}	91	/	ES
Barros et al. [36]	PCA-CCR (2, 2)	4-20-1	BPNN	50	no	ES
Sanei et al. [37]	SBM (3, 3)	5-6-1	three BPNNs	155 (46)	no	EF
Liu et al. [38]	BBC (4, 4)	7-8-4	BPNN	120 (20)	no	ES

Notes: *: CCR(2,1) denotes the used CCR-DEA model with two inputs and one output. **: the "2-3-1" BPNN model denotes the used BPNN model with two inputs, three neurons in the hidden layer, and one output; the "?" means that the article does not specify the number of neurons in the hidden layer. c: the "c" denotes the model used for classification. #: 181 (37, 36) denotes that the numbers of the total, validation, and test subsets are 181, 37, and 36, respectively; 250 (50) denotes the numbers of the total and test subsets are 250, and 50, respectively; 160 (4:1) denotes that the total data are divided into training and test subsets according to a 4:1 ratio.

It can be seen from Table 1 that there are more than 10 DEA combined models, including ANN (mainly BPNN and RBFNN models), SVR/SVM, RF, the bagging model, etc. There are three main functions (purposes) of establishing a DEA combined model. The first purpose is to reveal the function relationships between the input–output indicators of the DEA model; that is, the input and output indicators of the ANN model are exactly or partially the same as the DEA model [8,15]. The second purpose is to reveal the working principle of the DEA model [11,14,23,29]; that is, the efficiency score calculated via the DEA model is used as the output value of the machine learning model (ML) (e.g., BPNN, RF, SVR, etc.); input–output indicators of the DEA model are used as the inputs of the ML; to take all DMUs as modeling samples; and to divide the samples into training and test datasets (also called validation datasets in some studies, but they play the role of a test dataset, which is only used to evaluate the performance of the model instead of monitoring the training process to avoid over-training). This combined model is a subsystem of the DEA model, which can be used to calculate the value of input–output efficiency. Still, it cannot be used to simultaneously calculate input excesses and output shortages (or shortfalls). The third purpose is to construct the DEA-efficient frontier function of DEA obtained by establishing the DEA model, that is, to use the DEA-efficient (valid) DMUs as the modeling samples of the ML [9,15,16,23]. Since there are a relatively small number of valid DMUs, some scholars generate a certain number of virtual DMUs by uniformly taking values within the range of the indicator value of valid DMUs. Some scholars established a BPNN model for small samples. Therefore, building a combined DEA model is not a substitute for the DEA model but can augment rather than replace the DEA model. Na et al. [9] aimed to determine the total staff employed in government organizations in China; collected statistical data from 27 provinces, municipalities, and autonomous regions; set five input indicators (such as population) and two output indicators (administrative expenditure and the number of officially employed staff); and established the CCR and CGS-DEA models. There were 10 and 13 provinces that were valid for DEA. Then, according to the 10 and 13 valid DEA samples (all as training samples; no test and validation samples), the input layer had five neurons (that is, the indicators of the DEA model), and the output layer had two neurons (the output indicators of the DEA model); a DEA-efficient frontier BPNN model was established with a network structure of 5-3-2 and 26 connection weights, which is significantly larger than the number of training samples. It does not meet the requirements for BPNN modeling, as the number of training samples must be more than the number of network connection weights. Ren et al. [29] established a DEA model for a total of four DMUs (A, B, C, and D), with two input indicators and three output indicators; they generated uniformly distributed samples at five levels (which were all training samples; no test and validation samples), used a genetic algorithm to build a BPNN model with a 10-21-1 network topology, and constructed an efficient frontier of DEA with 253 network connection weights, which is 50 times the number of training samples; it does not meet the basic requirements for building a BPNN model. Na et al. [9] and Ren et al. [29] did not use validation samples to monitor the training process and could not judge whether over-training occurred in the training process. Because the training samples were less than the number of connection weights [9], and the number of training samples was even less than the number of hidden-layer neurons [29], over-training must occur during the training process. Therefore, it is impossible for BPNN models established in this way to have a generalization ability and practical value. In [15], the first case was to establish a DEA model based on the collected data of doctors and nurses (two input indicators) and the number of outpatients and inpatients (two output indicators) from 12 hospitals, and obtain three valid DEA samples (A, B, and D); then, they generated 125 virtual DEA-efficient DMUs using linear combinations of real DEA-efficient DMUs, A, B, and D. The researchers established a BPNN model to calculate the optimal numbers of doctors and nurses according to the number of inpatients and outpatients; that is, the output indicators of the DEA model were used as the input variables of the BPNN model, and the input indicator of the DEA-efficient DMUs was used as the output variable of BPNN. The second example studied in [15] is based on the borrowing data of 99 Indian microfinance institutions that had been collected; there were five input indicators (such as assets, etc.) and two output indicators (the number of borrowers and borrowings); according to the sample data of the six DEA-efficient DMUs, a total of 729 virtual DEA-valid samples were generated. The authors used the DEA model inputs as BPNN inputs and the DEA outputs as the BPNN outputs, established the BPNN model, and tried to obtain the optimal levels of inputs under a certain number of borrowers and borrowings—assets, offices, personnel, and expenses. For the two examples in Ref. [15], the authors did not explain whether the samples were divided into training, test, and validation subsets, nor how to determine a reasonable number of hidden-layer neurons and avoid over-training. Therefore, the validity and reliability of the results of Ref. [15] require further investigation. Tsolas et al. [30] studied the operational efficiency of 160 bank branches in Greece based on operational data; there were three inputs, namely personnel expenses, rents and depreciation, and operational expenses, and two outputs, namely net interest income and non-interest income (fee and trading income). The researchers established input-oriented DEA models, mainly including radial CCR, BCC, NIRS, NDRS, and non-radial FDH. Russell and others obtained the input–output efficiency score, and divided the efficiency score into four levels of 1–4; the authors took the input–output indicators of the DEA model as the inputs of the BPNN model (six in total) and the efficiency score level as the output (one), and randomly divided the 160 samples into training and test subsets at a ratio of 8:2. They established a BPNN model with certain results; however, since the test subset was not used to monitor the training process in real time, it was impossible to determine whether over-training occurred or not during the training process. Know et al. [23] employed ten years of longitudinal data from Japanese electronic manufacturing firms from 2003 to 2012, and each firm-year was treated as an individual DMU; the researchers eliminated non-normal data and obtained 1419 samples for modeling. There were three inputs for the DEA model (namely employees, total assets, and operating expenses) and two outputs (namely revenue and market value); they established output-oriented CCR and BCC models to produce input and output projections for each inefficient DMU to become DEA-efficient. Then, two BPNN models were established; the inputs of the BPNN1 model were the input–output indicators (five) of the DEA model, and the output variable was the efficiency score of the DEA model, revealing the production function of the DEA model. When building the BPNN1 model, the samples were randomly partitioned into training and test datasets at a 7:3 ratio, without a validation dataset; the BPNN1 model had a higher prediction accuracy, with only six DMUs beyond 10% error and with a maximum error of 17%. However, the input and output variables of the BPNN2 model were the input–output indicators of the DEA-valid DMUs, revealing the frontier function of the DEA model; the BPNN2 split 1419 DMUs into 772 training and 647 test samples, the established BPNN model demonstrated a high prediction accuracy of less than 10% error for 95% of the DMUs, but Ref. [23] had no validation samples to monitor the training process in real time, and it is impossible to determine whether over-training occurred or not during the training process. Similarly, Kwon (2014) also established two types of BPNN models that reveal the production function and frontier function of the DEA model based on the input–output indicators of the DEA model and their modeling results. The author split the data into training and test datasets at a ratio of 7:3; 54 samples were used to train the network to learn underlying patterns, and the remaining 24 samples were used to test the network. A total of five BPNN models were established, with network topologies of 8-11-1, 8-6-1, 8-4-1, 8-18-1, and 8-14-1, and connection weights of 111, 61, 41, 181, and 141, respectively; except for the third model, the numbers of network connection weights were larger than the number of training samples, and so they do not meet the requirements for building a BPNN model. A further review of the research on combining the DEA model with neural networks such as BPNN, RBFNN, PNN (probabilistic neural network), and SOM to establish DEA-efficient frontier functions or DEA production functions for efficiency prediction, efficiency classification, the screening of training data, and data processing refers to Kwon [16]. The combined DEA models with ANN are also used for risk management, feature selection, etc. In recent years, ML, such as SVR/SVM, RF, bagging, RSM, lightGBM, etc., has been introduced in combination with various DEA models, such as SBM, SESBM with undesirable output, and window analysis [4,11,22,31]. After removing outliers, Zhong et al. [13] obtained 710 sample data points from 95 rural commercial banks from 2011 to 2018 in Guangdong Province, China. The input indicators of the SESBM model were the number of employees, fixed assets, and intermediate business expenses, and the output indicator was the intermediate business income. The dataset was randomly split into training and test subsets with a proportion of 3:1. The authors used SESBM to measure the efficiency of DMUs in the training subset and then adjusted all the training datasets to the efficiency frontier based on projection values. The DEA model's input and output indicators were the machine learning model's input and output variables, respectively, and 15 machine learning models, such as BPNN and SVR, were established. Four performance metrics were used to evaluate the different ML algorithms, namely the mean square error (MSE), the mean absolute error (MAE), the root mean square error (RMSE), and the coefficient of correlation R^2. After comparing 15 MLs, the results show that the top four MLs for efficiency evaluation performance are BPNN, the extra trees regressor (ETR), RF, and the gradient boosting regressor (GBR). SVR is the worst, and BPNN is the best. Zhong et al. [13] also established combined models between SESBM with undesirable output and SVR, BPNN, etc.; the BPNN model is better than the other models.

To sum up, many scholars have conducted in-depth research on DEA combined models and achieved good results, which have played an essential role in management and other fields. However, the following problems still exist:

(1) The model's generalization ability and prediction accuracy are difficult to guarantee without a validation dataset. To establish MLs such as BPNN, except for [8], no other authors divided the samples into training, validation, and test datasets with similar properties. No study discusses how, according to the error changes of the validation dataset, to use the early-stopping or regularization methods to prevent over-training, and use the trial-and-error approach to determine a reasonable number of hidden-layer neurons to ensure the generalization ability and practical value [6,39–43].

Although, most studies clearly state (and some studies do not) to randomly divide the samples into training and test datasets (in some studies, the validation dataset is a test dataset) with similar properties according to the ratios of 8:2, 7:3, or 3:1 (or according to some algorithm). There are also studies stating that over-training should be avoided. Still, since no validation dataset is used to monitor the training process, it is impossible to judge whether over-training has occurred. When the number of training samples is less than three times the number of connection weights due to the extensive network topology, over-training is easy to occur during training. In the case of over-training, even if the error of the training dataset is minimal, and even if the RMSE of the test dataset is casually small, the established model has no generalization ability and practical value. Scholars should pay more attention to this problem.

(2) It is difficult to establish reliable and effective DEA-ML combined models for DEA modeling problems, usually with only small and medium-sized samples or frontier modeling with only a few DEA-valid samples. Most DEA efficiency modeling uses small or medium-sized samples. For modeling DEA-efficient frontier functions, small samples are usually used. To establish the DEA-BPNN combined model, most of the literature does not meet the basic requirement—the number of training samples must be more than the number of connection weights. Ref. [42] put forward a rule of thumb: you should aim to have at least five times as many cases (training samples) as connection weights in the network, and preferably ten times as many for establishing a reliable and effective BPNN model. According to the rule of thumb, you can determine the reasonable number of neurons in the hidden layer through trial-and-error. To establish DEA combined with SVR, RF, etc., you should avoid over-training and overfitting to carefully determine the model's parameters with small and medium-sized samples.

(3) It is not easy to judge which ML model is better than others. Among the ML models currently used for DEA combined modeling, some studies believe that BPNN performs better [10,12,30], while others believe that SVR and other models are better [13]. Which model has better performance is also a question worthy of study.

It can be seen that there are still several tricky problems that need to be improved in the establishment of reliable and effective DEA-combined MLs with a good generalization ability and production precision, particularly for frontier surface functions in the case of small and medium-sized samples.

2.2. Projection Pursuit Regression (PPR) Model

PPR is consistent with the BPNN model regarding the nonlinear approximation ability, especially for the nonlinear modeling of small and medium-sized samples that do not obey normal distribution [25–28,44,45]. The constraint condition of the PPR model is that the sum of the squares of the weights of all independent variables equals one. The modeling usually starts from linear to quadratic; cubic; polynomial ridge function (PRF); and one to two, three, etc., PRFs. So, there is a low possibility of over-training or overfitting during modeling [26,27,44,46]. It has been widely used in small and medium-sized samples such as agricultural engineering, water conservancy, demography, and earthquakes [26,44,46]. However, there are no studies on DEA-PPR combined modeling. Therefore, in this paper, the PPR model is introduced into efficiency evaluation research for the first time to construct the DEA-PPR combined model, hoping to improve the generalization ability, robustness, and reliability, mainly to solve the efficiency production function and frontier function of DEA combined modeling with small and medium-sized samples, and expand the DEA combined model and the new method of input–output efficiency research, as well as the application field of the PPR model.

The following are the innovations and contributions of this paper:

(1) We are the first to establish the DEA-PPR combined model to effectively and reliably solve the problem of input–output efficiency of small and medium-sized samples that

do not obey the normal distribution, to overcome the disadvantages of MLs, such as BPNN, SVR, etc., that can only be applied to large samples.

(2) We proposed the modeling principles and steps for establishing a BPNN model with good generalization ability. Through empirical research, the reliable and effective DEA-BPNN, DEA-SVR, and DEA-PPR combined models were found to verify the effectiveness of each other. The DEA-PPR model has a relatively better generalization ability and prediction accuracy among them.

(3) We established the DEA-PPR combined model to simulate the production function by setting the efficiency score to one, adopting the optimization technology to obtain the frontier function directly, and realizing the unification of the production and frontier surface functions.

(4) We established the DEA-PPR combined model of the frontier function of the DEA by generating virtual samples according to the valid DEA samples. According to the input-oriented DEA-PPR combined model, the optimal input quantity can be obtained; on the contrary, if the output-oriented DEA-PPR combined model is used, the optimal output quantity can be obtained, providing a decision-making basis and technical paths for DMUs to organize production, strengthen management, improve efficiency, and reduce costs.

3. Methodology and Principle

In this paper, for the prediction research of input–output efficiency, we establish combined models of DEA with BPNN, PPR, and SVR to improve the generalization ability, robustness, and practical value.

This paper briefly introduces some typical models to evaluate input–output efficiency.

3.1. DEA-CCR Model

Charnes et al. [1] first proposed the DEA-CCR model. There are n DMUs, and each DMU has m inputs and q outputs. We set x_i and y_r to represent the inputs and outputs of the DMU, $n \times m$ is the input matrix, and $n \times qs$ is the output matrix. For each DMU, we can obtain the ratio $\frac{\sum_{r=1}^{q} u_r y_{rk}}{\sum_{i=1}^{m} v_i x_{ik}}$ of all its outputs to inputs (where u_r is the output weight and v_i the input weight); then, the problem of the CCR model is converted to the problem of selecting the best weights. We take the input orientation as an example and obtain the specific planning equation, as in Equation (1):

$$\begin{aligned} max\theta_k &= \frac{\sum_{r=1}^{q} u_r y_{rk}}{\sum_{i=1}^{m} v_i x_{ik}} \\ &s.t. \frac{\sum_{r=1}^{q} u_r y_{rj}}{\sum_{i=1}^{m} v_i x_{ij}} \leq 1 \\ &u \geq 0, v \geq 0 \\ &(i = 1, 2, \ldots, m; r = 1, 2, \ldots, q; k, j = 1, 2, \ldots, n) \end{aligned} \tag{1}$$

The goal of the CCR model is to maximize the efficiency value of the rth DMU under the restriction that the other DMUs' efficiency score is less than or equal to 1. We can solve the problem by transforming it into linear programming. Thus, by constructing an effective frontier, all DMUs either fall within the effective frontier (CCR-effective, CCR-efficient, efficient, and valid) or outside the effective frontier (CCR-ineffective, CCR-efficient, invalid, and inefficient).

In practice, we would like to replace Equation (1) with the corresponding linear programming pairwise model and obtain Equation (2).

$$\begin{aligned} & min\theta \\ & \quad s.t. \sum_{j=1}^{n} \lambda_j x_{ij} \leq \theta x_{ik} \\ & \quad \sum_{j=1}^{n} \lambda_j y_{rj} \geq= y_{rk} \\ & \quad (i = 1,2,...,m; r = 1,2,...,q; k,j = 1,2,...,n; \lambda \geq 0) \end{aligned} \tag{2}$$

In Equation (2), λ is the coefficient of the DMU, and the optimal solution (θ) of Equation (2) represents the efficiency score. The range of the θ is (0,1]. For a specific DMU, the DMU is efficient when and only when $\theta = 1$, and if $\theta < 1$, the DMU is inefficient.

The CCR model is developed based on the constant return-to-scale (CRS) assumption. The DEA-BCC (or BCC) model, on the other hand, is based on a variable return-to-scale (VRS). The BCC and CCR models differ only in that the former, but not the latter, includes the convexity condition of $\sum_{j=1}^{n} \lambda_j = 1, \lambda_j \geq 0, \forall j$ in its constraints. Thus, as might be expected, they share properties in common and exhibit differences. They also share properties with the corresponding additive models [2,3,47].

Using the CCR and BCC models helps to determine the DMU's overall technical and scale efficiencies, and whether the data exhibit a VRS.

Furthermore, the CCR and BCC models fail to distinguish the DMUs with the highest efficiency scores of "1"; therefore, to overcome this problem, the super-efficiency CCR or BCC models were proposed in Ref. [2].

The CCR, BCC, and super-efficiency models are radial models and cannot fully consider the effect of slackness on efficiency. The slack-based-measured (SBM) DEA model (or SBM) and the super-efficiency SBM (called SE-SBM) were proposed to solve this problem [2]. Based on the slackness measurement, SBM is a non-radial method suitable for measuring efficiency when the input and output vary non-proportionally. SE-SBM is a model that combines super efficiency and the SBM.

In the super-efficiency evaluation method, the efficient DMUs are removed from the set, and the efficiency of the DMUs is re-evaluated; thus, the original non-efficient DMUs remain unchanged, and the original efficient DMUs can be greater than 1, then they can be compared [2,13].

According to Cooper et al. [3], the DEA models have a rule of thumb—the minimum number of DMUs should be equal to or greater than $max\{q \times m, 3 \times (q + m)\}$.

3.2. Machine Learning

Machine learning (ML) is a multi-disciplinary subject involving many disciplines, such as probability theory, statistics, approximation theory, convex analysis, and algorithm complexity theory. After decades of continuous development, ML is a well-known method that "uses algorithms to parse data, learn from it, and then make decisions or predictions about something unknown in the world" [10,40,42]. An important limitation of ML, such as BPNN, RF, SVR, etc., is that of over-training or overfitting. An ML trained to generalize well on new data (called the generalization ability) will produce a correct input–output mapping, even when the input differs slightly from the examples used to prepare the ML. However, when an ML model learns too many input–output examples, the ML model may memorize training examples that are not true of the underlying function that is to be modeled. Such a phenomenon is referred to as over-training or overfitting. During the training process or optimizing the parameters, the error of the training subset continuously decreases, even almost to zero. The error of the validation subset first decreases and then rises again. This is a sure sign that over-training or overfitting is occurring, and you should stop training once deterioration in the error of the validation subset is observed [40,42]. When an ML model is over-trained, it loses the ability to generalize between similar input–output patterns [6,40,42]. The most effective approach to ensure the generalization ability is to reserve data to cross-validate the ML's performance. First, the available dataset is randomly partitioned into training, validation, and test subsets. Second, the training subset is used to train the ML model, the validation subset is used to validate the ML model, and the generalization performance of the established ML model is measured on the test subset. Third, if the validation and test errors are close, the trained ML model will likely generalize well; otherwise, we must re-train the ML model [40,42]. If there is no validation or test subset, judging whether over-training has occurred is challenging. BPNN and SVR are the most widely used MLs, and the PPR is not used to combine with the DEA model. We briefly introduce the principles of the BPNN and PPR models as follows.

3.2.1. BPNN Model

The BPNN model, proposed by Rumelhart et al. [6,39,40,42,43,46], is currently the most widely used. The BPNN model consists of an input layer, a hidden layer (usually 1, but 2–3 layers are optional), and an output layer. Each layer consists of multiple neurons. It simulates the human processing process of external input information.

Corresponding to the BPNN with one hidden layer, the output of the jth output node of the pth training sample is

$$O_{pj} = f_2\left[\sum_{h=1}^{H} \omega_{jh} f_1\left(\sum_{i=1}^{n} \omega_{hi} x_{pi} + \theta_h\right) + \theta_j\right] \tag{3}$$

In Equation (3), ω_{hi} and ω_{jh} are the connection weights between the input layer and the hidden layer and between the hidden layer and the output layer; θ_j and θ_h are the thresholds of the hidden layer and the input layer; and $f_2(\cdot)$ and $f_1(\cdot)$ are the transfer functions of the output and hidden layers, respectively. n and H are the number of neurons in the input and hidden layers, respectively.

According to the principle of least squares, the objective function is obtained as

$$Q(\omega) = min\left\{\sum_{p=1}^{n} E_p^2\right\} = min\left\{\sum_{p=1}^{n} \sum_{j=1}^{m} \left(t_{pj} - O_{pj}\right)^2\right\} \tag{4}$$

In Equation (4), t_{pj} is the expected output value; m is the number of neurons in the output layer.

According to the existence theorem proposed by Hornik et al. [48], as long as sufficient neurons are in the hidden layer, a BPNN can always make the error of the training samples as small as possible (even close to 0). However, if the hidden neurons are too many, the BPNN will likely remember the pattern of the training samples and fail to generalize the training samples. Therefore, the network topology must be as compact as possible (the number of hidden layers and neurons should be as small as possible). Second, the collected data must be split into training, validation, and test subsets with similar properties. The validation subset is used to monitor the training process. At the beginning of the training process, the error of training and validation subsets monotonically decrease. The error of the validation subset reduces to a minimum and then rises as the training continues, indicating that over-training occurred. The connection weights before over-training are taken. Refs. [39,42,43,46] proposed the basic principles and steps to be followed when building a BPNN model.

3.2.2. PPR Model

The PPR model was proposed by Friedman and Stuetzle [25]. Because its constraint is that the sum of the squares of the weights of each independent variable is equal to 1, over-training is less likely to occur. Therefore, the PPR model has better reliability and robustness and is especially suitable for small and medium-sized (SM) samples that do not obey the normal distribution. In addition, in comparison with other techniques such as Jackknife, Bootstrap, Monte Carlo, Lasso, and Robust Regression, the PPR model does not sample from the total dataset and can reveal the natural structure characteristics of the data; of course, the modeling process is also reliable and unbiased [46,49,50]. The other techniques do sample to model and some bias exists. Furthermore, the distribution of the efficiency from DEA and the economies is very non-normal. Methods such as Jackknife, Lasso et al. are mainly suitable for normal distribution data.

Assuming the normalized data of the independent variable $x(i,j)$, we can obtain the sample projection value $z(i) = \sum_{j=1}^{m} a(j)x(i,j)$ and establish the PPR model based on the cubic polynomial ridge function (PRF) of the sample projection value, as follows [25,26,46]:

$$\begin{aligned} f[z(i)] &= c_0 + c_1 z(i) + c_2[z(i)]^2 + c_3[z(i)]^3 \\ &= c_0 + c_1 \sum_{j=1}^{m} a(j)x(i,j) + c_2\left[\sum_{j=1}^{m} a(j)x(i,j)\right]^2 + c_3\left[\sum_{j=1}^{m} a(j)x(i,j)\right]^3. \end{aligned} \quad (5)$$

In Equation (5), $c_0 \sim c_3$ is the coefficient of the cubic PRF, and m is the number of independent variables. We use the least squares method and take the minimum of the sum of squares fitting errors as the objective function:

$$Q(a,C) = min\sum_{i=1}^{n}\{y(i) - f[z(i)]\}^2 \quad (6)$$

Based on the parasitism–predation algorithm [51], we obtain the optimal global solution of Equation (6), the coefficient $(c_0 \sim c_3)$ of the cubic PRF of Equation (5), and the optimal projection vector coefficient $a(1), a(2), \ldots, a(m)$ as well.

Due to the length limitation of this article, more detailed information on MLs can be found in Refs. [52,53].

4. Empirical Researches

Whether the number of DMUs is small or large, the number of DEA-efficient DMUs is small. To establish the combined model of the DEA-efficient frontier function, we must generate sufficient virtual DEA-efficient samples.

4.1. Empirical Illustration Using Hospital Data

We use the data for 12 hospitals as an example from Cooper et al. [3] that depict the CCR efficiencies in Refs. [3,6]. The samples' data are shown in Table 2.

The primary purposes of establishing the two DEA-PPR combined models are as follows:

(1) To analyze the relationship between the efficiency and the DEA input–output indicators, judge the importance of the inputs and outputs, and predict the efficiency of new hospital data;

(2) To study how to improve the hospitals' managerial efficiency, and to provide the lower bounds for the inputs (the numbers of doctors and nurses) for each inefficient DMU to produce or service its current level of outputs (the numbers of inpatients and outpatients).

According to Refs. [3,6], we also run a input-oriented CCR under CRS, applying the MaxDEA 9.0 software [2]. Table 2 shows the DEA efficiencies of the twelve DMUs. The efficiency scores of the three DEA-efficient DMUs (A, B, and D) equal "1".

As aforementioned, we established the DEA-CCR model and obtained the efficiency of each DMU, determined whether one DMU is efficient or inefficient, constructed the efficient frontier of DEA, calculated the inputs' excesses of each decision unit, etc. However, the optimal solution is not single for the DEA model, is used to establish discontinuously changing frontiers and production functions (models for calculating input–output efficiency), and cannot predict the efficiency values of other (new) DMUs and their optimal input quantities (for input-oriented and specific outputs, the output-oriented is the optimal input). So, we should establish combined models such as DEA-PPR to overcome the disadvantages of DEA. This study will select the DEA-ANN, DEA-PPR, and DEA-SVR combined models and compare their performance.

Table 2. The DMUs' data and their DEA efficiency, as well as the predicted results.

DMUs	Original Data				ES	Efficient/ Inefficient	PPR Model			
	I1 *	I2	O1	O2			PPR-1	PPR-2	PPR	ESN
A	20	151	100	90	1	Efficient	0.996	0.018	1.014	1
B	19	131	150	50	1	Efficient	0.974	0.017	0.992	1
C	25	160	160	55	0.883	Inefficient	0.902	−0.012	0.891	0.874
D	27	168	180	72	1	Efficient	0.995	−0.008	0.987	0.941
E	22	158	94	66	0.764	Inefficient	0.787	−0.011	0.777	0.748
F	55	255	230	90	0.835	Inefficient	0.807	0.023	0.831	0.791
G	33	235	220	88	0.902	Inefficient	0.930	−0.002	0.928	0.902
H	31	206	152	80	0.796	Inefficient	0.806	−0.024	0.782	0.752
I	30	244	190	100	0.960	Inefficient	0.897	0.022	0.919	0.960
J	50	268	250	100	0.871	Inefficient	0.905	−0.028	0.877	0.819
K	53	306	260	147	0.955	Inefficient	1.002	−0.035	0.968	0.873
L	38	284	250	120	0.958	Inefficient	0.979	0.008	0.984	0.958
B-C **	20.9	141.2	160	55	1	Efficient	0.980	0.012	0.993	1(M) #
B-E	16.8	120.6	94	66	1	Efficient	0.982	0.015	0.997	0.963(N)
B-F	33.8	212.9	230	90	1	Efficient	1.002	−0.012	0.990	
B-G	29.8	208.6	220	88	1	Efficient	1.001	0.013	1.013	
B-H	24.7	164	152	80	1	Efficient	0.996	0.001	0.997	
B-I	28.8	207.1	190	100	1	Efficient	1.002	0.014	1.016	
B-J	37.5	233.3	250	100	1	Efficient	1.002	−0.018	0.985	
B-K	43.3	292.3	260	147	1	Efficient	0.984	−0.012	0.972	
B-L	36.4	259.5	250	120	1	Efficient	1.000	0.011	1.011	

Notes: *: I1 and I2 denote two inputs, the doctor and nurse, respectively; O1 and O2 denote two outpatient and inpatient outputs, respectively; ES denotes the efficiency score. **: B-C, B-E, . . ., and K-L represent the benchmark of DMU C, E, . . ., and K, respectively. #: the letters M and N in the bracket denote the new DMUs, and the column of ESN is the efficiency scores after adding new DMUs.

4.1.1. To Establish the DEA-PPR Combined Model of the DEA Production Function

According to the CCR model mentioned above, we obtain the efficiencies of 12 DMUs. Establishing a robust and reliable combined model is challenging due to the small samples. Therefore, we establish a PPR model with PRF by taking the efficiencies of 12 DMUs as the values of the output variable of PPR and the input–output indicators as the input variables of the PPR model. Because the CCR model describes a complex input–output relationship, the PPR model with a linear PRF cannot satisfy the accuracy. In this paper, we tried to establish a PPR model with a quadratic PRF. We input the above data into our compiled program based on the PPA [26,51] and obtain the optimal global solution, the first-dimension PPR model, and the DMUs' efficiency scores (ES),

$$ES_{PPR-1} = 1.00007 + 0.06054z_1 - 0.36947z_1^2 \quad (7)$$

where, $z_1 = -0.05135I_1 - 0.71259I_2 + 0.43926O_1 + 0.54464O_2$. The ES is shown in the ES_{PPR-1} column of Table 2. Since the coefficient of the quadratic term of the Formula (7) is less than 0, Formula (7) has a maximum value. When z_1 = 0.081928, the maximum value of ES_{PPR-1} is 1.00751.

It can be seen that, for the PPR model of the input-oriented DEA model, the two outputs are positive to the efficiency, and the two inputs are negative to the efficiency. If we take appropriate measures to reduce the input values of one DMU (or increase the output values), the z_1 will increase, and the ES rise; otherwise, the ES will decrease. The relationship between the ES and the input variables is entirely consistent with the production function of the DEA model. It can be seen from z_1 that the impact of I2 (nurse) on ES is more significant than that of I1 (doctor). Therefore, reducing the number of nurses is more conducive to improving the ES of these hospitals, while the two output indicators are not much different.

According to the PPR model (7), the mean absolute error (MAE) is calculated as MAE = 0.018, the mean absolute percentage error as MAPE = 2.2%, and the root mean squared error as RMSE = 0.024. That is to say, Equation (7) has revealed the production function of the DEA model well, but significant errors still exist in very few DMUs. For example, the ES of DEA-inefficient DMU K (1.002) is greater than DEA-efficient DMU B, which cannot accurately reflect the ES of the CCR. Then, we further build a PPR model with the second quadratic PRF to describe the ES:

$$ES_{PPR-2} = 0.00310 + 0.07887z_2 + 0.03985z_2^2 \tag{8}$$

where $z_2 = -0.87938I_1 + 0.31175I_2 + 0.31072O_1 + 0.18248O_2$. The ES in Equation (8) is shown in the ES_{PPR-2} column of Table 2. Since the coefficient of the quadratic term is greater than 0, Equation (8) has a minimum value. When $z_2 = -0.98959$, the minimum value ES_{PPR-2} is -0.03592. Thus, the PPR model with two quadratic PRFs reveals the production function of the DEA model as follows:

$$\begin{aligned} ES_{PPR} &= ES_{PPR-1} + ES_{PPR-2} \\ &= 1.00381 + 0.06054z_1 - 0.36947z_1^2 + 0.07887z_2 + 0.03985z_2^2 \end{aligned} \tag{9}$$

The ES calculated according to Equation (9) is shown in the ES_{PPR} column of Table 2. Therefore, for the DEA-PPR combined model of the original DEA model, if we take appropriate measures to reduce the input (or increase the output), the z_1 or z_2 increase, and the ES will rise; otherwise, the ES will decrease. Because the z_2 of all DMUs are greater than -0.98959, the larger the z_2 is, the higher the ES of the PPR model is. The relationship between the ES of the PPR model and the input variables is precisely the same as the production function of the original DEA model. Meanwhile, according to the combined DEA-PPR model in Equation (9) with two PRFs, the impact of I2 (nurse) on the hospital's ES is reduced because its coefficient in z_2 is greater than 0. In contrast, the impact of I1 (doctor) is significantly enhanced because the coefficient of z_2 reaches 0.87938, which is significantly larger than that of I1. Therefore, the impacts of the number of nurses and doctors on the hospital's ES are reduced, and the effects of the two output indicators are not much different, which is consistent with the actual operation of the hospital.

The linear relationship between the ES calculated with the DEA-PPR combined model and the ES of the original DEA model is $ES_{PPR} = 0.0303 + 0.968ES_{DEA}$, and the correlation coefficient is $R = 0.9737$, indicating that the ES values calculated according to the two models are significantly correlated at the 0.01 level. For the DEA-PPR combined model, MAE = 0.014, MAPE = 1.5%, and RMSE = 0.017. The performance metrics, such as MAE, MAPE, etc., are better than that of Equation (7). The ES of DMU K from Equation (9) equals 0.984 and is still more significant than that of benchmark DMU L (0.972), but smaller than that of other DEA-efficient and benchmark DMUs. Therefore, the performance metrics and the distribution of the ES of DMUs show that the established DEA-PPR combined model has a high prediction accuracy, reveals the production function characteristics of the CCR, and suggests good practical value.

We may suppose that, through investigation or other methods, we have obtained the numbers of the outpatients, inpatients, doctors, and nurses of two more hospitals, as shown in Tables 2 and 3, named M and N. If we rebuild the CCR, the ES of the 14 DMUs is shown in Table 2 (in the ESN column). Except for DMUs A, B, G, I, and L, the other DMUs' ES has changed. Furthermore, DMU D changes from being DEA-efficient to inefficient. So, we cannot rank the DMUs objectively when adding new DMUs.

Table 3. Comparison of the optimal number of doctors and nurses based on PPR and BPNN combined models.

DMUs	I_1	I_2	O_1	O_2	D_{PPR}	N_{PPR}	D_B	N_B	$D_{PPR\text{-}27}$	$N_{PPR\text{-}27}$	$D_{PPR\text{-}8}$	$N_{PPR\text{-}8}$	$D_{NN\text{-}729}$	$N_{NN\text{-}729}$
A	20	151	100	90	20.04	150.95	20.00	150.19	20.02	150.98	20.00	151.00	20.26	149.17
B	19	131	150	50	20.08	129.91	19.01	131.31	20.07	129.92	20.04	129.95	20.19	129.94
D	27	168	180	72	25.57	169.45	26.93	167.17	25.55	169.47	25.52	169.50	25.29	168.20
C	25	160	160	55	21.63	140.44	21.06	140.16	21.62	140.45	21.59	140.48	21.47	139.30
E	22	158	94	66	16.61	120.82	14.68	122.47	16.60	120.83	16.58	120.86	17.97	122.48
F	55	255	230	90	32.51	214.18	33.84	194.51	32.48	214.20	32.45	214.23	32.56	214.94
G	33	235	220	88	31.31	207.04	32.97	191.43	31.29	207.07	31.26	207.10	31.33	207.61
H	31	206	152	80	23.84	164.90	24.86	163.26	23.82	164.92	23.79	164.95	23.58	163.41
I	30	244	190	100	29.86	206.06	30.86	186.89	29.84	206.09	29.81	206.12	29.85	206.51
J	50	268	250	100	35.61	235.25	35.42	200.64	35.59	235.27	35.55	235.31	35.69	236.11
K	53	306	260	147	42.15	293.42	33.42	196.15	42.12	293.46	42.09	293.48	42.05	292.72
L	38	284	250	120	37.99	257.93	34.96	200.38	37.96	257.96	37.93	257.99	38.03	258.24
M	25	150	170	79	25.45	272.52	/	/	22.43	172.54	25.40	172.57	25.18	171.38
N	30	300	90	130	23.85	191.45	/	/	23.81	191.49	23.79	191.51	23.73	191.02

According to the above-established DEA-PPR combined model (9), we can quickly obtain the ES of DMU M and N to be 0.991 and 0.249, respectively.

For the above problem, since there are only 12 samples and four input–output indicators, even if we adopt the leave-one-out cross-validation method, the number of neurons in the hidden layer is one (there are seven connection weights); it does not meet the rule of thumb for building a BPNN model—the number of training samples should be more than five times the number of connection weights, so it is impossible to construct a DEA-BPNN, as well as DEA-SVR and DEA-RF models, with good generalization ability.

4.1.2. To Establish the DEA-PPR Combined Model of the DEA-Efficient Frontier Function

Because there is usually a relatively small number of DEA-efficient DMUs, we cannot directly build a DEA-BPNN combined model based on the efficient DMUs only. Ref. [15] constructed 125 virtual efficient DMUs based on three DEA-efficient DMUs, obtained 128 samples, and established a DEA-BPNN combined model. However, since Ref. [5] did not specify whether to divide the samples into training, validation, and test subsets with similar properties, nor how to prevent over-training (or overfitting, over-learning), the generalization ability and robustness of the established DEA-BPNN combined model is questionable and not explicit.

Based on Equation (9), the DEA-PPR combined model of the DEA production function, if we set the ES to be 1, we can obtain the frontier function model of the DEA-efficient DMUs. If I2 (I1), O1, and O2 are assumed, I1 (I2) can be obtained, and the rest can be obtained similarly. Of course, with the virtual samples of Ref. [15], we can also establish the DEA-PPR combined model of the DEA-efficient frontier functions. Referring to [15], we divided the closed interval [0,1] into five equal subintervals, each of the five values (0.2, 0.4, 0.6, 0.8, 1.0) for DEA-efficient A, B, and D, respectively. We generated $5^3 = 125$ combinations of DEA-efficient DMUs in all. We have a matrix of the above proportions, P, consisting of 125 rows and three columns, and matrix S of three DEA-efficient DMUs, as follows:

$$P = \begin{bmatrix} 0.2 & 0.4 & 0.6 & 0.8 & 1.0 & 0.2 & 0.4 & \ldots & 1.0 & 1.0 & 1.0 & 1.0 \\ 0.2 & 0.2 & 0.2 & 0.2 & 0.2 & 0.4 & 0.4 & \ldots & 1.0 & 1.0 & 1.0 & 1.0 \\ 0.2 & 0.2 & 0.2 & 0.2 & 0.2 & 0.2 & 0.2 & \ldots & 0.4 & 0.6 & 0.8 & 1.0 \end{bmatrix}^T$$

$$S = \begin{bmatrix} 20 & 151 & 100 & 90 \\ 19 & 131 & 150 & 50 \\ 27 & 168 & 180 & 72 \end{bmatrix}$$

Thus, we form the matrix $V = P_{125\times3}S_{3\times4}$ and generate 125 (5^3) virtual linear combinations of the DEA-efficient DMUs. Along with the matrix representing the DEA-efficient DMUs taken singly, we thus have 125 + 3 = 128 linear combinations.

We take the above 128 samples as modeling samples and the nine benchmark DMUs as test samples (or solve the optimal input of these DMUs), input the data into the PPA-based program, and establish the DEA-PPR combined model with one quadratic PRF of the DEA-efficient frontier function. For the input-oriented CCR model, referring to [14], we, respectively, establish, under the given output conditions, the combined DEA-PPR models for the optimal number of doctors and nurses (D_{PPR}, N_{PPR}) (represented by the letters D and N, respectively), as follows:

$$D_{PPR} = 38.6034 + 8.2998z_D - 0.00461z_D^2 \tag{10}$$

$$N_{PPR} = 263.057 + 54.931z_N + 0.0048z_N^2 \tag{11}$$

where $z_D = 0.851370_1 + 0.524570_2, z_N = 0.653700_1 + 0.756760_2; O_1, O_2$ has been centered on having a mean zero and scaled to have a standard deviation of 1. From the formulas of z_D and z_N, it can be seen that the number of outpatients significantly impacts the optimal number of doctors. In contrast, the number of inpatients significantly impacts the optimal number of nurses, but the difference between the two indicators is insignificant.

4.1.3. To Determine the Optimal Number of Doctors and Nurses

According to Equations (10) and (11), the frontier function of the DEA-PPR combined model, we can determine the optimal numbers of doctors and nurses. The optimal numbers of doctors and nurses for nine inefficient DMUs (verification samples) are shown in the D_{PPR} and N_{PPR} columns of Table 3.

4.1.4. To Compare the Performance of the Different Models

Based on the optimal numbers of doctors and nurses obtained from the DEA combined models, the performance metrics of nine verification samples, such as the MAE, MAPE, RMSE, correlation coefficient (R), maximum absolute error (E_{A-max}), and maximum relative (percentage) error (E_{R-max}) of the DEA-PPR combined model, were obtained, as shown in the D_{PPR} and N_{PPR} rows of Table 4. The performance metrics of the DEA-BPNN model established in Ref. [15] are shown in the D_B and N_B rows of Table 4.

From the performance metrics of different models shown in Table 4, it can be seen that the generalization ability of the DEA-PPR combined model with one quadratic PRF of the DEA-efficient frontier function is significantly better than that of the DEA-BPNN combined model in Ref. [15]. For the optimal number of doctors and nurses, the DEA-BPNN combined model has a relatively poor generalization ability, with the maximum relative errors reaching 29.6% and 49.0%, respectively, which shows that over-training occurs when the BPNN model is established (but Ref. [15] does not explicitly state whether they use the validation or test subsets to monitor the training process and how to avoid over-training).

Table 4. Comparison of performance between different models with different virtual combinations.

	MAE	MAPE (%)	RMSE	R	$E_{A\text{-}max}$	$E_{R\text{-}max}$ (%)
D_B	2.346	8.0	3.695	0.898	9.880	29.6
D_{PPR}	1.068	5.3	1.190	0.989	1.589	7.9
D_{SVR}	1.083	5.4	1.193	0.957	1.827	9.1
D_{NN}	1.163	4.2	1.281	0.983	2.262	6.8
D_{PPR-27}	1.068	3.7	1.188	0.989	1.882	5.6
D_{PPR-8}	1.065	3.7	1.182	0.989	1.820	5.3
$D_{PPR-729}$	1.070	3.66	1.192	0.989	1.894	5.8
D_{NN-729}	1.209	4.39	1.283	0.988	1.807	7.0
$D_{SVR-729}$	1.621	6.47	1.736	0.983	3.163	17.0
N_B	27.48	14.1	40.62	0.895	96.15	49.0
N_{PPR}	1.081	0.7	1.202	1	1.946	1.3
N_{SVR}	1.081	0.7	1.212	0.999	1.999	1.3
N_{NN}	1.111	0.6	1.253	0.999	2.379	1.0
N_{PPR-27}	1.078	0.6	1.199	1	1.939	0.9
N_{PPR-8}	1.075	0.6	1.193	1	1.672	0.9
$N_{PPR-729}$	1.083	0.56	1.204	1	1.952	0.9
N_{NN-729}	1.298	0.75	1.504	1	2.807	1.6
$N_{SVR-729}$	2.648	1.65	3.230	0.999	6.629	4.4

4.1.5. To Analyze the Robustness and Reliability of the DEA-PPR Combined Models

To further verify the robustness and reliability of the PPR model for small samples, we once divided the closed interval [0,1] into three and two equal subintervals, respectively, and generated virtual DEA-efficient samples. The former three values are 0.33, 0.66, and 1, and the latter two values are 0.5 and 1. We can generate $3^3 = 27$ and $2^3 = 8$ virtual samples, a total of 30 and 11 samples. We take the nine DEA-inefficient DMUs as the test subset and establish the DEA-PPR combined model with one quadratic PRF of the DEA-efficient frontier function based on the 27 and 8 virtual samples (respectively, represented with the subscripts 27 and 8), as follows:

$$D_{PPR-27} = 38.666 + 9.0173z_{D-27} - 0.0196z_{D-27}^2 \tag{12}$$

$$N_{PPR-27} = 263.554 + 59.576z_{N-27} + 0.0203z_{N-27}^2 \tag{13}$$

where $z_{D-27} = 0.8515O_1 + 0.5243O_2, z_{N-27} = 0.6543O_1 + 0.7562O_2$; O_1, O_2 has been normalized to have a mean zero and scaled to have a variance of 1, the same as follows:

$$D_{PPR-8} = 36.366 + 9.9389z_{D-8} - 0.0743z_{D-8}^2 \tag{14}$$

$$N_{PPR-8} = 247.515 + 65.574z_{N-8} + 0.0489z_{N-8}^2 \tag{15}$$

where $z_{D-8} = 0.8497O_1 + 0.5274O_2, z_{N-8} = 0.6526O_1 + 0.7577O_2$.

Since the virtual samples and quantities used to build DEA-PPR combined models differ, the normalized values (O_1, O_2) are also slightly different.

It can be seen from Table 4 that the performance metrics, such as the MAE, MAPE, and RMSE, of models with different virtual samples have a good consistency. Even if only eight virtual samples are generated, the established DEA-PPR combined models have a high fitting accuracy and generalization ability, indicating that the DEA-PPR combined models have good robustness and reliability.

4.1.6. To Establish the DEA-SVR Combined Model of the DEA-Efficient Frontier Function

Theoretically, an SVR model can minimize the structural risk, so it has a good nonlinear approximation ability, and has also been widely used in nonlinear data modeling, such as constructing DEA frontier functions [11,54,55]. However, in practice, the optimal values of the penalty factor and the width coefficient of the radial basis kernel function must be obtained via the multi-fold cross-validation method [54,55]; otherwise, there is a high possibility of overfitting during optimization. The performance metrics of the SVR model are directly related to whether the values of parameters such as the penalty factor and width coefficient are reasonable, so the model does not have strong stability [11]. Due to the length limitation, we only built the DEA-SVR combined model for the case of 125 virtual samples (taking nine inefficient DMUs and two new DMUs, M and N, to be test samples). If the value ranges of the penalty factor and the width coefficient are both $[10^{-2}, 10^5]$, we obtained the optimal solutions of the two parameters, 0.019 and 10^5 (the width coefficient is equal to the boundary value, which is not reasonable). The mean square error of the training samples is already less than 10^{-7}, but the error of the above nine benchmark verification samples is vast; the MAE of the optimal number of nurses is 21.01, the MAPE is 13.91%, the RMSE is 39.07, the E_{A-max} is 112.65, and the E_{R-max} is 74.63; clearly, overfitting occurred during the optimization process. Suppose the value ranges of the penalty factor and the width coefficient are adjusted to $[10^{-2}, 10^3]$. In that case, we obtain the optimal solutions of two parameters, 586.7 and 10^3 (the width coefficient is still equal to the boundary value, which is not reasonable), and the mean square error of the training sample is 0.36; the MAE, MAPE, RMSE, E_{A-max}, E_{R-max}, and other data on the optimal number of nurses in the nine benchmark verification samples of the DEA-SVR combined model are shown in the N_{SVR} row of Table 4. Similarly, when the DEA-SVR combined model of the optimal number of doctors is established when the value ranges of the penalty factor and the width coefficient are both $[10^{-2}, 10^5]$, the optimization process has overfitting; when the value ranges are adjusted to $[10^{-2}, 10^3]$, the optimal solutions are 484.0 and 10^3, respectively; the MAE, MAPE, RMSE, E_{A-max}, E_{R-max} and other data of the established DEA-SVR combined model are shown in the D_{SVR} row of Table 4. It can be seen that the DEA-SVR combined model has poorer performance than the DEA-PPR combined model.

It can be seen from the above modeling process that if the value of the penalty factor is too small, there is a high possibility of overfitting. Unfortunately, there is no commercial software for the SVR model that can, while optimizing the parameters, monitor the error changes of the test samples to prevent overfitting. It brings great difficulties and challenges to establish a DEA-SVR combined model.

4.1.7. To Establish the DEA-BPNN Combined Model of the DEA-Efficient Frontier Function

Although Ref. [15] generated 125 virtual samples and established a DEA-BPNN combined model of the frontier function, the errors of the nine verification samples were large (see Table 4), among which, for the relative errors of the optimal numbers of doctors and nurses in benchmark K, are as high as 22.8% and 32.9%, respectively. The relative error of the optimal number of nurses in benchmark L is also 22.8%, indicating that over-training may have occurred in the training process. So, we try to rebuild the DEA-BPNN combined model. Although many scholars have used the BPNN model to construct the frontier function and production function of the CCR model, few follow the basic principles and steps of establishing a reliable and robust BPNN model. According to the existence theorem proposed by Hornik et al. [48], as long as the hidden layer has sufficient neurons, the error of the training subset of a BPNN can reach as small a value as possible (even close to 0). However, the established BPNN model may have no generalization ability. Therefore, the core of building a BPNN model is to prevent over-training during the training process. We establish a DEA-BPNN combined model with good reliability, robustness, and generalization ability, and follow the basic principles and steps as follows:

(1) We randomly divide the samples into training, validation, and test subsets with similar properties according to the ratio of 2:1:1. (The number of the validation and test subsets should account for at least 15%, respectively.);
(2) We use the trial-and-error method and make the BPNN topology as compact as possible (usually one hidden layer and the number of neurons in the hidden layer is as small as possible). The ratio of the number of training subsets to the number of connection weights must be greater than one and should be more than five, preferably ten, according to the rule of thumb;
(3) We use the training subset to adjust the connection weights to reduce the sum of squares error (SSE) of the training subset and the validation subset to monitor the training process. Along with the training process, the SSE of the training subset gradually decreases, and the SSE of the validation subset first falls to a specific minimal value, and then begins to rise again, which is a sure sign that over-training is occurring. We stop training (called the early-stopping method). To take the network weights before the SSE begins to rise, we establish the BPNN model;
(4) We use the test subset to measure the prediction ability and performance of the BPNN model. If the SSE of the test subset is reasonably close to or slightly larger (generally less than 1.3 times) than the SSE of the training and validation subsets, the established BPNN model has a good generalization ability, reliability, robustness, and prediction ability, as well as practical value. Otherwise, we should restart the process from (3) until the BPNN model has good generalization and prediction abilities.

The above principles and steps must be wholly followed; otherwise, the generalization ability and practical value of the established BPNN cannot be guaranteed.

We randomly selected 68, 30, and 30 training, validation, and test samples from the above 128 samples. The three subsets had the same properties (the mean and standard deviation were almost the same; if they were inconsistent, we should resample). According to [15], the input layer had two neurons, outpatient and inpatient, and the output layer had one output, the optimal number of nurses or doctors. We established two BPNN models to predict them, respectively. According to the rule of thumb, the number of training samples must reach more than five times the number of connection weights, and the number of hidden-layer neurons cannot exceed three. Using the STATISTICA Neural Networks (SNNs) software [42], we used the logistic transformation function for the hidden layer and output layer; considering that the model must have a certain degree of extrapolation needs, the input and output data were linearly transformed into the range of [0.2, 0.8]. We used the quasi-Newton optimization algorithm and monitored the training process with the validation subset. We compared and studied the performance of the models with 1, 2, and 3 hidden neurons. Under the premise of following the above principles and no over-training, the model with one neuron in the hidden layer had a high enough performance; if the number of hidden-layer neurons was two or three, there is a high possibility of over-training, and the performance was not improved significantly. We established the DEA-BPNN combined models for the optimal numbers of nurses and doctors, respectively, and the performance metrics of nine benchmark verification samples are shown in the N_{NN} and D_{NN} rows of Table 4.

The performance metrics, such as the MAE, MAPE, RMSE, R, E_{A-max}, and E_{R-max} of the training, validation, and test subsets of the DEA-BPNN combined models for the optimal numbers of doctors and nurses, are shown in Table 5.

Table 5. Comparison of performance metrics of three subsets of DEA-BPNN combined models.

Model	Training Subset				Validation Subset				Test Subset			
	MAE	MAPE	$E_{A\text{-}max}$	$E_{R\text{-}max}$	MAE	MAPE	$E_{A\text{-}max}$	$E_{R\text{-}max}$	MAE	MAPE	$E_{A\text{-}max}$	$E_{R\text{-}max}$
D_{NN}	0.496	1.52	1.597	11.68	0.513	1.49	1.515	4.35	0.527	1.33	1.822	6.75
N_{NN}	2.170	2.06	3.245	8.60	2.417	2.17	6.13	9.47	2.649	2.66	10.06	10.29
D_{NN-729}	0.429	1.13	2.250	14.00	0.429	1.31	3.28	24.9	0.430	1.21	2.02	13.26
N_{NN-729}	0.833	0.37	15.75	15.40	0.787	0.30	9.32	2.13	0.880	0.39	8.55	6.37

It can be seen from Table 5 that the performance metrics of the training, validation, and test subsets are relatively close, indicating that the established BPNN models have a good generalization ability and reliability.

In cases where only 27 and 8 virtual samples are generated, even if we use the leave-one-out cross-validation method, it is impossible to establish DEA-BPNN and DEA-SVR combined models with generalization ability and reliability due to the small samples.

4.1.8. To Establish the Combined Models of the DEA-Efficient Frontier Function for Large Samples

To verify the suitability of the DEA-PPR combined model for large samples, we generated 729 (9^3) samples taking each of the nine values (0.2, 0.3, 0.4, 0.5, 0.6, 0.7, 0.8, 0.9, 1) for efficient DMUs A, B, and D, respectively. The minimal values of the samples were equal to that of 125 samples. We randomly divided the 731 samples into 371, 180, and 180 training, validation, and test datasets according to the ratio of 2:1:1. We established the DEA-PPR combined models for doctors and nurses, and the performance metrics of the nine benchmark verification samples are shown in the $D_{PPR\text{-}729}$ and $N_{PPR\text{-}729}$ rows of Table 4. By obeying the principles and steps, we established the DEA-BPNN combined models, and the performance metrics are shown in the $D_{NN\text{-}729}$ and $N_{NN\text{-}729}$ rows of Table 4. The performance metrics of the three subsets are shown in the $D_{NN\text{-}729}$ and $N_{NN\text{-}729}$ rows of Table 5. The optimal number of doctors and nurses is shown in the $D_{NN\text{-}729}$ and $N_{NN\text{-}729}$ columns of Table 3. We set the search range of parameters to be $[10^{-2}, 10^3]$, optimized the SVR models, and obtained the optimal values to be (7.762, 0.626) and (7.762, 3.206), respectively, which show that overfitting did not occur. The performance metrics of the nine benchmark verification samples are shown in the $D_{SVR\text{-}729}$ and $N_{SVR\text{-}729}$ rows of Table 4.

From the performance metrics of different models in Table 4, it can be seen that, for generating 125 and 729 virtual samples, respectively, the DEA-PPR, DEA-BPNN, and DEA-SVR combined models we established without over-training or overfitting have quite a high prediction accuracy and generalization ability. However, the DEA-PPR combined model slightly outperforms the DEA-BPNN and DEA-SVR for medium-sized samples, and the DEA-PPR combined model outperforms the DEA-BPNN and DEA-SVR for large samples. Furthermore, the DEA-BPNN combined model has a better generalization ability than the DEA-SVR combined models under medium-sized and large samples. More importantly, due to the strongly nonlinear approximation ability of the BPNN and SVR models, there is a high possibility of over-training and overfitting during training or optimizing the parameters within unreasonable search ranges. Cross-validation methods with validation subsets must be used, which require a cumbersome modeling process and cannot guarantee the generalization and prediction abilities of the model without following the principles and steps. Because the PPR model with a quadratic PRF is characterized by a relatively simple structure and moderate nonlinear approximation ability, it is especially suitable for modeling problems that are not very complex, such as the frontier surface, the production function, etc., of the DEA model. Whether for 729, 125, 27, or 8 virtual samples, the established DEA-PPR combined models have quite a good generalization ability and prediction ability. The PPR model is especially suitable for small and medium-sized samples that do not obey normal distribution. Furthermore, the DEA model is also ideal for analysis and modeling to research the input–output efficiency of small and medium-sized samples. Therefore, for the problem of small and medium-sized samples, the DEA-PPR combined model has unique advantages, such as good generalization, robustness, and reliability.

4.2. Empirical Illustration Using China's Provincial Carbon Dioxide Emission Quotas

As the world's top CO_2 emitter (accounting for 26.5% of the world's total emissions) and the second largest economy, China's achievement of peak CO_2 emissions and carbon neutrality (called the "dual carbon" target) will play a pivotal role in reducing the world's CO_2 emissions. Therefore, reasonably allocating China's provincial carbon emission quotas (CPCEQs) plays a vital role in promoting high-quality development and striving to achieve a "dual carbon" target. In this study, firstly, it is necessary to judge and evaluate the advantages and disadvantages of different allocation methods. Secondly, it is required to study the carbon emission efficiency of different periods (e.g., comparing efficiency between 2019 and 2030), and so on. DEA is the most used model to measure the efficiency of carbon emissions. It is a pity we cannot use the DEA model to compare the efficiency change between different periods and the advantages and disadvantages of other allocating methods.

The primary purposes of establishing the two DEA-PPR combined models are as follows:

(1) To build a relationship between the carbon emissions efficiency and the DEA inputs and outputs, judge the importance of the inputs and outputs, and predict the efficiency of new carbon-allocating methods and the quotas in 2030;

(2) To study ways to improve China's provincial carbon emissions efficiency and help to implement the "dual carbon" target, and provide lower bounds for the inputs (the carbon emissions) for each inefficient DMU to produce its current level of outputs (the GDP and population).

4.2.1. Data Resource

Referring to Gomes et al. [56], we take the gross domestic production (GDP) and the population of provinces as the output indicators and the actual carbon emissions as the input indicator. The data of the output indicators are from the *China Statistical Yearbook* (2020) (http://www.stats.gov.cn/sj/ndsj (accessed on 8 July 2023)), and the input data are from Chinese carbon emission accounts & datasets (https://www.ceads.net (accessed on 8 July 2023)). The data are shown in Table 6.

Table 6. The input–output data and ES of the DEA model and the predicted ES with the DEA-PPR combined model.

Provinces	CO_2 (Mt)	Population (10^4 Persons)	GDP (CNY 100 Million)	ES	Quotas	ES-q	Quotas-y	ES-y	Quotas-2030	ES-2030	SVR
Beijing	89.2	2154	35,371	1	89.2	0.6906	426.5	0.6516	84.3	0.9578	0.9877
Tianjin	158.5	1562	14,104	0.417	66.1	0.7636	179.2	0.6503	157.7	0.6094	0.4351
Hebei	914.2	7592	35,105	0.3133	286.4	0.3407	453.0	0.8803	910.0	0.4506	0.3410
Shanxi	564.9	3729	17,027	0.2562	144.7	0.5744	207.5	0.8229	727.0	0.3222	0.2874
Inner Mongolia	794.3	2540	17,213	0.1280	101.7	0.6701	547.6	0.2707	712.2	0.3484	0.1726
Liaoning	533.4	4352	24,909	0.3146	167.8	0.5214	361.3	0.6240	504.0	0.2491	0.3401
Jilin	203.7	2691	11,727	0.5245	106.8	0.6702	218.6	0.5558	213.5	0.5236	0.5353
Heilongjiang	278.2	3751	13,613	0.5231	145.5	0.5779	290.6	0.5776	262.9	0.4580	0.5302
Shanghai	192.9	2428	38,155	0.5581	107.7	0.6301	374.8	0.7978	173.0	0.7542	0.5641
Jiangsu	804.6	8070	99,632	0.6410	515.8	1.2084	787.1	0.9723	760.2	0.8652	0.6389
Zhejiang	381.4	5850	62,352	0.7201	274.7	0.4306	546.3	0.8834	379.7	0.8558	0.7108

Table 6. *Cont.*

Provinces	CO_2 (Mt)	Population (10^4 Persons)	GDP (CNY 100 Million)	ES	Quotas	ES-q	Quotas-y	ES-y	Quotas-2030	ES-2030	SVR
Anhui	408.1	6366	37,114	0.5923	241.7	0.4090	366.2	0.9428	427.8	0.4555	0.5950
Fujian	278.1	3973	42,395	0.5922	164.7	0.5287	352.6	0.9392	285.6	0.6615	0.5945
Jiangxi	242.3	4666	24,758	0.7393	179.1	0.4996	276.8	0.8492	267.5	0.6407	0.7402
Shandong	937.1	10,070	71,068	0.4823	452.0	0.4691	824.4	0.7480	885.4	0.5838	0.4947
Henan	460.6	9640	54,259	0.9059	417.3	0.3018	528.1	1	458.5	0.8236	0.8966
Hubei	354.8	5927	45,828	0.6636	235.4	0.4297	417.2	0.8935	371.9	0.5774	0.6600
Hunan	310.6	6918	39,752	0.8430	261.9	0.3874	385.3	0.9731	325.7	0.7527	0.8359
Guangdong	569.1	11,521	107,671	1	569.1	1.2367	826.4	1	510.3	1.4683	0.9877
Guangxi	246.7	4960	21,237	0.7689	189.7	0.4799	232.6	1	258.7	0.7177	0.7617
Hainan	43.1	945	5309	1	43.1	0.8542	74.5	0.6900	67.8	0.8221	0.9860
Chongqing	156.3	3124	23,606	0.7891	123.3	0.6135	238.1	0.8199	155.5	0.8515	0.7736
Sichuan	315.2	8375	46,616	1	315.2	0.3491	572.2	0.7915	313.7	0.9902	0.9877
Guizhou	261.1	3623	16,769	0.5394	140.9	0.5831	165.6	1	288.3	0.5035	0.5414
Yunnan	186.0	4858	23,224	1	186.0	0.4869	367.0	0.6421	216.1	0.7992	0.9881
Shaanxi	296.3	3876	25,793	0.5085	150.7	0.5553	308.7	0.7174	294.9	0.5161	0.5184
Gansu	164.5	2647	8718	0.6397	105.2	0.6820	108.8	1	172.4	0.6670	0.6323
Qinghai	51.8	608	2966	0.8322	43.1	0.8170	118.2	0.3213	48.9	0.8353	0.8346
Ningxia	212.4	695	3748	0.2028	43.1	0.8257	38.0	1	222.7	0.3959	0.2404
Xinjiang	455.3	2523	13,597	0.2214	100.8	0.6800	271.3	0.4695	408.2	0.2431	0.2574
Tianjin [b]		1562	14,104	1	60.7	0.9783					0.6401
Hebei [b]		7592	42,258	1	285.7	0.9669					0.7387
Shanxi [b]		3729	20,756	1	140.3	0.9747					0.6924
Inner Mongolia [b]		2540	17,213	1	96.7	0.9757					0.6219
Liaoning [b]		4352	24,909	1	164.0	0.9673					0.8048
Jilin [b]		2691	14,978	1	101.3	0.9869					0.6233
Heilongjiang [b]		3751	20,878	1	141.2	0.9745					0.6931
Shanghai [b]		2515	38,155	1	107.7	0.9107					0.8277
Jiangsu [b]		10,479	99,632	1	515.8	0.9770					0.622
Zhejiang [b]		5850	62,352	1	274.7	0.8479					0.6197
Average	/	/	/	0.6239	/	0.6089	/	0.7828	/	0.6567	/

Notes: [b] denotes the benchmark of the provinces of DEA-efficient DMUs. The quotas are obtained according to Refs. [56,57] using the ZSG-DEA model, and its ES is ES-q; we obtain the quotas (denoted in "Quotas-y" column of Table 6) according to an allocation system consisting of multiple variables. The quotas—2030 are obtained according to the GDP and carbon emission intensity of the "Outline" and "Plan".

We establish the input-oriented BCC model to measure the ES of the Chinese provincial carbon emissions in 2019, referring to Refs. [56,57] and the ES shown in Table 6. According to Refs. [56,57], we built the ZSG-DEA to allocate the provincial quotas and obtain the quotas, as shown in the "quotas" column of Table 6. If we rebuild the BCC model using the quotas and make the ES values all equal to 1, we cannot judge whether the ES will rise or decline. Meanwhile, we can allocate the provincial quotas by establishing an allocation system consisting of multiple variables [46] and obtain the quotas denoted in the "Quotas-y" column of Table 6. But how do we judge the ZSG-DEA's quotas and the quotas of the allocation system? We can obtain the provincial quotas in 2030 according to the GDP and carbon emission intensity in the Action Plan for Carbon Peaking before 2030 (namely "Plan"), as well as the 14th Five-Year Plan for Economic and Social Development and the Outline of the Vision for 2035 (namely "Outline") of provinces in China. We cannot compare and evaluate the ES of different allocation methods and periods solely through the DEA model because adding more DMUs changes the ES of the original DMUs. So, we must establish combined models to do so.

4.2.2. To Build the Combined Model Characterizing the DEA Production Function

(1) To build a DEA-PPR combined model

Referring to Section 4.1, we take the ES as the output variable and the three input–output indicators as the input variables, and build the DEA-PPR combined model with one quadratic PRF, and the 30 training and ten verification samples as shown in Table 6 (noted "b"), as follows:

$$ES_{PPR-1} = 0.7099 - 0.5571z_1 + 0.1613z_1^2 \tag{16}$$

where $z_1 = 0.8622I_1 - 0.4605O_1 - 0.2110O_2$, I_1 denotes CO_2 (Mt), O_1 denotes the population, and O_2 denotes the GDP, as shown in Table 6. The performance metrics of the training samples, such as the RMSE, AAE, MAPE, and $E_{A\text{-}max}$, are 0.1129, 0.0834, 17.48%, and 0.3305, respectively. The performance metrics of the verification samples are 0.0737, 0.0615, 6.15%, and 0.1172, respectively. The performance of the DEA-PPR model is not good enough, and we rebuild the DEA-PPR model with the second quadratic PRF as follows:

$$ES_{PPR-2} = -0.0557 + 0.0727z_2 + 0.0876z_2^2 \tag{17}$$

where $z_2 = -0.7195I_1 + 0.4261O_1 - 0.5484O_2$. The performance metrics of the training samples, such as the RMSE, AAE, MAPE, and $E_{A\text{-}max}$, are 0.0871, 0.0621, 14.83%, and 0.2448, respectively. The performance metrics of the verification sample are 0.0552, 0.0501, 5.00%, and 0.0819, respectively. Although we can build the third PRF to improve the performance, it is shown that the performance of the DEA-PPR model is good enough to describe the DEA production function. That is to say, the DEA-PPR combined model can approximately characterize the DEA production function as follows:

$$ES_{PPR} = ES_{PPR-1} + ES_{PPR-2} \tag{18}$$

The calculated ES of the provincial quotas, quotas-y, and quotas—2030 are all shown in the "ES-q", "ES-y", and "ES-2030" columns of Table 6.

According to the allocation system, the average ES of the quotas is greater than that of the actual carbon emissions, and shows that the allocated quotas are reasonable and improve the efficiency of carbon emissions. The average ES in 2030 is less than the actual carbon emissions in 2019 and shows that the carbon emissions efficiency in 2030 is lower than in 2019. To raise the carbon emissions efficiency in 2030, we must take measurements to reduce carbon emissions. If we can reduce carbon emissions by 5% in 2030, the average ES calculated with the DEA-PPR combined model is 0.6446, greater than the average ES in 2019.

The average ES of the quotas from the ZSG-DEA model is lower than the average ES of the actual carbon emissions, and shows that the ZSG-DEA model for carbon emission quotas can equalize the carbon emission efficiency of provinces, which does reduce carbon emission efficiency.

(2) To build the DEA-SVR and DEA-BPNN combined models

There are too few samples to build DEA-BPNN and DEA-RF combined models. We build an SVR model according to 4.6. We set the parameters' search range to be [0.01, 10] and obtain the optimal solution; the penalty factor equals 0.0258, and the width coefficient equals 10. The performance metrics of the training samples, such as the RMSE, AAE, and MAPE, are 0.0155, 0.0123, and 3.15, respectively. The performance metrics of the verification sample are 0.3205, 0.3116, and 31.16, respectively. Overfitting occurred during the optimization process.

4.2.3. To Build the Combined Models Characterizing the DEA Frontier Function

We cannot build the combined models of the DEA-efficient frontier function using the five DMUs only. Referring to Section 4.2, we also generate 2^5 or 3^5 virtual DEA-efficient DMUs by dividing the value interval [0, 1] into two (0.5, 1) or three (0.3333, 0.6667, 1) equal parts in a linear combination with the data of five DEA-efficient samples. We generated 32 and 243 virtual samples, respectively.

(1) To build a DEA-PPR combined model

We take the DEA model's input indicator as the PPR model's output variable and the two output indicators as the input variables. We take the ten samples in Table 6, denoted as "b", as the verification samples.

We build the PPR model with the first quadratic PRF as follows:

$$C_{PPR-32} = 677.96 + 272.23z_{32} + 0.9597z_{32}^2 \tag{19}$$

$$C_{PPR-243} = 765.99 + 172.26z_{243} + 0.4784z_{243}^2 \tag{20}$$

where $z_{32} = 0.8784O_1 + 0.4779O_2$ and $z_{243} = 0.8792O_1 + 0.4765O_2$. The performance metrics of the training and validation subsets, such as the RMSE, AAE, MAPE, and $E_{A\text{-}max}$, are shown in Table 7, respectively. The performance metrics show that the DEA-PPR combined model has good performance, generalization ability, and robustness for 32 or 243 virtual samples. The DEA-PPR combined model with one quadratic PRF already has a high enough prediction accuracy.

Table 7. Performance comparison of the PPR, BPNN, and SVR models in different virtual combinations.

	Training Subset					Validation Subset				
Model	MAE	MAPE (%)	RMSE	$E_{A\text{-}max}$	$E_{R\text{-}max}$ (%)	MAE	MAPE (%)	RMSE	$E_{A\text{-}max}$	$E_{R\text{-}max}$ (%)
C_{PPR-32}	12.30	3.29	15.53	43.20	36.41	8.37	5.05	14.66	35.00	27.37
C_{SVR-32}	10.92	5.03	15.00	45.78	106.3	42.54	27.43	53.41	132.2	49.06
$C_{PPR-243}$	12.08	1.89	14.69	45.33	41.81	9.52	5.73	14.86	37.51	22.82
$C_{SVR-243}$	11.40	2.10	14.67	51.29	119.1	36.45	24.33	44.44	106.2	44.77
$C_{BPNN-243}$	11.74	1.60	14.51	52.78	9.27	22.77	14.03	30.83	69.65	52.94
$C_{SVR-243}$	7.39	1.88	10.54	71.50	166.0	34.35	28.80	37.13	59.44	97.98

(2) To build the DEA-SVR and DEA-BPNN combined models

We build an SVR model according to Sections 4.1.6 and 4.1.7. When the parameters' search range is [0.01, 8.7096], the RMSE equals the RMSE of the DEA-PPR combined model. The optimal solution is that the penalty factor equals 0.2041, and the width coefficient equals 8.7096; that is, the optimal width coefficient equals the upper bound value. The performance metrics of the training and validation subsets are shown in Table 7.

It can be seen that the DEA-SVR combined model has almost the same performance for the training subset as DEA-PPR but has inferior performance for the validation subset compared to the DEA-PPR combined model. Overfitting occurred during the optimization process.

Furthermore, the main disadvantage of the SVR model is that its performance is directly dependent on the search range of the parameters. If we change the upper boundary limit, the performance will change, too. If we set the upper boundary limit to 100, the SVR has a poorer generalization ability.

(3) To build the DEA-BPNN and DEA-RF combined models

The 32 virtual samples are too few to build DEA-BPNN and DEA-RF combined models. The 243 virtual samples are barely suitable for building the DEA-BPNN and DEA-RF combined models.

We build a DEA-BPNN combined model. We randomly split the 248 (243 + 5) samples into training, validation (especially for the BPNN and RF model), and test subsets in the proportion of 6:2:2 [31,42,45] and obtain 148, 50, and 50 training, validation, and test samples. According to the rule of thumb, we should preferably have ten times as many training samples as the number of the connection weights. We determine the number of hidden neurons to be three; the connection weights are 13. The transfer functions in the hidden and output layers are Sigmoid applying the SNN [42]. We train the BPNN, monitor the error change in the validation subset, and build the BPNN model without over-training. The performance metrics of the DEA-BPNN combined model of the training and validation subset are shown in Table 7. The performance metrics of the validation and test subsets are 11.41, 1.61%, 14.64, 47.81, and 111.00, and 11.71, 3.78%, 13.83, 25.79, and 3.85, respectively. The performance metrics show that the built BPNN model has a good generation ability but is still poorer than the DEA-PPR combined model. According to the built model, we can conclude that GDP significantly impacts efficiency more than the population.

We build a DEA-RF combined model. We randomly split the 248 (243 + 5) samples into training and validation subsets in the proportion of 1:1 [46]. We take the default value of the parameters of the DPS software [46], the number of trees being 300, and build the DEA-RF combined model. The performance metrics of the DEA-RF combined model of the training and validation subsets are shown in Table 7. The validation subset's performance metrics are greater than the training subset's, indicating that overfitting has occurred in the optimization. The built DEA-RF combined model has a poor generalization ability.

The comparison of the performance metrics of different combined models shows that the DEA-PPR model has the best performance and generalization ability, robustness, and prediction accuracy, whether for large samples or small and medium-sized samples. The DEA-BPNN model without over-training has the second-best performance for medium-sized and large samples. The DEA-SVR combined model has a poorer generalization ability than the DEA-PPR and DEA-BPNN combined models, and overfitting occurs in optimizing the parameters under medium-sized samples. The DEA-RF combined model has the poorest generalization ability of the four combined models for large samples. We cannot build DEA-BPNN, DEA-SVR, and DEA-RF combined models for small samples.

5. Results and Concluding Remarks

5.1. The PPR and DEA Models Have Similarities in Frontier Morphology and Theoretical Consistency

According to the DEA modeling principle, the frontier surface of the input-oriented (or output-oriented) DEA is the top surface of a convex polyhedron that is convex (or concave) to the coordinate origin. The PPR model based on the quadratic PRFs comprises multiple (at least one) quadratic surfaces, which are convex polyhedra. It can be seen that the convex polyhedron of the PPR model and the top surface of the frontier surface of the DEA model have a similar morphology and consistent theory. Therefore, there is a low possibility of overfitting in building the DEA-PPR combined model, which thus has a good generalization ability and robustness. The widely used BPNN, RF, and SVR models with very complex curved surfaces have a low morphological similarity and theoretical consistency with the DEA model.

5.2. The Characterization Ability of the DEA-PPR Combined Model to the DEA Production Function

The main disadvantage of DEA is the static analysis of the existing DMUs. The production function constructed via DEA is discontinuous, so the newly added DMUs cannot be analyzed. If the DEA model is re-established with the newly added DMUs, the efficiency and ranking of the original DMUs will be changed, leading to a lack of comparability. For example, after adding new DMUs (M and N) and rebuilding the DEA model, the ES of the seven original DMUs changed, and DMU D changed from being DEA-efficient to inefficient.

To overcome its disadvantages, scholars have established combined models, such as DEA-BPNN, DEA-SVR, DEA-RF, etc., based on the input–output indicators and the ES of the DEA model that have played a good role. However, most combined models, such as BPNN, RF, etc., are mainly suitable for large samples and not suitable for small and medium-sized samples. Therefore, under the cases of small and medium-sized samples, we are the first to propose an innovative idea of establishing a DEA-PPR combined model and its modeling steps. For the two actual examples, we established reliable and effective combined models, describing the production function and the frontier surface function of the CCR, respectively. Especially in the cases of small and medium-sized samples that do not obey the normal distribution, it is challenging to establish a reliable DEA-BPNN, DEA-SVR, or DEA-RF combined model for the production or frontier surface functions of the CCR with generalization ability and prediction accuracy.

It can be seen from Tables 2 and 6 that the established DEA-PPR combined model can characterize the production function of the DEA model, has sufficient fitting accuracy, and predicts the ES of new DMUs. We input the data of the new DMUs (M and N) into the DEA-PPR combined model (Equation (10)), and obtained their ES values of 0.991 and 0.249, respectively.

Therefore, the DEA-PPR combined model can exploit the advantages of both nonparametric models and overcome their shortcomings.

5.3. The Characterization Ability of the DEA-PPR Combined Model to DEA-Efficient Frontier Function

We establish the DEA model to obtain the DEA-efficient frontier function consisting of a series of line segments of the polyline (approximately). There are only a few DEA-efficient DMUs, and we must fill the gap between the broken line and the actual situation. Of course, if new DMUs are added, it is impossible to analyze whether it is on the frontier surface or to calculate its optimal input amount. Therefore, according to the DEA-efficient DMUs, some scholars have divided each indicator into equal parts to generate virtual DEA-efficient DMUs. According to different equivalent fractions, we can establish combined models such as DEA-PPR, DEA-BPNN, DEA-SVR, etc., to describe the DEA-efficient frontier function. To establish the BPNN, RF, and SVR combined models, we must randomly divide the samples into training, validation, and test subsets with similar properties. We use cross-validation to train or optimize the parameters to avoid over-training.

We usually establish combined models for large samples, such as DEA-PPR, DEA-BPNN, DEA-SVR, etc., with a good generalization ability; otherwise, it is difficult to guarantee the generalization ability. The DEA-PPR combined model has a better generalization ability than the DEA-BPNN model, the DEA-SVR model, etc.

The PPR model is especially suitable for small and medium-sized samples of nonlinear and non-normal distribution. Our empirical research generates medium-sized and medium-sized virtual DMUs with a linear combination of DEA-efficient DMUs. We established reliable DEA-PPR combined models with a good generalization ability. Differently to the BPNN and SVR combined models, according to the DEA-PPR combined model, we can easily judge the importance and ranking of the indicators according to the coefficients of different inputs (or outputs). The DEA-PPR combined model outperforms the DEA-BPNN and DEA-SVR combined models.

In short, the DEA-PPR combined model outperforms DEA-BPNN, DEA-SVR, etc., for small, medium-sized, and large samples, and should be recommended for use regarding generalization ability, robustness, and accuracy.

According to the DEA-PPR combined model, we can obtain the optimal inputs (outputs) for new DMUs and take measures to improve efficiency and management.

6. Limitations and Future Research

This study mainly focuses on overcoming the weaknesses of the DEA model and the shortcomings of the DEA-BPNN and DEA-SVR combined models not being well applied to small and medium-sized samples. We introduce the PPR model and first establish a DEA-PPR combined model with a good generalization ability. However, there are some limitations. Future research should use more examples to verify the generalization ability and accuracy between DEA-BPNN, DEA-SVR, and DEA-PPR for establishing production and frontier functions. Second, we take the DEA-CCR model as an example, and in the next step, we should study other DEA models to verify the applicability of the DEA-PPR combined model. This paper is the first to focus on the DEA-PPR combined model, which requires more empirical research to confirm its universality and reliability.

Author Contributions: Conceptualization, W.L. and X.Y.; methodology, X.Y.; software, W.L.; validation, X.Y.; formal analysis, W.L.; investigation, X.Y.; resources, X.Y.; data curation, X.Y.; writing—original draft preparation, W.L. and X.Y.; visualization, X.Y.; supervision, W.L.; project administration, X.Y.; funding acquisition, X.Y. All authors have read and agreed to the published version of the manuscript.

Funding: This research is supported by Shanghai Business School Plateau Discipline Project of Business Administration, No. SHGDD-GSGL-2023-02-02.

Data Availability Statement: The data are available on request.

Conflicts of Interest: The authors declare no conflict of interest.

References

1. Charnes, A.; Cooper, W.W.; Rhodes, E. Measuring the efficiency of decision making units. *Eur. J. Oper. Res.* **1978**, *2*, 429–444. [CrossRef]
2. Cheng, G. *Data Envelopment Analysis: Methods and MaxDEA Software*; Intellectual Property Press: Beijing, China, 2014. Available online: http://www.maxdea.cn/ (accessed on 12 June 2023).
3. Cooper, W.; Seiford, L.; Tone, K. *Data Envelopment Analysis—A Comprehensive Text with Models, Applications, References and DEA-Solver Software*; Klumer Academic Publishers: Boston, MA, USA, 2007.
4. Panwar, A.; Olfati, M.; Pant, M.; Snasel, V. A review on the 40 years of existence of data envelopment analysis models: Historic development and current trends. *Arch. Comput. Methods Eng.* **2022**, *29*, 5397–5426. [CrossRef] [PubMed]
5. Wu, D.; Yang, Z.; Liang, L. Using dea-neural network approach to evaluate branch efficiency of a large Canadian bank. *Expert Syst. Appl.* **2006**, *31*, 108–115. [CrossRef]
6. Bishop, C. *Neural Networks for Pattern Recognition*; Oxford University Press: Oxford, UK, 1995.
7. Lewis, H.F.; Sexton, T.R. Network DEA: Efficiency analysis of organizations with complex internal structure. *Comput. Oper. Res.* **2004**, *31*, 1365–1410. [CrossRef]
8. Athanassopoulos, A.D.; Curram, S.P. A comparison of data envelopment analysis and artificial neural networks as tool for assessing the efficiency of decision making units. *J. Oper. Res. Soc.* **1996**, *47*, 1000–1016. [CrossRef]
9. Na, J.; Zhou, Z.; Zhou, H. A decision-analysis approach to determine the total staff employ in government organization—Establishment of local government organizations. *Chin. J. Manag. Sci.* **1997**, *5*, 7–17.
10. Ma, C.; Wu, C.; Zhang, S.; Yang, Y.; Li, H.; Han, Y. Decision making method for variable-rate fertilization based on data envelopment analysis and artificial neural network. *Trans. CSAE* **2004**, *20*, 152–155.
11. Zhu, N.; Zhu, C.; Emrouznejad, A. A combined machine learning algorithms and DEA method for measuring and predicting the efficiency of Chinese manufacturing listed companies. *J. Manag. Sci. Eng.* **2021**, *6*, 435–448. [CrossRef]
12. Olanrewaju, O. Integrated index decomposition analysis-artificial neural network-data envelopment analysis (IDA-ANN-DEA)—Implementation guide. *Energy Effic.* **2021**, *14*, 71–78. [CrossRef]
13. Zhong, K.; Wang, Y.; Pei, J. Super efficiency SBM-DEA and neural network for performance evaluation. *Inf. Process. Manag.* **2021**, *58*, 102728. [CrossRef]
14. Anouze, A.L.M.; Bou-Hamad, I. Data envelopment analysis and data mining to efficiency estimation and evaluation. *Int. J. Islam. Middle East. Financ. Manag.* **2019**, *12*, 169–190. [CrossRef]
15. Bose, A.; Patel, G.N. "NeuralDEA"—A framework using neural network to re-evaluate DEA benchmarks. *OPSearch* **2015**, *52*, 18–41. [CrossRef]
16. Kwon, H. Exploring the predictive potential of artificial neural networks in conjunction with DEA in railroad performance modeling. *Int. J. Prod. Econ.* **2017**, *183*, 159–170. [CrossRef]
17. Hong, H.K.; Ha, S.H.; Shin, C.K.; Park, S.C.; Kim, S.H. Evaluating the efficiency of system integration projects using data envelopment analysis (DEA) and machine learning. *Expert Syst. Appl.* **1999**, *16*, 283–296. [CrossRef]
18. Tsaples, G.; Papathanasiou, J.; Georgiou, A.C. An exploratory DEA and machine learning framework for the evaluation and analysis of sustainability composite indicators in the EU. *Mathematics* **2022**, *10*, 2277. [CrossRef]
19. Mirmozaffari, M.; Shadkam, E.; Khalili, S.M.; Kabirifar, K.; Yazdani, R.; Gashteroodkhani, T.A. A novel artificial intelligent approach: Comparison of machine learning tools and algorithms based on optimization DEA Malmquist productivity index for eco-efficiency evaluation. *Int. J. Energy Sect. Manag.* **2021**, *25*, 523–550. [CrossRef]

20. Yang, X.; Dimitrov, S. Data envelopment analysis may obfuscate corporate financial data: Using support vector machine and data envelopment analysis to predict corporate failure for nonmanufacturing firms. *INFOR Inf. Syst. Oper. Res.* **2017**, *55*, 295–311. [CrossRef]
21. Emrouznejad, A.; Shale, E. A combined neural network and DEA for measuring efficiency of large scale datasets. *Comput. Ind. Eng.* **2009**, *56*, 249–254. [CrossRef]
22. Saeidi, S.; Jouybanpour, P.; Mirvakilli, A.; Iranshahi, D.; Klemeš, J.J. A comparative study between modified data envelopment analysis and response surface methodology for optimisation of heterogeneous biodiesel production from waste cooking palm oil. *J. Clean. Prod.* **2016**, *136*, 23–30. [CrossRef]
23. Kwon, H.; Lee, J.; Roh, J. Best performance modeling using complementary DEA-ANN approach—Application to Japanese electronics manufacturing firms. *Benchmarking Int. J.* **2016**, *23*, 704–721. [CrossRef]
24. Farahmand, M.; Desa, M.; Nilashi, M. A combined data envelopment analysis and support vector regression for efficiency evaluation of large decision making units. *Int. J. Eng. Technol.* **2014**, *6*, 2310–2321.
25. Friedman, J.; Stuetzle, W. Projection pursuit regression. *J. Am. Stat. Assoc.* **1981**, *76*, 817–823. [CrossRef]
26. Lou, W. *The Projection Pursuit Theory Based on Swarm Intelligence Optimization Algorithms—New Developments, Applications, and Software*; Fudan University Press: Shanghai, China, 2021.
27. Hwang, T.; Lay, S.; Maechler, M. Regression modeling in back-propagation and projection pursuit learning. *IEEE Trans. Neural Netw.* **1994**, *5*, 342–353. [CrossRef] [PubMed]
28. Zhan, H.R.; Zhang, M.K.; Xia, Y.C. Ensemble projection pursuit for general nonparametric. *arXiv* **2022**, arXiv:2210.14467.
29. Ren, H.; Ma, X.R.; Li, H.B. Improvement of input evaluation for giant projects based on GA-BP neural network. *Syst. Eng.—Theory Pract.* **2015**, *35*, 1474–1481.
30. Tsolas, I.E.; Charles, V.; Gherman, T. Supporting better practice benchmarking: A DEA-ANN approach to bank branch performance assessment. *Expert Syst. Appl.* **2020**, *160*, 113599. [CrossRef]
31. Zhang, Z.; Xiao, Y.; Niu, H. DEA and Machine Learning for Performance Prediction. *Mathematics* **2022**, *10*, 1776. [CrossRef]
32. Fallahpour, A.; Kazemi, N.; Molani, M.; Nayyeri, S.; Ehsani, M. An Intelligence-Based Model for Supplier Selection Integrating Data Envelopment Analysis and Support Vector Machine. *Iran. J. Manag. Stud.* **2018**, *11*, 209–241.
33. Yazdanparast, R.; Tavakkoli-Moghaddam, R.; Heidari, R.; Aliabadi, L. A hybrid Z-number data envelopment analysis and neural network for assessment of supply chain resilience: A case study. *Cent. Eur. J. Oper. Res.* **2021**, *29*, 611–631. [CrossRef]
34. Sreekumar, S.; Mahapatra, S. Performance modeling of Indian business schools: A DEA-neural network approach. *Benchmarking Int. J.* **2011**, *18*, 221–239. [CrossRef]
35. Kao, H.; Huang, C.; Chen, J. Classification using DEA and SVM approaches: The empirical study of higher education. *Information* **2013**, *16*, 7801–7810.
36. Barros, C.; Wanke, P. Insurance companies in Mozambique: A two-stage DEA and neural networks on efficiency and capacity slacks. *Appl. Econ.* **2014**, *46*, 3591–3600.
37. Sanei, R.; Hosseinzadeh lotfi, F.; Fallah, M.; Sobhani, F. An estimation of an acceptable efficiency frontier having an optimum resource management approach, with a combination of the DEA-ANN-GA technique (A case study of branches of an insurance company). *Mathematics* **2022**, *10*, 4503. [CrossRef]
38. Liu, Q.; Shang, J.; Wang, J.; Niu, W.; Qiao, W. Evaluation and prediction of the safety management efficiency of coal enterprises based on a DEA-BP neural network. *Resour. Policy* **2023**, *83*, 103611. [CrossRef]
39. Lou, W.G. Evaluation and prediction of soil quality based on artificial neural network in the Sanjiang Plain. *Chin. J. Manag. Sci.* **2002**, *10*, 79–83.
40. Haykin, S. *Neural Networks and Learning Machines*; China Machine Press: Beijing, China, 2009.
41. Zhang, G.; Patuwo, E.; Hu, M. Forecasting with artificial neural networks: The state of the art. *Int. J. Forecast.* **1998**, *14*, 35–62. [CrossRef]
42. StatSoft, Inc. *STATISTICA Neural Networks*; StatSoft, Inc.: Tulsa, OK, USA, 2011.
43. Zhong, C.; Lou, W.G.; Wang, C. Neural Network-Based Modeling for Risk Evaluation and Early Warning for Large-Scale Sports Events. *Mathematics* **2022**, *10*, 3228. [CrossRef]
44. Singhee, A. SiLVR: Projection Pursuit for Response Surface Modeling. In *Machine Learning in VLSI Computer-Aided Design*; Elfadel, I., Boning, D., Li, X., Eds.; Springer Nature: Cham, Switzerland, 2019.
45. Hall, P. On projection pursuit regression. *Ann. Stat.* **1989**, *17*, 573–588. [CrossRef]
46. Yu, X.H.; Xu, H.Y.; Lou, W.G.; Xu, X.; Shi, V. Examining energy eco-efficiency in China's logistics industry. *Int. J. Prod. Econ.* **2023**, *258*, 108797. [CrossRef]
47. Banker, R.D.; Charnes, A.; Cooper, W.W. Some models for estimating technical and scale inefficiencies in data envelopment analysis. *Manag. Sci.* **1984**, *30*, 1078–1092. [CrossRef]
48. Hornik, K.; Stinchcombe, M.; White, H. Multilayer feedforward networks are universal approximators. *Neural Netw.* **1989**, *2*, 359–366. [CrossRef]
49. Chen, C.; Tuo, R. Projection Pursuit Gaussian Process Regression. *IISE Trans.* **2023**, *55*, 901–911. [CrossRef]
50. Tao, H.; Tao, J.; Li, Q.; Aihemaiti, M.; Jiang, Y.; Yang, W.; Wei, J. Average relative flow of single-wing labyrinth drip irrigation tape based on projection pursuit regression. *Sci. Rep.* **2022**, *12*, 8543. [CrossRef] [PubMed]

51. Mohamed, A.A.A.; Hassan, S.A.; Hemeida, A.M.; Alkhalaf, S.; Mahmoud, M.; Eldin, A.M.B. Parasitism–Predation algorithm (PPA): A novel approach for feature selection. *Ain Shams Eng. J.* **2020**, *11*, 293–308. [CrossRef]
52. Athey, S.; Imbens, G. Machine learning methods that economists should know about. *Annu. Rev. Econ.* **2019**, *11*, 685–725. [CrossRef]
53. Marsland, S. *Machine Learning: An Algorithmic Perspective*; CRC Press: Boca Raton, FL, USA; Taylor & Francis Group: Abingdon, UK, 2015.
54. Valero-Carreras, D.; Aparicio, J.; Guerrero, N.G. Support vector frontiers: A new approach for estimating production functions through support vector machines. *Omega* **2021**, *104*, 102490. [CrossRef]
55. Gomes, E.G.; Lins, M.P.E. Modelling undesirable outputs with zero sum gains data envelopment analysis models. *J. Oper. Res. Soc.* **2008**, *59*, 616–623. [CrossRef]
56. Zhou, X.; Niu, A.Y.; Lin, C.X. Optimizing carbon emission forecast for modelling China's 2030 provincial carbon emission quota allocation. *J. Environ. Manag.* **2023**, *325*, 116523. [CrossRef]
57. Tang, Q.; Zhang, C. Data processing system (DPS) software with experimental design, statistical analysis and data mining developed for use in entomological research. *Insect Sci.* **2013**, *20*, 254–260. [CrossRef]

mathematics

Article

A New Multi-Target Three-Way Threat Assessment Method with Heterogeneous Information and Attribute Relevance

Yang Gao * **and Na Lyu**

Information and Navigation College, Air Force Engineering University, Xi'an 710077, China
* Correspondence: gao_yang_mail@126.com

Abstract: Target threat assessment provides support for combat decision making. The multi-target threat assessment method based on a three-way decision can obtain threat classification while receiving threat ranking, thus avoiding the limitation of traditional two-way decisions. However, the heterogeneous situation information, attribute relevance, and adaptive information processing needs in complex battlefield environment bring challenges to existing methods. Therefore, this paper proposes a new multi-target three-way threat assessment method with heterogeneous information and attribute relevance. Firstly, dynamic assessment information is represented by heterogeneous information, and attribute weights are calculated by heterogeneous Criteria Importance Through Intercriteria Correlation (CRITIC). Then, the conditional probability is calculated by the heterogeneous weighted Technique for Order Preference by Similarity to Ideal Solution (TOPSIS), and the adaptive risk avoidance coefficients are constructed by calculating the uncertainty of the assessment value, and then the relative loss function matrices are constructed. Finally, the comprehensive loss function matrices are obtained by the weighted Heronian mean (HM) operator, and the comprehensive thresholds are calculated to obtain the three-way rules. The case study shows that compared with the existing methods, the proposed method can effectively handle the heterogeneous information and attribute relevance, and obtain the risk avoidance coefficients without presetting or field subjective settings, which is more suitable for the complex mission environment.

Keywords: heterogeneous information; three-way decision; threat assessment; attribute relevance; risk avoidance coefficient

MSC: 90B50; 90C70

1. Introduction

The research and application of military technology is an important part of the development of science and technology. Modern war is a highly informationised and even intelligent systems' confrontation [1,2]. It involves many operational elements. Typically, to gain an operational advantage, both sides need to focus their superior forces on the other side's high-value targets in combat decision making [3]. In the course of combat, targets with a higher threat degree are usually considered to be high-value targets, which need to allocate more resources to attack or interfere first [4,5]. Therefore, target threat assessment is an important issue in modern military combat decision making [6,7].

A typical implementation process of target threat assessment is shown in Figure 1. Briefly, in complex mission scenarios, first determine the threat assessment attributes, then normalize the assessment data, select appropriate threat assessment methods, and finally obtain the threat ranking of the targets. Often, it is desirable to minimize human involvement in the above processes in order to improve timeliness.

With the increasing complexity of the combat environment and the increasing variety of combat forms, the study object of target threat assessment gradually includes air targets, ground targets, radiation source targets, group targets, and so on. The selection of target threat assessment attributes needs to consider the scenarios, and usually there are discrete attributes, such as the target type; continuous attributes, such as the target location; etc., and the target information comes from the situation information base (historical data), various types of sensors, and so on. Due to the different sources and accuracy of target situation information, different forms of information representation are justified, i.e., the assessment information is heterogeneous. The choice of evaluation methods is crucial. Target threat assessment methods include methods based on multi-attribute decision making (MADM), neural-network-based assessment, Bayesian-network-based estimation, and methods based on fuzzy set theory [8–12]. The characteristics and deficiencies of these methods are listed in Table 1.

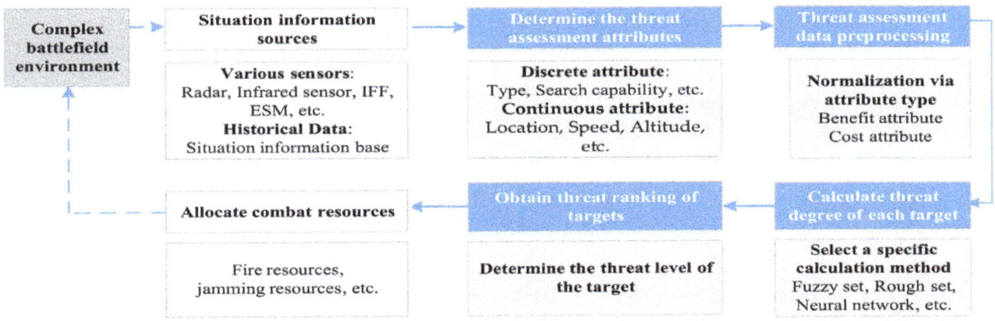

Figure 1. The implemented process of target threat assessment.

Table 1. The comparison among target threat assessment methods.

Method	Theoretical Basis	Characteristics	Deficiencies
Zhang et al. [8]	Bayesian inference and evidence theory	The method is based on Bayesian network, which has good interpretability.	Bayesian network structure and conditional probability are difficult to effectively determine.
Luo et al. [9]	MADM, information entropy, and AHP	The method can explicitly represent assessment indicators and their importance, and can calculate the threat degree.	The method is static and subjective parameters are difficult to determine.
Ma et al. [10]	Cloud model	The method can effectively handle ambiguity and randomness of situation information.	The membership function is hard to determine.
Wang et al. [11]	Intuitionistic fuzzy set and fuzzy reasoning	The method can handle the uncertainty of target situation information.	The reasoning rules grow rapidly with the increase in evaluation indicators, leading to decision-making difficulties.
Yu et al. [12]	Long short-term memory network	The method has learning capacities and generalization ability.	The network needs dataset to train and computations are relatively complex.

Although the research objects of each method in Table 1 are not the same, the methods are universal. The advantages and disadvantages of each method are caused by its theoretical basis, which is detailed in Table 1. Among them, dynamic target threat assessment methods based on fuzzy MADM, which have a good ability to represent uncertainty and to directly calculate the threat degree, have received extensive attention [13–17]. Follow-up studies are also based on this model. However, there are common problems with these methods or models:

(1) Ranking results vary between different assessment methods. Different methods have different focuses and often give different results in threat ranking. This increases the difficulty of selecting high-threat targets reasonably.

(2) These methods usually are two-way decisions and can cause misjudgment. For the threat value that is higher than a certain threshold value, take the priority of the combat strategy, and for the lower value than the threshold value, take the strategy of abandoning the combat. The result of such a decision is an either/or, and if the information is insufficient to support the decision, false judgements are often made, leading to an irrational allocation of combat resources.

To address the above problems, a target threat assessment method that objectively realizes multi-target threat classification is needed. Three-way decision, based on the decision–theoretic rough sets model, was proposed by Yao et al. [18]. It succeeds in rationally assigning semantic interpretations to the positive, negative, and boundary domains of rough sets, which correspond to acceptance, rejection, and delay decisions, respectively, in practical decision making. Since the proposal of the three-way decisions, many scholars have refined and extended it, and it has been widely used in many fields [19–25].

We first introduced three-way decision into multi-target dynamic threat assessment under an intuitionistic fuzzy MADM environment [26,27], which can obtain threat classification while receiving threat ranking. The application of three-way decision in the field of target threat assessment can be notated as multi-target three-way threat assessment. Subsequently, the literature [28–32] conducted improvement studies, whose main concerns are the optimization calculation of conditional probability and decision thresholds. However, the above methods still cannot meet the practical needs in complex combat scenarios well, which are manifested in the following aspects:

(1) The evaluation information coming from different sources is usually heterogeneous and uncertain. For example, different sensors provide information with different accuracies and it is not reasonable to use the same representation. Discrete and continuous attributes should be represented differently. The representation and processing of assessment information are relatively simple in [26–29]. A single form of fuzzy numbers, such as intuitionistic fuzzy set, is usually used, which ignores the differences of assessment attributes and is inconsistent with the actual situation.

(2) The attribute relevance is often ignored in existing methods of multi-target three-way threat assessment. Among the existing methods, many weight calculation methods are used, including subjective, objective, and comprehensive weight methods, but the influence of attribute relevance on weight calculation is rarely considered. At the same time, when aggregating information, its influence is usually also ignored.

(3) In combat, it is usually necessary to minimize the influence of humans in the decision process to improve timeliness. Therefore, it is desirable to reduce or avoid subjective settings of parameters in the evaluation process. In existing methods, the risk avoidance coefficients are usually presetting or field subjective settings. On the one hand, the reasonableness depends on the subjective experience, and how to select causing problems, and on the other hand, it may affect the timeliness of combat decision making.

To address the above problems, this paper proposes a new multi-target three-way threat assessment method with heterogeneous information and attribute relevance. In the study of the application of three-way decision, other scholars have considered heterogeneous information processing [33–36] and the attribute relevance [37,38] separately. Good attempts were indeed made, despite problems such as the potential loss of information during the conversion of heterogeneous information to a single format in some studies. However, in target threat assessment, there has not been a systematic study.

The main contributions are as follows:

(1) The dynamic threat assessment information is represented by heterogeneous forms, such as real numbers, interval numbers, three-parameter intervals, and four-parameter intervals. Which form to use is determined by the source and type of assessment information in a specific mission scenario.

(2) The conditional probabilities are estimated based on heterogeneous weighted TOPSIS, where the attribute weights are calculated by the heterogeneous CRITIC. Obviously, the CRITIC considers both the variability and relevance of attributes [39]. At the same time, the above calculations are performed directly on the heterogeneous assessment information without the need for information conversion.

(3) The adaptive risk avoidance coefficients are calculated by the uncertainty of the assessment value. Then, the relative loss function matrices can be obtained quickly. There is no need to set them subjectively or in advance, and they can objectively reflect the acquisition of situation information.

(4) The comprehensive loss function matrix is constructed by aggregating the relative loss function matrix under each attribute via the weighted HM operator, which can effectively reflect the correlation between the aggregated data.

(5) The proposed method can directly obtain threat ranking and threat classification based on the assessment information without additional parameter settings. It can meet the timeliness need for combat and can even be used directly in autonomous intelligent systems.

Through the heterogeneous representation and correlation processing of evaluation information, as well as the design of adaptive risk avoidance coefficients, the three-way threat assessment method proposed in this paper is more suitable for complex combat environment. The specific structure of this paper is as follows: Section 2 introduces the analytical ground. Section 3 introduces the proposed method. In Section 4, the case study and comparison analysis show that the proposed method is effective. Section 5 concludes this paper.

2. Analytical Ground

2.1. Fuzzy MADM

We first introduce the fuzzy MADM and its application to target threat assessment. In the implementation process, the assessment attributes are used as decision attributes and the targets are used as alternatives. The specific process is expressed as follows:

Assume that alternatives (targets) set $T = \{T_1, T_2, \cdots, T_m\}$ consists of m elements, $A = \{A_1, A_2, \cdots, A_n\}$ consists of n assessment attributes, $t = \{t_1, t_2, \cdots, t_K\}$ is a set of assessment moments, and $\mathbf{W} = (w_1, w_2, \cdots, w_n)$ is the attribute weight vector, where $\sum_{j=1}^{n} w_j = 1$. The assessment matrix can be denoted as $\mathbf{Z}(t_k) = \left(z_{ij}(t_k)\right)_{m\times n}$, where $z_{ij}(t_k)$ is the assessed value for the attribute $A_j, j \in \{1, 2, \cdots, n\}$ of the target T_i $i \in \{1, 2, \cdots, m\}$ at moment t_k $k \in \{1, 2, \cdots, K\}$. The value can be expressed in the form of fuzzy numbers, etc.

Remark 1. *As mentioned in the Introduction, what the symbols represent is determined by the specific mission scenario. More specifically, the targets can be air targets, ground targets, radiation source targets, group targets, and so on. The assessment attributes could be the type, distance, course angle, speed, height, interface ability, etc. As for the form of the assessed value, it depends on the source and type. In this paper, multi-parameter intervals are used to represent heterogeneous information, which does not mean that other forms, such as linguistic variables, are not allowed. If other types of data are used, a modification of the proposed method is sufficient.*

Then, there are usually the following methods to receive ranking results:

(1) Arithmetic weighting method. The weight of each decision attribute is obtained through subjective expert experience or objective data methods. Using operators such as the arithmetic mean, the multi-attribute information is aggregated, and the multiple alternatives are ranked [40].

(2) Method based on ideal solutions. By calculating the distance of each option to the positive and negative ideal solutions, the closeness and other related indexes can be calculated, and the multiple alternatives are ranked by the closeness, such as the TOPSIS method and its improvement methods, the VIKOR method, and so on [41,42].

(3) Dominance decision method. Using decision attributes, a series of dominance relations are constructed, and the set of alternatives is narrowed down by the dominance relations to make the judgement of the superiority or inferiority of the alternatives [43].

As for heterogeneous fuzzy MADM, it usually means that the assessment attributes are not represented in the same form, i.e., some attributes are real numbers, some are interval numbers, and so on. Accordingly, there are two types of processing methods; one is to transform heterogeneous information into the same form, and the other is to extend the last two methods to heterogeneous information environments. The first type of transformation process involves information loss [33]. Therefore, in this paper, we use the second type of method, constructing heterogeneous weighted TOPSIS to estimate conditional probability.

2.2. Three-Way Decision

The study of three-way decision can be divided into three main categories: connotation, extension, and application. The connotation study focuses on the computation of conditional probability, loss function, and decision thresholds; the extension study focuses on the combination of three-way decision with other uncertainty theories, decision methods, etc.; and the application study focuses on the application to specific problems [44–47].

The three-way decision based on decision–theoretic rough sets is as follows:

Definition 1 ([48]). *Let U be a finite and non-empty set, $R \subseteq U \times U$ be an equivalence relation, $apr = (U, R)$ be a rough approximation space. U can be parted by R, expressed as $U/R = \{[x] |\ x \in U\}$, and thresholds are set as $0 \le \beta < \alpha \le 1$. For $\forall A \subseteq U$, the lower and upper approximation sets of the probabilistic rough set can be defined:*

$$\underline{apr}_{(\alpha,\beta)}(A) = \{x \in U |\ \Pr(A|\ [x]) \ge \alpha\}, \tag{1}$$

$$\overline{apr}_{(\alpha,\beta)}(A) = \{x \in U |\ \Pr(A|\ [x]) > \beta\}, \tag{2}$$

where $\Pr(A|\ [x])$ is the conditional probability, expressed as $\Pr(A|\ [x]) = |[x] \cap A| / |[x]|$.

The universe U can be parted into three regions by thresholds, expressed as

$$POS(A) = \{x \in U \mid \Pr(A|[x]) \ge \alpha\}, \tag{3}$$

$$BND(A) = \{x \in U \mid \beta < \Pr(A|[x]) < \alpha\}, \tag{4}$$

$$NEG(A) = \{x \in U \mid \Pr(A|[x]) < \beta\}. \tag{5}$$

Let $\Omega = \{A, \neg A\}$ be the state sets of targets, which means a target belongs to A or not; actions are set as $AC = \{a_P, a_B, a_N\}$, where a_P, a_B, and a_N denote $x \in POS(A)$, $x \in BND(A)$, and $x \in NEG(A)$, respectively. The loss function regarding the risk of different actions is exhibited in Table 2. Usually, the loss functions satisfy $0 \le \lambda_{PP} \le \lambda_{BP} < \lambda_{NP}$ and $0 \le \lambda_{NN} \le \lambda_{BN} < \lambda_{PN}$.

Table 2. Loss function matrix.

	$A(P)$	$\neg A(N)$
a_P	λ_{PP}	λ_{PN}
a_B	λ_{BP}	λ_{BN}
a_N	λ_{NP}	λ_{NN}

Then, the three-way decision rules can be expressed as

(P) If $\Pr(A|[x]) \ge \alpha$ and $\Pr(A|[x]) \ge \gamma$, decide $x \in POS(A)$;

(B) If $\Pr(A|[x]) \le \alpha$ and $\Pr(A|[x]) \ge \beta$, decide $x \in BND(A)$;

(N) If $\Pr(A|[x]) \le \beta$ and$\Pr(A|[x]) \le \gamma$, decide$x \in NEG(A)$;

where the thresholds are defined as

$$\alpha = \frac{(\lambda_{PN} - \lambda_{BN})}{(\lambda_{PN} - \lambda_{BN}) + (\lambda_{BP} - \lambda_{PP})} \tag{6}$$

$$\beta = \frac{(\lambda_{BN} - \lambda_{NN})}{(\lambda_{BN} - \lambda_{NN}) + (\lambda_{NP} - \lambda_{BP})} \tag{7}$$

$$\gamma = \frac{(\lambda_{PN} - \lambda_{NN})}{(\lambda_{PN} - \lambda_{NN}) + (\lambda_{NP} - \lambda_{PP})} \tag{8}$$

Further assuming that $\beta < \alpha$, we can obtain the following:

(P) If $\Pr(A|[x]) \geq \alpha$, decide $x \in POS(A)$;

(B) If $\beta < \Pr(A|[x]) < \alpha$, decide $x \in BND(A)$;

(N) If $\Pr(A|[x]) \leq \beta$, decide $x \in NEG(A)$.

For this study, its goal was to construct a three-way decision about whether the target should be attacked first or not. If the target is classified into the positive domain, it means that the threat level is high and the attack or interference needs to be given priority. If it is divided into negative areas, it means that the threat level is low and there is no need to attack or interfere first. Otherwise, more information is needed to make a judgement.

2.3. Multi-Target Three-Way Threat Assessment Method

Multi-target three-way threat assessment is an application of three-way decision to the area of multi-target threat assessment. The key problems are how to obtain conditional probability and comprehensive decision thresholds via assessment information.

First, the conditional probability usually can be calculated by ideal solutions, being inspired by the literature [49]. More specifically, decision methods based on ideal solutions, such as TOPSIS and VIKOR, can be used to compute conditional probability. Relative closeness and compromise ranking values are used to represent the conditional probability of a target being prioritized for attack, respectively.

Then, the relative loss function matrix of each target under each attribute is constructed by the fuzzy evaluation information [50,51] and risk avoidance coefficient. The comprehensive loss function matrix of the target under the attribute set is aggregated based on the attribute weight and arithmetic mean operator.

Finally, the thresholds are calculated by the comprehensive loss function matrix. The three-way decision rules can also be obtained by conditional probability and decision thresholds.

The above process is analyzed and improved as follows:

(1) In the above process, assessment information is usually represented by a class of fuzzy numbers, such as intuitionistic fuzzy numbers, and conditional probabilities are obtained through intuitionistic fuzzy TOPSIS. In this paper, we consider the heterogeneous representation of situation information and extend the calculation of conditional probability to heterogeneous weighted TOPSIS, whose attribute weights are calculated by the CRITIC method.

(2) In the construction of the relative loss function matrix, the risk avoidance coefficients usually need to be set subjectively and set to the same value, which is not only difficult to determine, but also inconsistent with the actual situation. In this paper, adaptive risk avoidance coefficients are designed based on the uncertainty of the evaluation values.

(3) In the aggregation of relative loss function matrices, the correlation among the assessment attributes is ignored. In this paper, the HM operator is used to aggregate the relative loss function matrices.

Remark 2. *Although there can be many ways to calculate conditional probability and decision thresholds, such as the dominance relation in study* [52]*, the method based on an ideal solution is widely used due to its simplicity and ease of implementation in the existing target threat assessment based on three-way decision.*

3. The Proposed Multi-Target Three-Way Threat Assessment Method

In this section, we will describe how the proposed method is realized. The evaluation process is shown in Figure 2. It should be noted that the representation of targets, assessment attributes, etc., is the same as above. Table 3 gives a quick reference to the symbols and acronyms.

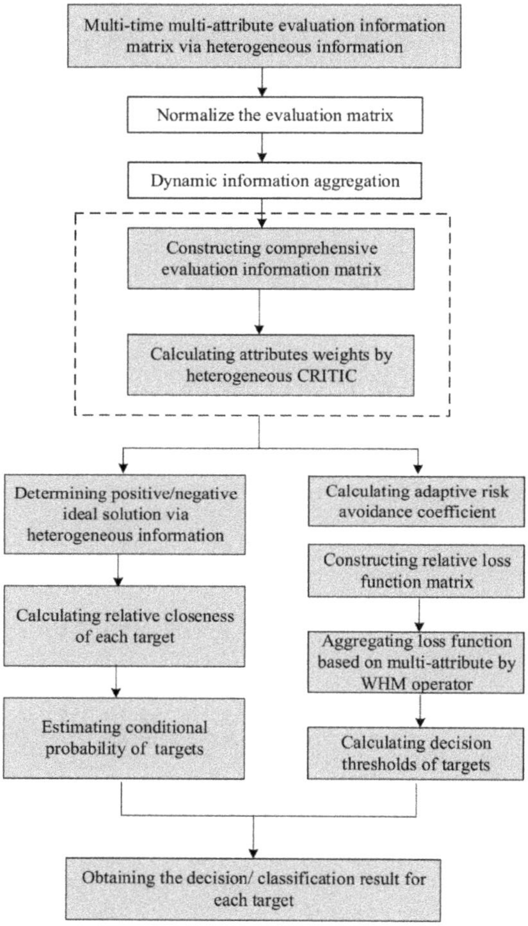

Figure 2. The evaluation process of proposed method.

Table 3. Main nomenclature.

Symbol/Acronym	Description
CRITIC	Criteria Importance Through Intercriteria Correlation
TOPSIS	Technique for Order Preference by Similarity to an Ideal Solution
HM	Heronian mean
$WHM^{p,q}$	Weighted HM operator with parameter
VIKOR	VIšekriterijumsko KOmpromisno Rangiranje
MADM	Multi-attribute decision making
T	Target set
T_i	The i-th target
m	Total number of targets
A	Evaluation attributes set

Table 3. *Cont.*

Symbol/Acronym	Description
A_j	The j-th evaluation attribute
n	Total number of attributes
t	Evaluation moments set
t_k	The k-th moment
$\mathbf{Z}(t_k)$	Evaluation matrix of targets at moment t_k
$z_{ij}(t_k)$	The element of $\mathbf{Z}(t_k)$, j-th evaluation attribute value of i-th target
$\mathbf{W} = (w_1, w_2, \cdots, w_n)$	Attribute weight vector
$F_i, i = \{1,2,3,4\}$	Different forms set of assessment values
$\widetilde{\mathbf{Z}}(t_k)$	Normalized evaluation matrix of targets at moment t_k
$\widetilde{z}_{ij}(t_k)$	Normalized form of $z_{ij}(t_k)$
$\eta = (\eta_1, \eta_2, \cdots, \eta_K)$	Weights of multi-time
$\widetilde{\mathbf{Z}}$	Multi-time integration matrix of $\widetilde{\mathbf{Z}}(t_k)$

3.1. *Multi-Time Multi-Attribute Evaluation Information Matrix via Heterogeneous Information*

Due to the complexity of the combat environment, it is difficult to effectively interpret the uncertainty using a single fuzzy number form. Therefore, the value of the assessment attribute is expressed by heterogeneous fuzzy information. More specifically, $z_{ij}(t_k)$ is mainly based on four different forms of information: (1) real numbers (F_1); (2) interval numbers (F_2); (3) three-parameter interval numbers (F_3); and (4) four-parameter interval numbers (F_4). F_i denotes the set of the assessment values, and $F_i \cap F_j = \varnothing$ $(i \neq j)$, where $\varnothing$ is an empty set.

Remark 3 ([53]). *A non-negative three-parameter interval number* α *is expressed by* $\alpha = [\alpha^{\mathrm{L}}, \alpha^{\mathrm{M}}, \alpha^{\mathrm{U}}]$, *where* $0 \leq \alpha^{\mathrm{L}} \leq \alpha^{\mathrm{M}} \leq \alpha^{\mathrm{U}}$. *And a non-negative four-parameter interval number* α *is expressed by* $\alpha = [\alpha^{\mathrm{L}}, \alpha^{\mathrm{M}_1}, \alpha^{\mathrm{M}_2}, \alpha^{\mathrm{U}}]$, *where* $0 \leq \alpha^{\mathrm{L}} \leq \alpha^{\mathrm{M}_1} \leq \alpha^{\mathrm{M}_2} \leq \alpha^{\mathrm{U}}$.

The basic operations on multi-parameter interval numbers are added here. Given two non-negative multi-parameter interval numbers $\alpha = [\alpha_1, \alpha_2, \cdots, \alpha_n]$ and $\beta = [\beta_1, \beta_2, \cdots, \beta_n]$ and a positive real number θ, some operations are as follows:

(1) $\alpha + \beta = [\alpha_1 + \beta_1, \alpha_2 + \beta_2, \cdots, \alpha_n + \beta_n]$;
(2) $\alpha \times \beta = [\alpha_1\beta_1, \alpha_2\beta_2, \cdots, \alpha_n\beta_n]$;
(3) $\theta\alpha = [\theta\alpha_1, \theta\alpha_2, \cdots, \theta\alpha_n]$;
(4) $d(\alpha, \beta) = \sum\limits_{i=1}^{n} |\alpha_i - \beta_i|$.

3.2. *Constructing Comprehensive Evaluation Information Matrix*

3.2.1. Normalize the Evaluation Information Matrix

In the target threat assessment, the attributes may be usually given by benefit or cost criteria. The magnitude and type of evaluation attributes affect subsequent calculations. Usually, this impact is eliminated by standardizing the evaluation information matrix. There are a number of standardized and normalized calculation methods that will not be repeated here; please refer to [6,36]. Through the normalized process of [6,36], we can obtain a normalized evaluation information matrix, denoted as $\widetilde{z}_{ij}(t_k)$.

3.2.2. Dynamic Information Aggregation

The target threat assessment should combine the multi-time heterogeneous fuzzy evaluation information. We can obtain the time series weight based on the Poisson distribution method with an inverse form. The closer to the current time, the more important the situation information is. Thus, the series weight vector $\boldsymbol{\eta} = (\eta_1, \eta_2, \cdots, \eta_K)$ of K times can be calculated by

$$\begin{cases} \eta_k = \dfrac{k!}{\phi^k} \Big/ \sum\limits_{j=1}^{K} \frac{j!}{\phi^j} \\ \sum\limits_{k=1}^{K} \eta_k = 1 \end{cases}, \tag{9}$$

where $\eta_k \geq 0$ and $0 < \phi < 2$. Usually, the setting of ϕ is 1.5.

Then, combined with time series weights, we can obtain a comprehensive evaluation information matrix, denoted as $\widetilde{\mathbf{Z}} = \left(\widetilde{z}_{ij}\right)_{m \times n}$, where $\widetilde{\mathbf{Z}} = \sum\limits_{k=1}^{K} \eta_k \widetilde{\mathbf{Z}}(t_k)$ and $\widetilde{z}_{ij} = \sum\limits_{k=1}^{K} \eta_k \widetilde{z}_{ij}(t_k)$.

3.3. Calculating Assessment Attribute Weights

In order to take into account the variability and relevance of the attributes, the CRITIC model is applied to calculate the attribute weights of a heterogeneous comprehensive evaluation information matrix, which can be called heterogeneous CRITIC. The main difference with other CRITIC methods [38] is that, through the definition of the distance function, it is suitable for heterogeneous information environments with multi-parameter interval numbers. The main steps are as follows:

(1) Calculate the standard deviation of A_j, where

$$\begin{cases} \overline{z}_j = \frac{1}{m} \sum\limits_{i=1}^{m} \widetilde{z}_{ij} \\ S_j = \sqrt{\dfrac{\sum\limits_{i=1}^{m} \left[d(\widetilde{z}_{ij} - \overline{z}_j)\right]^2}{m-1}} \end{cases}. \tag{10}$$

(2) Calculate the correlation coefficient r_{ij} between A_i and A_j, where

$$r_{ij} = \frac{\sum\limits_{k=1}^{m} d(\widetilde{z}_{ki} - \overline{z}_i) \cdot d\left(\widetilde{z}_{kj} - \overline{z}_j\right)}{\sqrt{\sum\limits_{k=1}^{m} [d(\widetilde{z}_{ki} - \overline{z}_i)]^2} \cdot \sqrt{\sum\limits_{k=1}^{m} \left[d\left(\widetilde{z}_{kj} - \overline{z}_j\right)\right]^2}}. \tag{11}$$

(3) Further, the conflictual relationship R_j of A_j can be expressed by

$$R_j = \sum_{i=1}^{m} (1 - r_{ij}) \tag{12}$$

(4) Calculate the information load C_j of A_j, where

$$C_j = S_j \sum_{i=1}^{m} (1 - r_{ij}) = S_j \times R_j \tag{13}$$

(5) Finally, the weight w_j of A_j can be expressed by

$$w_j = \frac{C_j}{\sum\limits_{j=1}^{n} C_j} \tag{14}$$

3.4. Estimating Conditional Probability by Heterogeneous Weighted TOPSIS

The conditional probability of each target can be estimated by heterogeneous weighted TOPSIS.

(1) For the comprehensive information matrix, define the heterogeneous positive ideal solution (HPIS) z^+ and heterogeneous negative ideal solution (HNIS) z^- as follows:

$$z_j^+ = \max_i \widetilde{z}_{ij},\ z_j^- = \min_i \widetilde{z}_{ij},\ \text{where } z_{ij} \in F_1 \tag{15}$$

$$z_j^+ = \left[\max_i \widetilde{z}_{ij}^L, \max_i \widetilde{z}_{ij}^R\right],\ z_j^- = \left[\min_i \widetilde{z}_{ij}^L, \min_i \widetilde{z}_{ij}^R\right],\ \text{where}, z_{ij} \in F_2 \tag{16}$$

$$z_j^+ = \left[\max_i \widetilde{z}_{ij}^L, \max_i \widetilde{z}_{ij}^M, \max_i \widetilde{z}_{ij}^R\right],\ z_j^- = \left[\min_i \widetilde{z}_{ij}^L, \min_i \widetilde{z}_{ij}^M, \min_i \widetilde{z}_{ij}^R\right],\ \text{where}, z_{ij} \in F_3 \tag{17}$$

$$z_j^+ = \left[\max_i \widetilde{z}_{ij}^L, \max_i \widetilde{z}_{ij}^{M_1}, \max_i \widetilde{z}_{ij}^{M_2}, \max_i \widetilde{z}_{ij}^R\right],\ z_j^- = \left[\min_i \widetilde{z}_{ij}^L, \min_i \widetilde{z}_{ij}^{M_1}, \min_i \widetilde{z}_{ij}^{M_2}, \min_i \widetilde{z}_{ij}^R\right],\ \text{where } z_{ij} \in F_4. \tag{18}$$

z^+ implies the evaluation of the state A, and z^- implies the evaluation of the state $\neg A$.
(2) Calculate relative closeness of each target.
The distance between the target T_i and HPIS is calculated by

$$D(T_i, z^+) = \sum_{j=1}^{n} w_j d\left(\widetilde{z}_{ij}, z_j^+\right), \tag{19}$$

The distance between the target T_i and HNIS is calculated by

$$D(T_i, z^-) = \sum_{j=1}^{n} w_j d\left(\widetilde{z}_{ij}, z_j^-\right), \tag{20}$$

The relative closeness of each target is expressed as

$$RC(T_i) = \frac{D(T_i, z^-)}{D(T_i, z^-) + D(T_i, z^+)} = \frac{\sum_{j=1}^{n} w_j d\left(\widetilde{z}_{ij}, z_j^-\right)}{\sum_{j=1}^{n} w_j d\left(\widetilde{z}_{ij}, z_j^-\right) + \sum_{j=1}^{n} w_j d\left(\widetilde{z}_{ij}, z_j^+\right)}. \tag{21}$$

(3) Estimate conditional probabilities of targets.
Obviously, $RC(T_i)$ represents the probability of the target T_i being in the state A [26,49]. Thus,

$$\Pr(A|T_i) = RC(T_i). \tag{22}$$

3.5. Calculating Decision Thresholds by Evaluation Values

Firstly, define the absolutely maximum value $z^j_{\max}$ and minimum value $z^j_{\min}$ for the attribute, where

$$z^j_{\max} = \begin{cases} 1 & , z_{ij} \in F_1 \\ [1,1] & , z_{ij} \in F_2 \\ [1,1,1] & , z_{ij} \in F_3 \\ [1,1,1,1] & , z_{ij} \in F_4 \end{cases} \text{ or } z^j_{\max} = z_j^+,\ z^j_{\min} = \begin{cases} 0 & , z_{ij} \in F_1 \\ [0,0] & , z_{ij} \in F_2 \\ [0,0,0] & , z_{ij} \in F_3 \\ [0,0,0,0] & , z_{ij} \in F_4 \end{cases} \text{ or } z^j_{\min} = z_j^-.$$

(1) Calculate the uncertainty of multi-parameter interval numbers, expressed as

$$\delta(\widetilde{z}_{ij}) = \begin{cases} 0 & , z_{ij} \in F_1 \\ \left(\widetilde{z}_{ij}^R - \widetilde{z}_{ij}^L\right) & , z_{ij} \notin F_1 \end{cases} \tag{23}$$

(2) Calculate the adaptive risk avoidance coefficient $\sigma(\widetilde{z}_{ij})$ of $\widetilde{z}_{ij}$, expressed as

$$\sigma(\widetilde{z}_{ij}) = \begin{cases} \frac{-0.5}{\max_i \delta(\widetilde{z}_{ij}) - \min_i \delta(\widetilde{z}_{ij})} \left[\delta(\widetilde{z}_{ij}) - \max_i \delta(\widetilde{z}_{ij})\right] & , z_{ij} \notin F_1 \\ 0.5 & , z_{ij} \in F_1 \end{cases} \tag{24}$$

where $0 \leq \sigma(\widetilde{z}_{ij}) \leq 0.5$.

Remark 4. *The risk avoidance coefficients reflect the acquisition of situation information. The more sufficient the situation information that can be obtained, the bigger the value of the risk avoidance coefficient* [26,50]. *We use the uncertainty of the assessed value to indicate the extent of acquisition. The greater the uncertainty, the less adequate the access. The relative magnitude of uncertainty in the assessed value is measured by the range of the upper and lower limits of the interval.*

(3) Construct relative loss function matrices via the adaptive risk avoidance coefficient. The relative loss function matrix of each target under each attribute is expressed as

$$\lambda(\widetilde{z}_{ij}) = \begin{pmatrix} \lambda_{PP}^{ij} & \lambda_{PN}^{ij} \\ \lambda_{BP}^{ij} & \lambda_{BN}^{ij} \\ \lambda_{NP}^{ij} & \lambda_{NN}^{ij} \end{pmatrix} = \begin{pmatrix} 0 & d\left(\widetilde{z}_{ij}, z_{\max}^{j}\right) \\ \sigma(\widetilde{z}_{ij})d\left(\widetilde{z}_{ij}, z_{\min}^{j}\right) & \sigma(\widetilde{z}_{ij})d\left(\widetilde{z}_{ij}, z_{\max}^{j}\right) \\ d\left(\widetilde{z}_{ij}, z_{\min}^{j}\right) & 0 \end{pmatrix} \tag{25}$$

(4) Aggregate loss function based on multi-attribute information by weighted HM operator.

Since there is correlation in the assessment attributes, there is also correlation in the relative loss function constructed from the attribute values. Therefore, the relative loss function matrices of the target across attributes are aggregated using the weighted HM operator.

Definition 2 ([54]). *Let $z_i (i = 1, 2, \cdots, n)$ be a collection of non-negative numbers, $W = (w_1, w_2, \cdots, w_n)$ is the weight vector of $z_i (i = 1, 2, \cdots, n)$, $p \geq 0$, $q \geq 0$, and p, q do not take the value 0 simultaneously. If $WHM^{p,q}$ satisfies*

$$WHM^{p,q}(z_1, z_2, \cdots, z_n) = \left(\frac{2}{n(n+1)} \sum_{i=1}^{n} \sum_{j=1}^{n} (nw_i z_i)^p (nw_j z_j)^q \right)^{\frac{1}{p+q}}, \tag{26}$$

then $WHM^{p,q}$ is the weighted HM operator with a parameter. Usually, we can set $p = q = 1$.

The comprehensive loss function matrix of each target under multi-attribute information via the weighted HM operator is expressed as

$$\lambda_i = \begin{pmatrix} \lambda_{PP}^{i} & \lambda_{PN}^{i} \\ \lambda_{BP}^{i} & \lambda_{BN}^{i} \\ \lambda_{NP}^{i} & \lambda_{NN}^{i} \end{pmatrix} = \begin{pmatrix} 0 & WHM^{1,1}(\lambda_{PN}^{i1}, \lambda_{PN}^{i2}, \cdots, \lambda_{PN}^{in}) \\ WHM^{1,1}(\lambda_{BP}^{i1}, \lambda_{BP}^{i2}, \cdots, \lambda_{BP}^{in}) & WHM^{1,1}(\lambda_{BN}^{i1}, \lambda_{BN}^{i2}, \cdots, \lambda_{BN}^{in}) \\ WHM^{1,1}(\lambda_{NP}^{i1}, \lambda_{NP}^{i2}, \cdots, \lambda_{NP}^{in}) & 0 \end{pmatrix}. \tag{27}$$

(5) Calculate the comprehensive decision threshold.

The corresponding decision thresholds of each target are calculated by

$$\alpha_i = \frac{(\lambda_{PN}^{i} - \lambda_{BN}^{i})}{(\lambda_{PN}^{i} - \lambda_{BN}^{i}) + (\lambda_{BP}^{i} - \lambda_{PP}^{i})}, \tag{28}$$

$$\beta_i = \frac{(\lambda_{BN}^{i} - \lambda_{NN}^{i})}{(\lambda_{BN}^{i} - \lambda_{NN}^{i}) + (\lambda_{NP}^{i} - \lambda_{BP}^{i})}. \tag{29}$$

3.6. Obtaining Three-Way Decision Rules

The three-way decisions rules are as follows:

(P1) If $\Pr(A|T_i) \geq \alpha_i$, decide $T_i \in POS(A)$, which means that the target threat level is high and there is a need to attack or interfere first;

(B1) If $\beta_i < \Pr(A|T_i) < \alpha_i$, decide $T_i \in BND(A)$, which means that the target needs more situation information to be analyzed;

(N1) If $\Pr(A|T_i) \leq \beta_i$, decide $T_i \in NEG(A)$, which means that the target threat level is low and there is not a need to attack or interfere first.

4. Case Study

The case is from [6], which is about dynamic threat assessment of an unmanned aerial vehicle (UAV) swarm against ground targets. Assume there are three UAVs in the swarm, four ground targets in the combat area, six evaluation attributes, and three moments' information, i.e., $T = \{T_1, T_2, T_3, T_4\}$, $A = \{A_1, A_2, A_3, A_4, A_5, A_6\}$, and $t = \{t_1, t_2, t_3\}$. More specifically, the six evaluation attributes are the number of fire units, reliability, viability, searching ability, damage ability, and anti-jamming ability.

The evaluation process in study [6] is divided into consensus process and selection process. For the multiple UAV consensus-reaching process, we do not need to pay attention to it. We just compare the threat assessment method used in its selection process with our method. As for dynamic heterogeneous information processing, the two papers are similar.

4.1. Three-Way Threat Assessment Based on Heterogeneous Information Processing

The comprehensive evaluation matrix $MS\widetilde{V}$ of study [6] is exactly the comprehensive evaluation information matrix $\widetilde{\mathbf{Z}} = \left(\widetilde{z}_{ij}\right)_{m\times n}$ of this paper, as listed in Table 4.

Table 4. Comprehensive evaluation information matrix.

	A_1	A_2	A_3	A_4	A_5	A_6
T_1	0.473	0.961	[0.836, 0.862]	[0.830, 0.846]	[0.802, 0.828, 0.862]	[0.913, 0.936, 0.948, 0.970]
T_2	1.000	0.973	[0.841, 0.870]	[0.941, 0.956]	[0.949, 0.971, 0.983]	[0.838, 0.861, 0.877, 0.900]
T_3	0.622	0.950	[0.871, 0.910]	[0.984, 0.988]	[0.833, 0.850, 0.855]	[0.798, 0.828, 0.856, 0.869]
T_4	0.342	0.983	[0.921, 0.941]	[0.811, 0.847]	[0.814, 0.855, 0.863]	[0.842, 0.863, 0.879, 0.887]

The key steps are as follows:

(1) Based on Formulas (10)–(14), we can obtain the conflictual relationship of evaluation attributes.

$$R = (1.5956\ 1.0910\ 1.4764\ 0.9230\ 0.9377\ \ 1.2690);$$

then, the weight vector of evaluation attributes is calculated as $W = (0.3745\ 0.0127\ 0.0939\ 0.1382\ 0.1715\ 0.2092)$.

(2) Based on Formulas (15)–(22) of heterogeneous weighted TOPSIS, we can obtain z^+ and z^-, which are calculated as

$$Z^+ = (1.000\ 0.9830\ [0.9210\ 0.9410][0.9840\ 0.9880]\ [0.9490\ 0.9710\ 0.9830]\lfloor 0.9130\ 0.9360\ 0.9480\ 0.9700\rfloor),$$

$$Z^- = (0.3420\ 0.9500\ [0.8360\ 0.8620][0.8110\ 0.8460][0.8020\ 0.8280\ 0.8550][0.7980\ 0.8280\ 0.8560\ 0.8690]).$$

Then, the conditional probabilities of the target are calculated as

$$\Pr(A|T) = (0.3015\ 0.8149\ 0.3558\ 0.1057).$$

(3) The adaptive risk avoidance coefficients can be calculated by Formula (24), expressed as

$$\sigma\left(\widetilde{Z}\right) = \begin{pmatrix} 0.5000 & 0.5000 & 0.3421 & 0.3125 & 0 & 0.2692 \\ 0.5000 & 0.5000 & 0.2632 & 0.3281 & 0.3421 & 0.1346 \\ 0.5000 & 0.5000 & 0 & 0.5000 & 0.5000 & 0 \\ 0.5000 & 0.5000 & 0.5000 & 0 & 0.1447 & 0.5000 \end{pmatrix}.$$

(4) Based on Formulas (25)–(27), the comprehensive loss function matrix of each target based on multi-attribute information is listed in Table 5.

Table 5. Comprehensive loss function matrix.

		A	$\neg A$
	a_P	0	0.1656
T_1	a_B	0.1881	0.0184
	a_N	3.2251	0
	a_P	0	0.0284
T_2	a_B	0.3480	0.0011
	a_N	4.1386	0
	a_P	0	0.1456
T_3	a_B	0.2270	0.0127
	a_N	3.3164	0
	a_P	0	0.2478
T_4	a_B	0.3405	0.0387
	a_N	2.9068	0

(5) We can further calculate the decision thresholds based on Table 5 via Formulas (28) and (29). The results of each target are shown in Table 6.

Table 6. Conditional probability and decision thresholds.

	T_1	T_2	T_3	T_4
α_i	0.4389	0.0727	0.3693	0.3804
β_i	0.0060	0.0003	0.0041	0.0149
$\Pr(A\vert T_i)$	0.3015	0.8149	0.3558	0.1057

(6) From Table 6, we can obtain the ranking results based on conditional probability, i.e., $T_2 \succ T_3 \succ T_1 \succ T_4$. We can further obtain the classification results based on decision rules P(1)–N(1): $POS(A) = \{T_2\}$ and $BND(A) = \{T_1, T_3, T_4\}$. They imply that we should attack or interfere with T_2 first and need more information to analyze T_1, T_3, and T_4.

Here, the comparison of the proposed method with study [6] is shown in Figure 3. For comparison, the threat degree of our method is represented by $\Pr(A|T_i)$ and the threat degree from study [6] is converted to

$$Threat(T_i) = 1 - Q_i \tag{30}$$

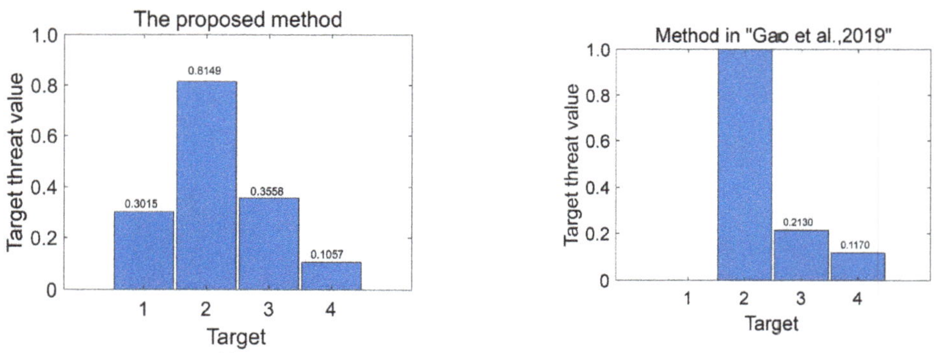

(a) (b)

Figure 3. The results of the proposed method (**a**) and method in [6] (**b**).

From Figure 3, we can see that the ranking results are basically the same. They all agree that T_2 and T_3 have the highest and second highest threat levels. For the difference between T_1 and T_4, which is due to the difference in the calculation of attribute weights, this paper uses a data-based objective weighting method, whereas the AHP used in [6] is based on subjective judgement. The weight vector in [6] is $\boldsymbol{W} = (0.041\ 0.138\ 0.227\ 0.158\ 0.347\ 0.089)$.

In order to avoid the decision conflict caused by different methods, this paper introduces heterogeneous weighted TOPSIS into the multi-target threat assessment method, which can obtain the threat ranking along with the threat classification. With the above heterogeneous information processing, T_1 and T_4 are in the $BND(A)$, which means that more information is needed to assess whether priority strikes or interference are required.

4.2. Analysis of Attribute Relevance

The attribute relevance is considered by the weight calculation and the aggregation of relative loss functions under multiple attributes, i.e., the heterogeneous CRITIC and weighted HM operators are used, respectively.

In order to analyze the advantages of the correlation processing, we compare the classification results of the proposed method with those methods without an HM operator, without CRITIC, etc. The specific methods are denoted and described as follows:

TH1: Instead of using the HM operator in our method, the weighted average operator is used.

TH2: Instead of using the HM operator and CRITIC in our method, the weighted average operator and equal weights are used.

TH3: Instead of using the HM operator and CRITIC in our method, the weighted average operator and weights from [6] are used.

TH4: Instead of using the CRITIC in our method, the weights from [6] are used.

The differences among these methods are listed in Table 7.

Table 7. The differences among evaluation methods.

Method	Weight Calculation	Information Aggregation
Proposed method	Heterogeneous CRITIC	Weighted HM
TH1	Heterogeneous CRITIC	Weighted average operator
TH2	Equal weights	Weighted average operator
TH3	AHP in [6]	Weighted average operator
TH4	AHP in [6]	Weighted HM

The conditional probability of targets under different methods is listed in Table 8.

Table 8. The conditional probability.

Method/Pr($A \mid T_i$)	T_1	T_2	T_3	T_4
Proposed method	0.3015	0.8149	0.3558	0.1057
TH1	0.3015	0.8149	0.3558	0.1057
TH2	0.2914	0.7360	0.3648	0.1821
TH3	0.1642	0.7552	0.3276	0.2293
TH4	0.1642	0.7552	0.3276	0.2293

The threat degree of TH1 is the same as the proposed method, and the threat degrees of TH2 and TH3 (TH4) are shown in Figure 4.

The comprehensive decision thresholds of different methods are listed in Table 9.

Accordingly, the $POS(A)$, $BND(A)$, and $NEG(A)$ of different methods are listed in Table 10.

In order to analyze the effectiveness of the attribute relevance, the above methods are discussed through the ranking results, the classification results, the relative magnitude of the threat degree of the target, and the relative magnitude of the decision domains. The ranking and classification results directly reflect the output of the methods. The relative magnitude of the threat degree of the targets can help analyze the reasonableness of target classification. The relative magnitude of the decision domains is obtained through the decision thresholds, representing the probability of classifying the target into this domain.

For a target with a large threat degree, the larger the positive domain is, the more beneficial it is to divide it into the positive region, and the more reasonable it is.

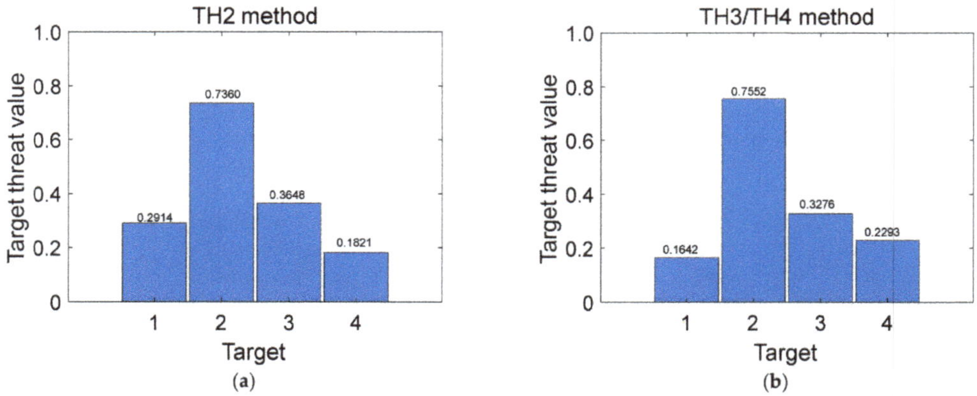

Figure 4. The results of TH2 method (**a**) and TH3/TH4 method (**b**).

Table 9. Comprehensive decision thresholds.

Method	Thresholds	T_1	T_2	T_3	T_4
Proposed method	α_i	0.4389	0.0727	0.3693	0.3804
	β_i	0.0060	0.0003	0.0041	0.0149
TH1	α_i	0.3847	0.1872	0.3608	0.3403
	β_i	0.0906	0.0220	0.0773	0.1003
TH2	α_i	0.3278	0.1879	0.3038	0.2766
	β_i	0.0627	0.0262	0.0532	0.0936
TH3	α_i	0.4282	0.1551	0.2296	0.3030
	β_i	0.0348	0.0273	0.0636	0.0505
TH4	α_i	0.4416	0.0537	0.1464	0.2351
	β_i	0.0009	0.0004	0.0023	0.0016

Table 10. Decision domains of different methods.

Method	*POS*(*A*)	*BND*(*A*)	*NEG*(*A*)
Proposed method	$\{T_2\}$	$\{T_1, T_3, T_4\}$	$\varnothing$
TH1	$\{T_2\}$	$\{T_1, T_3, T_4\}$	$\varnothing$
TH2	$\{T_2, T_3\}$	$\{T_1, T_4\}$	$\varnothing$
TH3	$\{T_2, T_3\}$	$\{T_1, T_4\}$	$\varnothing$
TH4	$\{T_2, T_3\}$	$\{T_1, T_4\}$	$\varnothing$

More specifically, combining the analysis in Section 4.1 and the results in Table 9, we can see that

(1) T_1 and T_4 receive different ranking results under different decision-making methods; thus, they need more information to analyze. All the methods in Table 8 put them in the boundary domain. The results of the calculations are compatible with the theoretical analyses.

(2) From Figures 3 and 4, in terms of the relative magnitude of the target's threat degree, the relative difference between T_1 and T_3, expressed as $\Delta_{Threat}(T_3, T_1)$, is smaller than the relative difference between T_1 and T_4, expressed as $\Delta_{Threat}(T_1, T_4)$. When both T_1 and T_4 belong to the boundary domain, it is more reasonable that T_1 also belongs to the boundary domain. Therefore, the proposed method and the results of TH1 are more

reasonable. The reason why their methods are more reasonable is that their attribute weights are calculated by heterogeneous CRITIC.

(3) The relative magnitudes of decision domains under different methods are shown in Figure 5. We can further compare the proposed method with the TH1 method, both of which use heterogeneous CRITIC. However, for T_2, which has the significantly highest threat degree, the proposed method has a smaller boundary domain and a higher discrimination degree. This makes it easier to determine T_2 as the priority target in the proposed method. This effect is due to the further use of the HM operator in the proposed method.

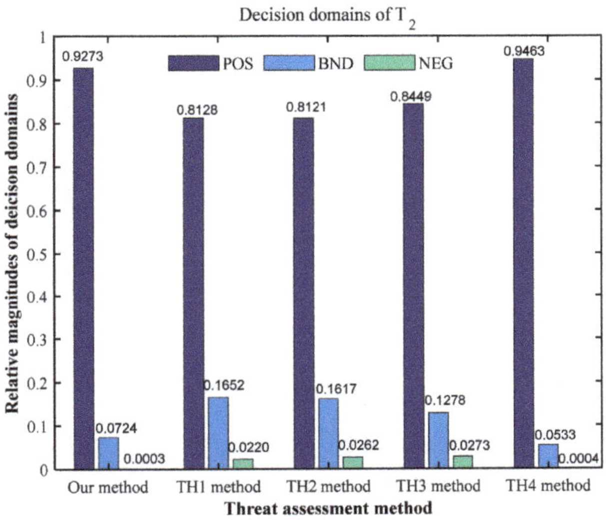

Figure 5. The relative magnitude of decision domains of T_2.

Combining the above analyses, the proposed method can achieve more reasonable three-way classification results by considering the attribute relevance.

4.3. Analysis of Risk Avoidance Coefficient

The risk avoidance coefficients are used in the construction of the relative loss function matrices. Compared with existing methods, this paper exploits adaptive risk avoidance coefficients, which can directly be calculated according to the uncertainty of the assessed values.

Based on the related content of Section 3.5, the adaptive risk avoidance coefficient curves via Formula (24) for different attributes are shown in Figure 6. Since the values of A_1 and A_2 are real numbers, we consider that there is no uncertainty. Thus, the curves in Figure 6 are only for A_3, A_4, A_5, and A_6.

As we can see from Figure 6, the corresponding risk avoidance coefficients can be obtained directly based on the uncertainty of the assessed attribute values. Table 11 gives the comparison between the proposed method and the existing threat assessment methods on setting the risk avoidance coefficient.

It can be seen from Table 10 that, in the construction of risk avoidance coefficients, existing methods usually preset or subjectively set the risk avoidance coefficients. And for the convenience of calculation, they set the same coefficient for each evaluation value. This is not in line with the actual situation. Accordingly, the advantages of constructing it objectively in this paper are as follows:

(1) As mentioned in Remark 4, the risk avoidance coefficient reflects the acquisition of information. Each assessment attribute has a different source, and it is not reasonable to set its risk avoidance coefficient as the same value. We set it via the uncertainty of information, which is more in line with the reality.

(2) The risk avoidance coefficients are calculated directly and do not need to be set by a human. Timeliness and objectivity of target threat assessment can be guaranteed.

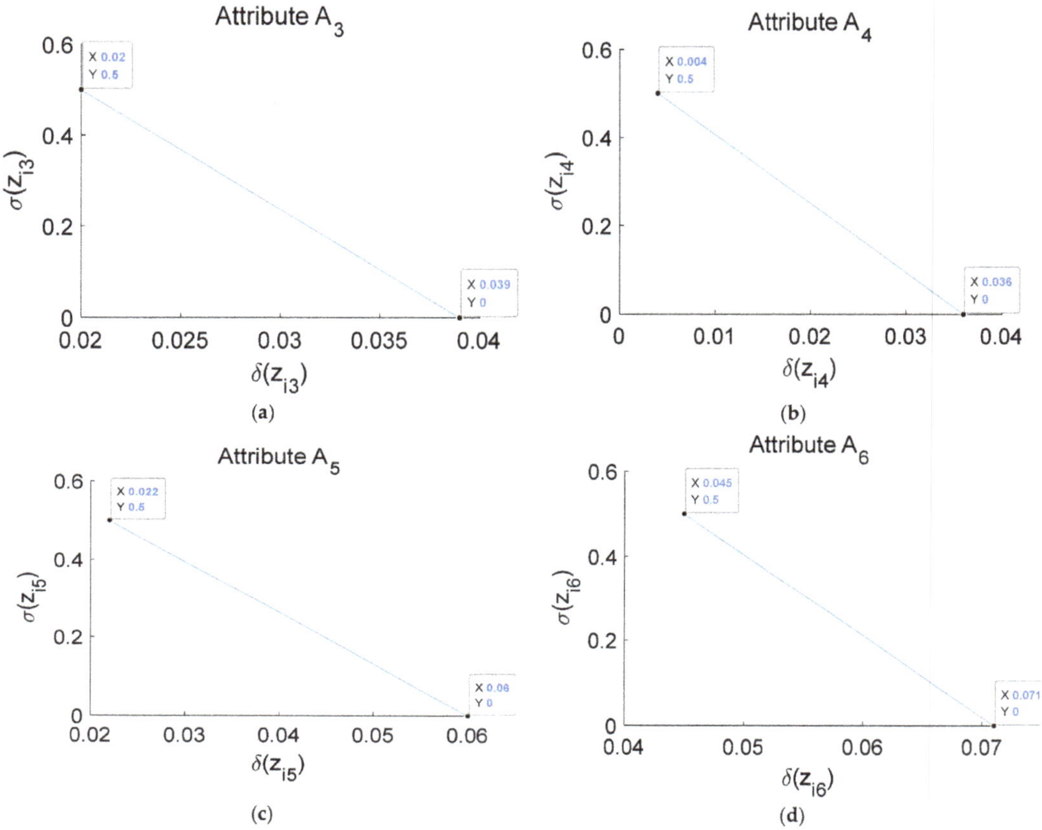

Figure 6. Adaptive risk avoidance threshold curves of A_3 (**a**), A_4 (**b**), A_5 (**c**), and A_6 (**d**).

Table 11. The comparison among different methods.

Method	Are There Risk Avoidance Coefficients Involved?	How to Calculate Risk Avoidance Coefficients?	Are All Risk Avoidance Coefficients the Same?
Gao et al. [25]	Yes	Subjectively setting	Yes
Gao et al. [26]	Yes	Subjectively setting	Yes
Yin et al. [27]	Yes	Subjectively setting	Yes
Li et al. [29]	Yes	Subjectively setting	Yes
Peng et al. [31]	Yes	Subjectively setting	Yes
Our method	Yes	Objectively calculating	No

Remark 5. *The mainstream trend of multi-target three-way threat assessment methods is to construct a relative loss function matrix through risk avoidance coefficients and ultimately calculate decision thresholds. However, there are still some methods that do not use risk avoidance coefficients, such as subjectively constructing loss function matrices, which are not within the scope of discussion.*

5. Conclusions

For the problem of multi-target threat assessment with heterogeneous information and attribute relevance, we propose a new multi-target three-way threat assessment method. First, the dynamic assessment information is represented by heterogenous forms. The comprehensive evaluation information matrix can be obtained by the normalization and aggregation. Based on the comprehensive evaluation information matrix, attribute weights are calculated by heterogeneous CRITIC. The conditional probability is calculated by the heterogeneous weighted TOPSIS. Then, the adaptive risk avoidance coefficients are constructed by the uncertainty of the assessment value, and the relative loss function matrices are constructed. Subsequently, the comprehensive loss function matrices are obtained by the weighted HM operator. The three-way decision rules are obtained via decision thresholds. The case study shows that the proposed method can effectively handle the heterogeneous information and attribute relevance, which is more suitable for the combat environment. Compared with existing methods, this study has the following features and benefits:

(1) It expands the research of three-way decision and target threat assessment. In particular, both the heterogeneity and the relevance of information have been considered in target threat assessment. This is rare in the study of existing multi-target threat assessment based on three-way decision. Therefore, this study is more in line with an actual combat mission environment.

(2) For the representation and processing of heterogeneous information, there is no loss of information. Neither the calculation of conditional probabilities by heterogeneous weighted TOPSIS nor the calculation of weights by heterogeneous CRICTIC involves the conversion of heterogeneous information formats. Whereas in some of the existing studies, the conversion of heterogeneous information into the same format may result in information loss.

(3) The treatment of attribute correlation is relatively comprehensive and includes both weight calculation and information aggregation. The proposed method considers the attribute relevance in terms of both in the weights' calculation and in the aggregation of relative loss function matrices, which is rarely considered by the other three-way threat assessment methods. The consideration of attribute relevance makes the results of target threat classification more reasonable and credible.

(4) The adaptive risk avoidance coefficients can be calculated based on the uncertainty of the attribute information. Compared with other methods in which the risk avoidance coefficients are set subjectively, it is more reasonable and effective, and avoids subjective experience limitations. It can meet the timeliness need for a wartime decision.

However, despite the above-mentioned advantages, there are still some issues that need further investigation. First, the diversity of heterogeneous information representations deserves further study. For example, heterogeneous information is represented by linguistic variables, hesitant fuzzy numbers, etc. Second, when the decision makers are groups, how can the method of this paper be generalized to three-way group decision making [55]? Finally, adaptive risk aversion function curves can be optimized in conjunction with human psychological decision theories such as regret theory [56].

Author Contributions: Conceptualization, Methodology, Writing—original draft, Writing—review and editing, Y.G.; Investigation, N.L. All authors have read and agreed to the published version of the manuscript.

Funding: This work was supported by Project funded by China Postdoctoral Science Foundation (2021M693940).

Data Availability Statement: Data are contained within the article.

Acknowledgments: Support by colleagues and the university is acknowledged.

Conflicts of Interest: The authors declare no conflicts of interest.

References

1. Lives, I.V.; Joseph, M.; Robbins, M.J.; Lunday, B.J. Solving nonstationary Markov decision processes via contextual decomposition: A military air battle management application. *Eng. Appl. Artif. Intell.* **2023**, *233*, 120949.
2. Nalin, A.; Tripodi, P. Future warfare and responsibility management in the AI-based military decision-making process. *J. Adv. Mil. Stud.* **2023**, *14*, 83–97. [CrossRef]
3. Li, Y.F.; Shi, J.P.; Jiang, W.; Zhang, W.G.; Lyu, Y.X. Autonomous maneuver decision-making for a UCAV in short-range aerial combat based on an MS-DDQN algorithm. *Def. Technol.* **2022**, *18*, 1697–1714. [CrossRef]
4. Li, S.; He, X.Y.; Xu, X.; Zhao, T.; Song, C.Y.; Li, J.B. Weapon-target assignment strategy in joint combat decision-making based on multi-head deep reinforcement learning. *IEEE Access* **2023**, *11*, 113740–113751. [CrossRef]
5. Barlas, M.A.; Gözde, H.; Özden, S. Modelling and optimization of UHF-band EW base WTA problem within the scope of threat assessment. *Def. Sci. J.* **2022**, *72*, 73–82. [CrossRef]
6. Gao, Y.; Li, D.S. Consensus evaluation method of multi-ground-target threat for unmanned aerial vehicle swarm based on heterogeneous group decision making. *Comput. Electr. Eng.* **2019**, *74*, 223–232. [CrossRef]
7. Li, S.Y.; Chen, M.; Wang, Y.H.; Wu, Q.X. Air combat decision-making of multiple UCAVs based on constraint strategy games. *Def. Technol.* **2022**, *18*, 368–383. [CrossRef]
8. Zhang, Y.; Xiao, Q.L.; Deng, X.Y.; Jiang, W. A multi-source information fusion method for ship target recognition based on Bayesian inference and evidence theory. *J. Intell. Fuzzy Syst.* **2022**, *42*, 2331–2346. [CrossRef]
9. Luo, R.N.; Huang, S.C.; Zhao, Y.; Song, Y.F. Threat assessment method of low altitude slow small (LSS) targets based on information entropy and AHP. *Entropy* **2021**, *23*, 1292. [CrossRef]
10. Ma, S.D.; Zhang, H.Z.; Yang, G.Q. Target threat level assessment based on cloud model under fuzzy and uncertain conditions air combat simulation. *Aerosp. Sci. Technol.* **2017**, *67*, 49–53. [CrossRef]
11. Wang, Y.; Liu, S.Y.; Niu, W. Threat assessment method based on intuitionistic fuzzy similarity measurement reasoning with orientation. *China Commun.* **2014**, *6*, 119–128.
12. Yu, X.J.; Wei, S.; Fang, Y.Y.; Sheng, J.L.; Zhang, L. Low altitude slow small target threat assessment algorithm by exploiting sequential multifeatured with long short-term memory. *IEEE Sens. J.* **2023**, *23*, 21524–21533. [CrossRef]
13. Fan, C.; Fu, Q.; Song, Y.; Lu, Y.; Li, W.; Zhu, X. A newmodel of interval-valued intuitionistic fuzzy weighted operators and their application in dynamic fusion target threat assessment. *Entropy* **2022**, *24*, 1825. [CrossRef] [PubMed]
14. Zhao, R.J.; Yang, F.B.; Ji, L.N. An extended fuzzy CPT-TODIM model based on possibility theory and its application to air target dynamic threat assessment. *IEEE Access* **2022**, *10*, 21655–21669. [CrossRef]
15. Kong, D.P.; Chang, T.Q.; Wang, Q.D. A threat assessment method of group targets based on interval-valued intuitionistic fuzzy multi-attribute group decision-making. *Appl. Soft. Comput.* **2018**, *67*, 350–369. [CrossRef]
16. Zhang, Q.; Hu, J.H.; Feng, J.F. Air multi-target threat assessment method based on improved GGIFSS. *J. Intell. Fuzzy Syst.* **2019**, *36*, 4127–4139.
17. Zhang, K.; Kong, W.R.; Liu, P.P. Assessment and sequencing of air target threat based on intuitionistic fuzzy entropy and dynamic VIKOR. *J. Syst. Eng. Electron.* **2018**, *29*, 305–310. [CrossRef]
18. Yao, Y.Y. The Dao of three-way decision and three-world thinking. *Int. J. Approx. Reason.* **2023**, *162*, 109032. [CrossRef]
19. Zhang, C.; Zhang, M.; Yang, G.; Xue, T.; Zhang, Z.; Liu, L.; Wang, L.; Hou, W.; Chen, Z. Three-way selection random forest optimization model for anomaly traffic Detection. *Electronics* **2023**, *12*, 1788. [CrossRef]
20. Chen, X.; Zou, L. Three-way decision models based on ideal relations in multi-attribute decision-making. *Entropy* **2022**, *24*, 986. [CrossRef] [PubMed]
21. Wang, H.; Cui, Z.W.; Liu, R.G.; Fang, L.; Sha, Y. A multi-type transferable method for missing link prediction in Heterogeneous social networks. *IEEE Trans. Knowl. Data Eng.* **2023**, *35*, 10981. [CrossRef]
22. Song, J.K.; He, Z.G.; Jiang, L.N.; Liu, Z.C.; Leng, X.L. Research on hybrid multi-attribute three-way group decision making based on improved VIKOR model. *Mathematics* **2022**, *10*, 2783. [CrossRef]
23. Jia, F.; Wang, Y.; Su, Y. GRA-based dynamic hybrid multi-attribute decision-making for the performance evaluation of elderly-care services. *Mathematics* **2023**, *11*, 3176. [CrossRef]
24. Ali, A.; Rehman, N.; Ali, M.; Hila, K. A novel approach to three-way decision model under fuzzy soft dominance degree relations and emergency situation. *Expert Syst. Appl.* **2024**, *239*, 122369. [CrossRef]
25. Blundo, C.; Loia, V.; Orciuoli, F. A time-aware approach for MOOC dropout prediction based on rule induction and sequential three-way decisions. *IEEE Access* **2023**, *11*, 113189–113198. [CrossRef]
26. Gao, Y.; Li, D.S.; Zhong, H. A novel target threat assessment method based on three-way decisions under intuitionistic fuzzy multi-attribute decision making environment. *Eng. Appl. Artif. Intell.* **2020**, *87*, 103276. [CrossRef]
27. Gao, Y.; Huang, Y.C.; Cheng, G.B.; Duan, L. Multi-target threat assessment method based on VIKOR and three-way decisions under intuitionistic fuzzy information. *Acta Electron. Sin.* **2021**, *49*, 542–549.
28. Yin, T.F.; Zhang, R.T.; Su, Q.R. Threat assessment of aerial targets based on improved GRA-TOPSIS method and three-way decisions. *Math. Biosci. Eng.* **2023**, *20*, 13250–13266. [CrossRef] [PubMed]
29. Li, B.; Tian, L.Y.; Han, Y.; Chen, D.Q. Three-way decision of target threat decision making based on adaptive threshold algorithms. *J. Eng.* **2020**, *13*, 293–297. [CrossRef]

30. Li, W.H.; Gao, P.X.; Chen, J.; Lu, Y.Q. Unmanned aerial vehicle situation assessment based on cumulative prospect theory and three-way decision. *J. ShangHai Jiao Tong Univ.* **2022**, *56*, 1479–1490.
31. Li, T.C.; Ye, J.; Lu, K. Air combat threat assessment based on improved three-way decision-Topsis method. *J. Gun Launch Control* **2022**, *43*, 1–8, 20.
32. Peng, L.; Zhang, T.; Zhang, X. Threat assessment for aerial targets based on three-way multi-criteria decision making. In Proceedings of the 2021 IEEE International Conference on Networking, Sensing and Control (ICNSC), Xiamen, China, 3–5 December 2021; Volume 1, pp. 1–6.
33. Tan, R.P.; Zhang, W.D.; Yang, L.H. Decision-making method based on set pair analysis and VIKOR under heterogeneous information environment and application to typhoon disaster assessment. *Soft. Comput.* **2023**, *27*, 8289–8314. [CrossRef]
34. Wu, J.M.; Huang, Z.H.; Li, J.J.; Liu, D.Y. Three-way multi-attribute decision-making with multiple decision makers in heterogeneous incomplete decision systems. *J. Amb. Intel. Hum. Com.* **2023**, *14*, 7469–7483. [CrossRef]
35. Liang, D.C.; Wang, M.W.; Xu, Z.S. Heterogeneous multi-attribute nonadditivity fusion for behavioral three-way decisions in interval type-2 fuzzy environment. *Inf. Sci.* **2019**, *496*, 242–263. [CrossRef]
36. Li, G.; Kcou, G.; Peng, Y. A group decision making model for integrating heterogeneous information. *IEEE Trans. Syst. Man Cybern. Syst.* **2018**, *48*, 982–992. [CrossRef]
37. Mondal, A.; Roy, S.K. Behavioural three-way decision making with Fermatean fuzzy Mahalanobis distance: Application to the supply chain management problems. *Appl. Soft Comput.* **2024**, *151*, 111182. [CrossRef]
38. Modal, A.; Roy, S.K.; Deveci, M. Regret-based domination and prospect-based scoring in three-way decision making using q-rung orthopair fuzzy Mahalanobis distance. *Artif. Intell. Rev.* **2023**, *56*, S2311–S2348. [CrossRef]
39. Diakoulaki, D.; Mavrotas, G.; Papayannakis, L. Determining objective weights in multiple criteria problems: The CRITIC method. *Comput. Oper. Res.* **1995**, *22*, 763–770. [CrossRef]
40. Kumar, K.; Chen, S.M. Multi-attribute decision making based on q-ring orthopair fuzzy Yager prioritized weighted arithmetic aggregation operator of q-rung orthopair fuzzy numbers. *Inf. Sci.* **2024**, *657*, 119984. [CrossRef]
41. Chakraborty, S. TOPSIS and modified TOPSIS: A comprehensive analysis. *Decis. Anal. J.* **2022**, *2*, 100021. [CrossRef]
42. Zhu, G.N.; Ma, J.; Hu, J. A fuzzy rough number extended AHP and VIKOR for failure mode and effects analysis under uncertainty. *Adv. Eng. Inform.* **2022**, *51*, 101454. [CrossRef]
43. Qin, H.W.; Peng, Q.W.; Ma, X.Q. A novel interval-valued Fermatean fuzzy three-way decision making method with probability dominance relations. *Expert Syst. Appl.* **2024**, *242*, 122727. [CrossRef]
44. Ye, J.; Sun, B.Z.; Bai, J.C.; Bao, Q.; Chu, X.L.; Bao, K. A preference-approval structure-based non-additive three-way group consensus decision-making approach for medical diagnosis. *Inform. Fusion* **2024**, *101*, 102008. [CrossRef]
45. Xu, T.; Luo, F. Interval-value based movement strategy of three-way decisions. *Inform. Sci.* **2024**, *652*, 119761. [CrossRef]
46. Yang, X.; Li, Y.H.; Li, T.R. A review of sequential three-way decision and multi-granularity learning. *Int. J. Approx. Reason.* **2023**, *152*, 414–433. [CrossRef]
47. Yin, L.J.; Zhang, Q.H.; Zhao, F.; Liu, D.; Wang, G.Y. Superiority of three-way decisions from the perspective of probability. *Artif. Intell. Rev.* **2023**, *56*, 1263–1295. [CrossRef]
48. Yang, Y.Y. Three-way decision: An interpretation of rules in rough set theory. *Rough Sets Knowl. Technol.* **2009**, *5589*, 642–649.
49. Liang, D.C.; Xu, Z.S.; Liu, D. Method for three-way decisions using ideal TOPSIS solutions at Pythagorean fuzzy information. *Inf. Sci.* **2018**, *435*, 282–295. [CrossRef]
50. Jia, F.; Liu, P.D. A novel three-way decision model under multiple-criteria environment. *Inf. Sci.* **2019**, *471*, 29–51. [CrossRef]
51. Jia, F.; Liu, P.D. Multi-attribute three-way decisions based on ideal solutions under interval-valued intuitionistic fuzzy environment. *Int. J. Approx. Reason.* **2021**, *138*, 12–37. [CrossRef]
52. Modal, A.; Roy, S.K.; Pamucar, D. Regret-based three-way decision making with possibility dominance and SPA theory in incomplete information system. *Expert Syst. Appl.* **2023**, *211*, 118688. [CrossRef]
53. He, X.; Chang, J.; Liu, W. Harmonic averaging operators of four parameters interval numbers and applications to decision making. *Fuzzy Syst. Math.* **2017**, *31*, 135–143.
54. Gao, Y.; Li, D.S. UAV swarm cooperative situation perception consensus evaluation method based on three-parameter interval number and Heronian mean operator. *IEEE Access* **2018**, *6*, 73328–73340. [CrossRef]
55. Modal, A.; Roy, S.K.; Zhan, J.M. A reliability-based consensus model and regret theory-based selection process for linguistic hesitant-Z multi-attribute group decision making. *Expert Syst. Appl.* **2023**, *228*, 120431. [CrossRef]
56. Lei, W.J.; Ma, W.M.; Li, X.N.; Sun, B.Z. Three-way group decision based on regret theory under dual hesitant fuzzy environment: An application in water supply alternatives selection. *Expert Syst. Appl.* **2024**, *237*, 121249. [CrossRef]

mathematics

MDPI

Article

Consistency Improvement in the Analytic Hierarchy Process

Valerio Antonio Pamplona Salomon [1,*] and Luiz Flavio Autran Monteiro Gomes [2]

[1] Department of Production, UNESP–Sao Paulo State University, Guaratingueta 12516-410, SP, Brazil
[2] Pro-Rectory of Research, Federal University of ABC, Santo André 09210-580, SP, Brazil; luiz.autran@ufabc.edu.br
* Correspondence: valerio.salomon@unesp.br

Abstract: Consistency checking is one of the reasons for the Analytic Hierarchy Process (AHP) leadership in publications on multiple criteria decision-making (MCDM). Consistency is a measure of the quality of data input in the AHP. The theory of AHP provides indicators for the consistency of data. When an indicator is out of the desired interval, the data must be reviewed. This article presents a method for improving the consistency of reviewing the data input in an AHP application. First, a conventional literature review is presented on the theme. Then, an innovative tool of artificial intelligence is shown to confirm the main result of the conventional review: this topic is still attracting interest from AHP and MCDM researchers. Finally, a simple technique for consistency improvement is presented and illustrated with a practical case of MCDM: supplier selection by a company.

Keywords: AHP; consistency; MCDM; supplier selection

MSC: 91B06

1. Introduction

The multiple criteria decision-making (MCDM) approach contributes to decision-making in situations where multiple alternatives must be evaluated considering multiple criteria [1]. The MCDM is a methodology, a collection of methods developed from the 1960s to solve decision problems [2]. This article is focused on the Analytic Hierarchy Process (AHP), a leading MCDM method for decades [3–5]. One main reason for the AHP's leadership in publications on MCDM is its solid mathematical foundation [6]. The AHP's fundamentals provide a ground for research and development of this MCDM method. The AHP theory and practice have "seven pillars", which include the following [7]:

1. Ratio scales derived from reciprocal pairwise comparisons.
2. Pairwise comparisons and the 1–9 Saaty Scale.
3. Sensitivity of the eigenvector to judgments.
4. Extending the scale from 1 to 9 to 1–$\mathbb{R}$.
5. Additive synthesis of priorities.
6. Rank preservation or rank reversal.
7. Group decision-making with an aggregation of individual judgments or priorities.

Another main reason for the great number of AHP publications is the need to solve practical problems with a handy tool. AHP applications include the following [6,8]:

- Educational decisions: Admitting students and faculty selection.
- Financial and marketing decisions: Advertising, credit analysis, downsizing, project management, and resource allocation.
- Governmental or social decisions: Affirmative action, energy and fuel regulations, food and drug, and smoking policies.
- Human resources and personal decisions: Career choices, entrepreneurial development, performance evaluations, and human tracking.

- Sports decisions: Drafts, predictions, and salary cap.
- Supply chain decisions: Information technology, logistics, outsourcing, and supplier and vendor selection.

The pairwise comparison matrix A of a set of n objects is a central element in the AHP. Components of $A = [a_{ij}]$ represent w_i/w_j [9], where w is the vector of the weights for the compared objects $i = 1,2,3\ldots n$. Equation (1) presents one way to generate w from A:

$$A\,w = \lambda_{\max} w \quad (1)$$

where w is the right eigenvector of A, and $\lambda_{\max}$ is its maximum eigenvalue.

To be consistent means no change of mind. Consistency is "conformity with previous practice" [8]. A 100%-consistent pairwise comparison matrix A satisfies Equation (2):

$$a_{ij} = w_i/w_j \quad (2)$$

$\forall i,j = 1,2,3\ldots n$.

A consequence of the consistency of A is presented in Equation (3) [10]:

$$a_{ij} = \frac{w_i}{w_j} = \frac{w_i/w_k}{w_j/w_k} = \frac{a_{ik}}{a_{jk}}$$

$$a_{ik} = a_{ij}a_{jk} \quad (3)$$

$\forall i,j,k = 1,2,3\ldots n$.

In the AHP, pairwise comparisons are usually performed regarding a linear 1–9 scale, which is named the Saaty Scale here but is also named "The Scale" [11] or "Fundamental Scale of Absolute Numbers" [8]. With the Saaty Scale, A becomes a positive reciprocal matrix, satisfying conditions $a_{ij} > 0$ and $a_{ij} = 1/a_{ji}$, $\forall i,j = 1,2,3\ldots n$. A consequence of this positiveness and reciprocity is that $\lambda_{\max} \geq n$. A corollary from consistency is $\lambda_{\max} = n$ [11].

Despite some criticism and the proposal of different scales [12,13], the Saaty Scale prevails in AHP applications [14]. After all, the Saaty Scale allows for comparisons concerning weight dispersion and weight uncertainty [15]. Nevertheless, the use of the Saaty Scale does not guarantee that A will be a consistent matrix, satisfying Equations (2) and (3). In the example below, A, B, and C are all pairwise comparison matrices obtained with the Saaty Scale. However, only A is 100% consistent; B and C are not:

$$A = \begin{bmatrix} 1 & 3 & 9 \\ 1/3 & 1 & 3 \\ 1/9 & 1/3 & 1 \end{bmatrix},\quad B = \begin{bmatrix} 1 & 3 & 5 \\ 1/3 & 1 & 3 \\ 1/5 & 1/3 & 1 \end{bmatrix},\quad C = \begin{bmatrix} 1 & 7 & 3 \\ 1/7 & 1 & 3 \\ 1/3 & 1/3 & 1 \end{bmatrix}$$

The consistency of A is noted with $a_{12}a_{23} = 3 \times 3 = 9 = a_{13}$. The inconsistency of B and C is noted with $b_{12}b_{23} = 3 \times 3 \neq 5 = b_{13}$ and $c_{12}c_{23} = 7 \times 3 \neq 3 = c_{13}$. The eigenvalues for A, B, and C are $\lambda_{\max_A} = 3$, $\lambda_{\max_B} \approx 3.04$, and $\lambda_{\max_C} \approx 3.99$, respectively. The eigenvectors are $w_A \approx [0.69, 0.23, 0.08]$, $w_B \approx [0.64, 0.26, 0.10]$ and $w_C \approx [0.69, 0.19, 0.12]$. As A is 100% consistent, one question arises: By how much are B and C inconsistent matrices? Since $\lambda_{\max_B}$ and w_B are closer to $\lambda_{\max_A}$ and w_A than $\lambda_{\max_C}$ and w_C, it seems that B is less inconsistent than C. Therefore, Q1 and Q2 are two research questions:

Q1: How can we measure the consistency of a pairwise comparison matrix?
Q2: How can we improve the consistency of a pairwise comparison matrix?

To answer Q1 and Q2, this article presents a literature review on consistency measurement and consistency improvement (Section 2), with innovative support from artificial intelligence (AI) in Section 2.2. Then, a simple technique for consistency improvement is presented (Section 3) with a practical case of MCDM: a supplier selection by a manufacturing company (Section 4). Finally, Section 5 presents this article's conclusions and proposal for future research.

2. Literature Review

2.1. Background

Consistency and the Saaty Scale have been major subjects in AHP theory since the presentation of the seminal works [11,16,17]. The first document published on the AHP [16] introduced the Saaty Scale, with the former name "The Scale" but starting with zero being defined for "not comparable" when "there is no meaning to compare two objects". The document does not address the consistency measurement, focusing on obtaining the weights with the eigenvector.

The subsequent documents published on the AHP [11,18–22] updated the Saaty Scale, deleting the zero, as presented in Table 1:

Table 1. Saaty Scale [8,11,17–22].

Intensity of Importance	Definition	Explanation
1 [1]	Equal importance	The two compared objectives have the same importance
3	Moderate importance	Experience and judgment slightly favor one object over another
5	Strong importance	Experience and judgment strongly favor one object over another
7	Demonstrated importance	One object is very strongly favored and this dominance is demonstrated in practice
9	Absolute importance	The evidence favoring one object over another is of the highest possible order of affirmation
Reciprocals of above		If object i has one of the above nonzero numbers when compared to j, then j has the reciprocal value when compared to i
Rationals	Ratios arising from the scale	If consistency were to be forced or when measurements are available

[1] In two-object problems, one may use $1+\epsilon,\ 0<\epsilon\leq 1/2$, to indicate very slight dominance between two nearly equal objects.

Documents published previously, other than the AHP's eponymous book (Saaty, 1980) [17], average 65.3 citations, as presented in Table 2. The outlier is Saaty (1977) [11] with over 6000 citations, the most cited document on MCDM [6].

Table 2. Citations of the first published documents on the AHP.

Authorship (Year) [Reference]	Title of the Document	Citations
Saaty (1974) [16]	Measuring the fuzziness of sets	100
Saaty and Khouja (1976) [18]	A measure of world influence	41
Saaty and Rodgers (1976) [19]	Higher education in the United States (1985–2000). Scenario construction using a hierarchical framework with eigenvector weighting	76
Alexander and Saaty (1977) [20]	The forward and backward processes of conflict analysis	45
Saaty (1977) [11]	A scaling method for priorities in hierarchical structures	6636
Saaty (1977) [21]	Scenarios and priorities in transport planning: Application to Sudan	58
Saaty and Bennett (1977) [22]	A theory of analytical hierarchies applied to political candidacy	72

Source: www.scopus.com (accessed on 6 December 2023).

Saaty (1977) [11] introduced the consistency measurement, proposing the consistency index CI as in Equation (4):

$$CI = \frac{\lambda_{\max} - n}{n - 1} \tag{4}$$

If $\boldsymbol{A}$ is 100% consistent, then $\lambda_{\max} = n$ and $CI = 0$. In this case, Equations (2) and (3) are satisfied.

The consistency ratio CR is a better measure for the consistency of a comparison matrix since it compares CI with a random index RI obtained with the simulation of positive reciprocal matrices [23–25], as presented in Equation (5):

$$CR = \frac{CI}{RI} \tag{5}$$

Table 3 presents values for RI as a function of the matrix order n.

Table 3. Random consistency indexes.

n	Original	ORNL–PITT (1982)	EC–GWU (1990)	UU (1991)	Usual
3	0.416	0.58	0.52333	0.4887	0.52
4	0.851	0.90	0.88604	0.8045	0.89
5	1.115	1.12	1.10983	1.0591	1.11
6	1.150	1.24	1.25390	1.1797	1.25
7	1.345	1.32	1.34516	1.2519	1.35
8	1.334	1.41		1.3171	1.40
9	1.315	1.45		1.3733	1.45
10	1.420	1.49		1.4055	1.49
11	1.395	1.51		1.4213	1.51
12	1.482	1.48		1.4497	1.54
13	1.491	1.56		1.4643	1.56
14	1.470	1.57		1.4822	1.57
15	1.466	1.59		1.4969	1.58

In the AHP literature, RI values vary because they were obtained with different numbers of randomly simulated matrices. Originally, RI was obtained with 50 matrices for each n [11]. A study performed at the University of Pittsburgh (PITT) with support from the Oak Ridge National Laboratory (ORNL) increased the number of matrices to 500 [26]. A statistical experiment conducted at the George Washington University (GWU) with the Software Expert Choice (EC) experimented with incomplete matrices [27], increasing the number of simulated matrices to thousands. Perhaps the most accurate estimation for RI was performed in the University of Ulster (UU), Northern Ireland [28]. However, the usual values for RI are presented in the last column of Table 3. The usual values combine the ORNL–PITT values with EC–GWU: for $n \leq 7$, the usual values are the EC-CWU values rounded to hundredths; for $n > 7$, the usual values are the same for ORNL–PITT [8].

Table 4 presents values of CR for matrices $\mathbf{A}$, $\mathbf{B}$, and $\mathbf{C}$ (Section 1) for RI presented in Table 3.

Table 4. Consistency ratio values with different random consistency indexes.

Matrix	Original	ORNL–PITT	EC–GWU	UU	Usual
A	0	0	0	0	0
B	0.05	0.03	0.04	0.04	0.04
C	1.62	1.16	1.28	1.37	1.29

As $\lambda_{\max_A} = 3$, then $CI_A = 0$, resulting in $CR_A = 0$ for all RI values. This result is expected since $\mathbf{A}$ is a 100%-consistent matrix, satisfying Equations (2) and (3).

As $\lambda_{\max_B} \approx 3.04$, then $CI_B \approx 0.02$, making CR_B vary from 0.03 to 0.04. As $\lambda_{\max_C} \approx 3.99$, then $CI_C \approx 0.22$, making CR_C vary from 0.38 to 0.53. CR_B and CR_C are expected to be greater than zero, since $\mathbf{B}$ and $\mathbf{C}$ are not 100%-consistent matrices. However, $CR_C > CR_B$, indicating that $\mathbf{C}$ is more inconsistent than $\mathbf{B}$. The question is as follows: is the inconsistency of $\mathbf{B}$ or $\mathbf{C}$ acceptable? To answer this question, the 0.1 threshold was proposed [11].

The 0.1 threshold considers that the normalized values for w_i are from 0 to 1; the required order for RI was as small as 10% but not smaller than 1% because inconsistency itself is important, since "without it new knowledge that changes preferences cannot be admitted" [9]. Saaty [17] further suggested that for matrices of orders three and four, the thresholds could be 0.5 and 0.8, respectively [29]. For larger matrices, even a CR = 0.2 could be tolerated, but no more [30]. Other consistency indices were proposed, such as the geometrical consistency index [31]. In this article, the usual CI, CR, and its 0.1 threshold are adopted. This adoption is for an alignment with the original AHP theory and its usual practice.

Considering the 0.1 threshold, $\mathbf{B}$ is not 100% consistent, but it is an acceptable matrix, and $\mathbf{C}$ is an inconsistent unacceptable matrix. Then, the c_{ij} components of $\mathbf{C}$ must be revised to improve its consistency, or simply to increase CR_C.

One simple way to increase the CR of a comparison matrix is by comparing the differences between its components and the components of a 100%-consistent matrix. As the components with greater differences are more inconsistent with the others, these components are first suggested to be revised. The differences compose the deviations matrix $\bar{\mathbf{C}}$ as in Equation (6):

$$\bar{c}_{ij} = \left|c_{ij} - w_{C_i}/w_{C_j}\right| \tag{6}$$

$\forall i, j = 1, 2, 3 \ldots n$.

In our case, $\bar{C}$ is as follows:

$$\bar{C} = \begin{bmatrix} 0 & 3.34 & 2.74 \\ 0.13 & 0 & 1.43 \\ 0.16 & 0.30 & 0 \end{bmatrix}$$

As $c_{12} = 3.34$ is the greatest component of $\bar{C}$, it is suggested that it should be revised from $c_{12} = 7$ to $c'_{12} = w_1/w_2 \approx 0.69/0.19 \approx 3.66$, resulting in C':

$$C' = \begin{bmatrix} 1 & 3.66 & 3 \\ 0.27 & 1 & 3 \\ 1/3 & 1/3 & 1 \end{bmatrix}$$

$\lambda_{max_{C'}} \approx 3.64$ and $CR_{C'} \approx 0.17$. C' is less inconsistent than C, but the inconsistency of both matrices is unacceptable since CR_C and $CR_{C'}$ are greater than the 0.1 threshold.

With one more iteration, C'' is found:

$$C'' = \begin{bmatrix} 1 & 2.66 & 3 \\ 0.38 & 1 & 3 \\ 1/3 & 1/3 & 1 \end{bmatrix}$$

$\lambda_{max_{C''}} \approx 3.04$ and $CR_{C''} \approx 0.04$. Now, C'' is an acceptable pairwise comparison matrix with $CR_{C''} \approx 0.096$. The changes from C to C'' result in $w''_C \approx [0.58, 0.27, 0.16]$, different than the former w_c. Of course, this would need approval by the decision-maker or by whoever is in charge of making the comparisons.

The simple *A–B–C* example illustrates the concepts and variables of consistency as *CI* and *CR*. Section 3 presents a technique for consistency improvement in more complex cases with $n > 3$. Before it, the next subsection presents how consistency has been measured and analyzed in the more recent AHP literature.

2.2. Recent Literature on Consistency Measurement and Improvement

The literature on consistency measurement of pairwise comparison matrix is a major part of the AHP literature. Therefore, it has also been prolific in the literature since the 1970s. This section focuses on the last ten years: documents published from 2013. This is the focus of the new Scopus Database tool, its artificial intelligence (AI) tool.

Most literature reviews are based on two databases: Clarivate's Web of Science or Elsevier's Scopus [32]. Despite both databases having similar contents, Scopus was selected for this research because it is free through institutional access [5]. Despite expected similar contents between Scopus and Web of Science, a second reason to exclusively search Scopus was the uniformity of search characteristics, such as search strings. Finally, the third reason for choosing Scopus was its new AI tool (https://www.elsevier.com/products/scopus/scopus-ai, accessed on 6 December 2023). Still in a beta phase, this tool allows for focusing on publications from recent years.

The question of "How to measure the consistency for a pairwise comparison matrix?" in the Scopus AI tool resulted in four key insights from the abstracts:

1. Inconsistency reduction: Various iterative and non-iterative algorithms have been developed to reduce inconsistency in pairwise comparison matrices [33].
2. Inconsistency indices: Different inconsistency indices have been proposed to measure the deviation from a consistent matrix, such as Koczkodaj's inconsistency index, Saaty's inconsistency index, geometric inconsistency index, and logarithmic Manhattan distance [34–36].
3. New measures: Some studies have introduced new inconsistency measures for incomplete pairwise comparison matrices and interval pairwise comparison matrices [36,37].
4. Comparative analysis: Comparative analyses have been conducted to evaluate the performance of different inconsistency indices using Monte Carlo simulations [33,37].

Scopus AI concludes that "there are several methods and indices available to measure the consistency of pairwise comparison matrices, and their effectiveness can be evaluated through comparative analyses and simulations" (https://www.scopus.com/search/form.uri?display=basic#scopus-ai, accessed on 29 December 2023).

Figure 1 presents a "conceptual map" generated by Scopus AI. This map groups the keywords into three branches, separating pairwise comparisons from the pairwise comparison matrix.

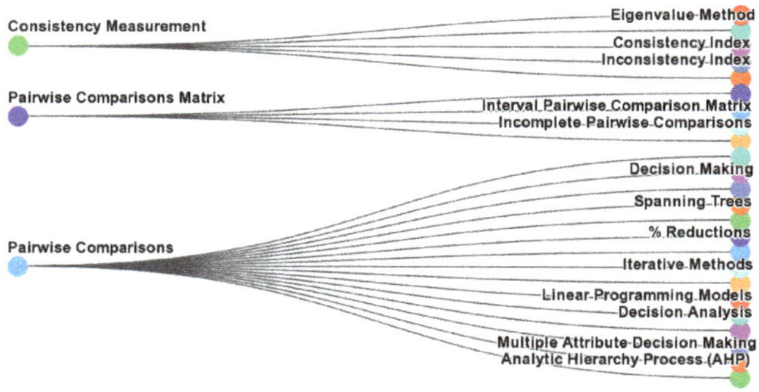

Figure 1. Conceptual map for "How to measure the consistency for a pairwise comparison matrix?". Source: Scopus AI.

Scopus AI concludes by highlighting three topics for expert research:

- What are the mathematical methods used to measure consistency in pairwise comparison matrices?
- How does the *CR* help in evaluating the reliability of a pairwise comparison matrix?
- Can inconsistency in a pairwise comparison matrix affect the accuracy of decision-making processes?

These three points are connected, indeed. For instance, if the *CR* helps in evaluating the reliability of a pairwise comparison matrix, it affects the accuracy of the decision-making process.

The literature review concludes that *CR* and the 0.1 threshold have been accepted for the consistency measurements and analyses of pairwise comparison matrices.

3. Consistency Improvement

Sections 1 and 2.1 present the ***A–B–C*** example with three 3-n pairwise comparison matrices. Real problems certainly involve more matrices with $n > 3$. Therefore, consistency improvement becomes more complex.

With $n = 2$, there is no possibility for inconsistency, since $k = i$ or $k = j$, always satisfying Equation (3), $\forall i, j, k = 1, 2$. With $n = 3$, and, for instance, $i = 1$, $j = 2$, and $k = 3$, Equation (3) may not be satisfied, as it occurrs with $b_{13} \neq b_{12}b_{23}$ and $c_{13} \neq c_{12}c_{23}$. With $n \geq 4$, the possibility for inconsistency increases with three combinations of $n(n-1)/2$ comparisons.

Iterations with just one change in an inconsistent comparison matrix may not be effective. On the other hand, replacing all comparisons seems to be unfair or illogical. Therefore, we propose to change only the a_{ij} comparisons, which brings significant deviation to $a_{ik}a_{kj}$, initially computing the expected value γ_{ij} as in Equation (7):

$$\gamma_{ij} = \frac{\sum_{k=1}^{n} a_{ik}a_{jk}}{n-2} \tag{7}$$

$\forall i, j, k = 1, 2, 3 \ldots n$ and $j > i$.

The absolute deviation between the value provided in the comparison matrix and the expected value for consistency satisfying Equation (3) is $\psi_{ij} = |a_{ij} - \gamma_{ij}|$. For inconsistent comparison matrices, we suggest that the a_{ij} with ψ_{ij} between the average $\bar{\psi}$ plus or less one-third of its standard deviation must be replaced by w_i/w_j.

For instance, let us consider the 4-n pairwise comparison matrix $\boldsymbol{D}$:

$$\boldsymbol{D} = \begin{bmatrix} 1 & 1/2 & 1/3 & 9 \\ 2 & 1 & 8 & 3 \\ 3 & 1/8 & 1 & 2 \\ 1/9 & 1/3 & 1/2 & 1 \end{bmatrix}$$

$\lambda_{\max_D} \approx 5.28$ and $CR_D \approx 0.43$. As $CR_D > 0.1$, $\boldsymbol{D}$ is an inconsistent pairwise comparisons matrix, and its inconsistency is unacceptable. For $\boldsymbol{D}$, $\psi_{12} \approx 1.02$, $\psi_{13} \approx 3.92$, $\psi_{14} \approx 7.92$, $\psi_{23} \approx 6.92$, $\psi_{24} = 14$, and $\psi_{34} \approx 11.69$. The average value is $\bar{\psi} \approx 7.58$, and its standard deviation is approximately 4.80. Only ψ_{14} and ψ_{23} are in the interval [5.98, 9, 18]. Then, $\boldsymbol{D'}$ is obtained by replacing d_{14} and d_{23} by γ_{14} and γ_{23}, respectively:

$$\boldsymbol{D'} = \begin{bmatrix} 1 & 1/2 & 1/3 & 1 \\ 2 & 1 & 1 & 3 \\ 3 & 1 & 1 & 2 \\ 1 & 1/3 & 1/2 & 1 \end{bmatrix}$$

$\lambda_{\max_{D'}} \approx 4.40$ and $CR_{D'} \approx 0.016$. As $CR_{D'} < 0.1$, $\boldsymbol{D'}$ is also an inconsistent pairwise comparison matrix, but its inconsistency is acceptable. However, the eigenvector also changes from $w_D = [0.22, 0.52, 0.18, 0.07]$ to $w_{D'} = [0.14, 0.36, 0.36, 014]$ to be validated by the decision-maker.

It is important to note that our proposed technique for consistency improvement resulted in individual significant changes in the comparison matrix $\boldsymbol{D}$ to $\boldsymbol{D'}$. Therefore, replacing $d_{14} = 9$ and $d_{23} = 8$ by $d'_{14} = d'_{23} = 1$ are big changes that result in a new vector of weights. These must all be validated by the decision-maker. Furthermore, this is a major limitation of our proposal. If the decision maker does not agree with the changes, then he (she or they) must review the comparisons by himself (herself or themselves). However, our proposal is not solely based on mathematics. The comparisons are connected, and the mathematics may capture the connection as presented in the next section, with a case of consistency improvement from the real world.

4. A Case of Consistency Improvement in Supply Chain Decision-Making

Supplier selection is one of the decision-making problems mostly solved by AHP applications [4]. This problem consists of choosing a single alternative (supplier) from a set of alternatives (suppliers). Table 5 presents an example of data for supplier selection considering three criteria (Delivery, Price, and Quality) and four alternatives (Suppliers 1, 2, 3, and 4):

Table 5. Example of data for a supplier selection problem.

Supplier	Delivery	Price [USD]	Quality
1	Slow	200,000	Acceptable
2	Regular	400,000	Excellent
3	Quick	300,000	Good
4	Regular	300,000	Very Good

In this case, it is clear that Quality is the most important criterion, but it is not clear by how much it is more important than others. Furthermore, it is not clear which one is more important: Delivery or Price. Then, a pairwise comparison matrix is a good tool to figure out the relative importance of the criteria. Table 6 presents a comparison matrix among the criteria.

Table 6. Pairwise comparison of the criteria for a supplier selection problem.

Criterion	Delivery	Price	Quality
Delivery	1	1/3	1/5
Price	3	1	1/3
Quality	5	3	1

The comparison matrix of the criteria has the same components of matrix $\boldsymbol{B}$ presented in Section 1. This matrix is equal to $\boldsymbol{B}^T$. Then, both matrices have the same $\lambda_{\max} \approx 3.04$ and $CR \approx 0.04$. Therefore, this matrix is inconsistent but acceptable, since its $CR < 0.1$. The decision-maker who provided the comparison matrix of the criteria understood the concepts of the Saaty Scale.

The eigenvector for the comparison matrix of the criteria has the same components of w_B, but in reverse order: [0.10, 0.26, 0.64]. It results in Quality being the most important criterion with 64% of weight, followed by Price and Delivery with 26% and 10%, respectively.

Table 7 presents a comparison matrix among Suppliers 1 to 4 regarding criterion Delivery. According to Table 5, Supplier 3 has the best performance in delivering quickly; Suppliers 2 and 4 deliver regularly, and Supplier 3 delivers slowly.

Table 7. Pairwise comparison of suppliers regarding their deliveries.

	Supplier 1	Supplier 2	Supplier 3	Supplier 4
Supplier 1	1	1/3	1/5	1/3
Supplier 2	3	1	1/3	1
Supplier 3	5	3	1	3
Supplier 4	3	1	1/3	1

The comparison matrix of suppliers on their deliveries has $\lambda_{\max} \approx 4.064$ and $CR \approx 0.024$. Therefore, this matrix is inconsistent but acceptable, since its $CR < 0.1$. The eigenvector for the comparison matrix is [0.08, 0.20, 0.52, 0.20]. It results in Supplier 3 being the best in Delivery with 52% of weight, followed by Suppliers 2 and 4 tied at 20%, and Supplier 1 being the worst with 8%.

For Price, there are available data as presented in Table 5. Weights for suppliers on Price are obtained by normalizing their reciprocals, as presented in Table 8.

Table 8. Weights for suppliers regarding their prices.

Supplier	Price [USD 1000]	Reciprocal	Weight
1	200	1/200	35%
2	400	1/400	18%
3	300	1/300	24%
4	300	1/300	24%

Table 9 presents a comparison matrix for suppliers regarding the Quality criterion. According to Table 5, suppliers' performances vary greatly: from Acceptable (Supplier 1) to Excellent (Supplier 2), including Good (Supplier 3) and Very Good (Supplier 4).

Table 9. Pairwise comparisons of suppliers regarding their quality.

	Supplier 1	Supplier 2	Supplier 3	Supplier 4
Supplier 1	1	1/7	1/3	1/5
Supplier 2	7	1	5	3
Supplier 3	3	1/5	1	3
Supplier 4	5	1/3	1/3	1

The comparison matrix of suppliers on their quality Q, has $\lambda_{\max_Q} \approx 4.39$ and $CR_Q \approx 0.146$. Therefore, this matrix is inconsistent and unacceptable, since $CR_Q > 0.1$. The eigenvector for the comparison matrix is [0.06, 0.58, 0.21, 0.16]. It results in Supplier 2 as the best in Quality with 58% of weight, followed by Suppliers 3, 4, and 1 with 21%, 16%, and 5%, respectively. The weights for Suppliers 1 and 2 are expected to be the lowest and the highest ones. However, there is a clear inversion between Good Supplier 3 and Very Good Supplier 4. Then, Q must be revised.

For Q, $\psi_{12} \approx 0.076$, $\psi_{13} \approx 0.057$, $\psi_{14} \approx 0.514$, $\psi_{23} \approx 3.333$, $\psi_{24} = 5.2$, and $\psi_{34} = 2.4$. The average value is $\bar{\psi} \approx 1.93$, and its standard deviation is approximately 2.08. Only ψ_{34} is in the interval [1.23,2.62]. Then, Q' is obtained by replacing $d_{24} = 3$ with $\gamma_{14} = 3/5$ as presented in Table 10:

Table 10. Revised pairwise comparison of suppliers regarding their quality.

	Supplier 1	Supplier 2	Supplier 3	Supplier 4
Supplier 1	1	1/7	1/3	1/5
Supplier 2	7	1	5	3
Supplier 3	3	1/5	1	3/5
Supplier 4	5	1/3	5/3	1

$\lambda_{\max_{Q'}} \approx 4.08$ and $CR_{Q'} \approx 0.030$. As $CR_{Q'} < 0.1$, Q' is also an inconsistent pairwise comparison matrix, but its inconsistency is acceptable. The eigenvector changes from $w_Q = [0.05, 0.58, 0.21, 0.16]$ to $w_{Q'} = [0.06, 0.57, 0.14, 0.23]$, which makes much more sense considering the initial data in Table 5 with more weight for Very Good Supplier 4 than for Good Supplier 3.

Table 11 presents, again, the weights for the suppliers regarding each criterion (decision matrix), and it also presents their overall weights (decision vector).

Table 11. Decision matrix and decision vector for the case of supplier selection.

Supplier	Delivery 10%	Price 26%	Quality 64%	Overall
1	0.08	0.35	0.06	0.14
2	0.20	0.18	0.57	0.43
3	0.52	0.24	0.14	0.20
4	0.20	0.24	0.23	0.23

Keeping the comparisons of Q (Table 9) and its eigenvector in the decision matrix would result in a different decision vector: [0.14,0.43,0.25,0.18]. Both decision vectors are close to each other, indicating that Supplier 2 has the highest overall performance. However, there are significant changes in the second and third-best suppliers. Therefore, the consistency improvement in this case results in a more reliable decision. Astoundingly, the decision-maker recognized that he caused slight confusion in the last comparison, comparing Supplier 3 to Supplier 4 regarding their quality. Instead of "3", the decision-maker was thinking of "1/3", since the quality of Supplier 4 is better than Supplier 3's. The mathematics of the proposed technique quickly identified this comparison as most divergent among all. Then, the decision-maker agreed with the new comparison matrix (Table 11) and its eigenvector.

Complimentary procedures such as Sensitivity Analysis or Robustness Tests are not conducted in this case because they are out of the scope of this work.

5. Conclusions

Consistency measurement and improvement is still an attractive subject of research in the AHP literature. This is evidenced by the literature review presented in Section 2. After all, consistency checking is an advantage of applying AHP instead of other MCDM methods, which do not include this check. However, when the consistency test fails, the decision process stalls.

This article presents a procedure for the improvement of consistency of pairwise comparison matrices. The simple procedure considers the means and the standard deviations to a consistent matrix. Besides being simple, it is a highly efficient procedure requiring few changes in the pairwise comparison matrix.

The first proposal for future research is the test of the proposed procedure with more cases other than in supply chain management. This proposal is very reliable due to the applicability of the AHP in many fields of decision-making, from computer science and engineering to health and medical applications. Mathematical simulations of inconsistent matrices, for instance, with Monte Carlo experiments or similar algorithms of randomness, could also be interesting.

Finally, some important advances in the AHP not included in this work may be considered in future research, such as the adoption of Fuzzy Sets Theory (FST) or the study of Group Decision-Making. Much older than the AHP literature, FST gained attention earlier this century with the proposal of Fuzzy Hesitant and Fuzzy Intuitionistic Sets. The study on consistency measurements and improvements in hybrid AHP–FST, especially with the new types of fuzzy sets, has not yet been studied.

Author Contributions: Conceptualization, L.F.A.M.G. and V.A.P.S.; methodology, L.F.A.M.G. and V.A.P.S.; software, L.F.A.M.G. and V.A.P.S.; validation, L.F.A.M.G. and V.A.P.S.; formal analysis, L.F.A.M.G. and V.A.P.S.; investigation, L.F.A.M.G. and V.A.P.S.; resources, L.F.A.M.G. and V.A.P.S.; data curation, L.F.A.M.G. and V.A.P.S.; writing—original draft preparation, L.F.A.M.G. and V.A.P.S.; writing—review and editing, L.F.A.M.G. and V.A.P.S.; visualization, L.F.A.M.G. and V.A.P.S.; supervision, L.F.A.M.G. and V.A.P.S.; project administration, L.F.A.M.G. and V.A.P.S.; funding acquisition, L.F.A.M.G. and V.A.P.S. All authors have read and agreed to the published version of the manuscript.

Funding: Grant No. 2023/14761-5, Sao Paulo Research Foundation (FAPESP).

Data Availability Statement: No new data were created or analyzed in this study. Data sharing is not applicable to this article.

Acknowledgments: We acknowledge the anonymous reviewers whose comments and suggestions improved the quality of this article.

Conflicts of Interest: The authors declare no conflicts of interest.

Abbreviations

The following abbreviations and notations, alphabetically sorted, are used in this manuscript:

A, B, C, D, and Q	Pairwise comparison matrices
AHP	Analytic Hierarchy Process
AI	Artificial intelligence
$\bar{C}$	Deviations matrix
C', C'', D', and Q'	Revised pairwise comparison matrices
CI	Consistency index
CR	Consistency ratio
EC	Software Expert Choice

FST	Fuzzy Sets Theory
GWU	George Washington University
MCDM	Multiple criteria decision-making
n	Matrix order
ORNL	Oak Ridge National Laboratory
PITT	University of Pittsburgh
Q1	Research question 1
Q2	Research question 2
UU	University of Ulster
$\mathbf{R}$	Set of real numbers
RI	Random index
w	Right eigenvector of a pairwise comparison matrix
γ_{ij}	Expected value for consistent pairwise comparison
$\lambda_{\max}$	Maximum eigenvalue of a pairwise comparison matrix
ψ_{ij}	Deviation between a pairwise comparison and its expected consistent value

References

1. Alvarez Gallo, S.; Maheut, J. Multi-criteria analysis for the evaluation of urban freight logistics solutions: A systematic literature review. *Mathematics* **2023**, *11*, 4089. [CrossRef]
2. Greco, S.; Ehrgott, M.; Figueira, J.F. *Multiple Criteria Decision Analysis: State of the Art Surveys*, 2nd ed.; Springer: New York, NY, USA, 2016.
3. Wallenius, J.; Dyer, J.S.; Fishburn, P.C.; Steuer, R.E.; Zionts, S.; Deb, K. Multiple criteria decision making, multiattribute utility theory: Recent accomplishments and what lies ahead. *Manag. Sci.* **2008**, *54*, 1336–1349. [CrossRef]
4. Khan, S.A.; Chaabane, A.; Dweiri, F.T. Multi-criteria decision-making methods application in supply chain management: A systematic literature review. In *Multi-Criteria Methods and Techniques Applied to Supply Chain Management*; Salomon, V., Ed.; Tech Open: London, UK, 2018; pp. 3–31. [CrossRef]
5. Ruiz Bargueno, D.; Salomon, V.A.P.; Marins, F.A.S.; Palominos, P.; Marrone, L.A. State of the art review on the Analytic Hierarchy Process and urban mobility. *Mathematics* **2021**, *9*, 3179. [CrossRef]
6. Petrillo, A.; Salomon, V.A.P.; Tramarico, C.L. State-of-the-art review on the Analytic Hierarchy Process with benefits, opportunities, costs, and risks. *J. Risk Financ. Manag.* **2023**, *16*, 372. [CrossRef]
7. Saaty, T.L. Fundamentals of the Analytic Hierarchy Process. In *The Analytic Hierarchy Process in Natural Resource and Environmental Decision Making: Managing Forest Ecosystems*; Schmoldt, D.L., Kangas, J., Mendoza, G.A., Pesonen, M., Eds.; Springer: Dordrecht, the Netherlands, 2001; Volume 3, pp. 15–35. [CrossRef]
8. Saaty, T.L. *Theory and Applications of the Analytic Network Process: Decision Making with Benefits, Opportunities, Costs, and Risks*; RWS: Pittsburgh, PA, USA, 2013.
9. Saaty, T.L. *Fundamentals of Decision Making and Priority Theory with the Analytic Hierarchy Process*, 1st ed.; RWS: Pittsburgh, PA, USA, 1994.
10. Saaty, T.L.; Vargas, L.G. Inconsistency and rank preservation. *J. Math. Psychol.* **1984**, *28*, 205–214. [CrossRef]
11. Saaty, T.L. A scaling method for priorities in hierarchical structures. *J. Math. Psychol.* **1977**, *15*, 234–281. [CrossRef]
12. Ma, D.; Zheng, X. 9/9-9/1 scale method of AHP. In Proceedings of the 2nd International Symposium on the AHP, Pittsburgh, PA, USA, 11–14 August 1991; Volume 1, pp. 197–202. [CrossRef]
13. Salo, A.A.; Hamalainen, R.P. On the measurement of preferences in the Analytic Hierarchy Process. *J. Multi-Criteria Decis. Anal.* **1997**, *6*, 309–319. [CrossRef]
14. Asadabadi, M.R.; Chang, E.; Saberi, M.; Zhou, Z. Are MCDM methods useful? A critical review of Analytic Hierarchy Process (AHP) and Analytic Network Process (ANP). *Cogent Eng.* **2019**, *6*, 1–11. [CrossRef]
15. Goepel, K.D. Comparison of judgment scales of the Analytical Hierarchy Process: A new approach. *Int. J. Inf. Technol. Dec.* **2019**, *18*, 445–463. [CrossRef]
16. Saaty, T.L. Measuring the fuzziness of sets. *J. Cybern.* **1974**, *4*, 53–61. [CrossRef]
17. Saaty, T.L. *The Analytic Hierarchy Process*; McGraw-Hill: New York, NY, USA, 1980.
18. Saaty, T.L.; Khouja, M. A measure of world influence. *Confl. Mana. Peace* **1976**, *2*, 31–48. [CrossRef]
19. Saaty, T.L.; Rodgers, P.C. Higher education in the United States (1985–2000): Scenario construction using a hierarchical framework with eigenvector weighting. *Socio Econ. Plan. Sci.* **1976**, *10*, 251–263. [CrossRef]
20. Alexander, J.M.; Saaty, T.L. The forward and backward processes of conflict analysis. *Syst. Res.* **1977**, *22*, 87–98. [CrossRef]
21. Saaty, T.L. Scenarios and priorities in transport planning: Application to the Sudan. *Transport. Res.* **1977**, *11*, 343–350. [CrossRef]
22. Saaty, T.L.; Bennett, J.P. A theory of analytical hierarchies applied to political candidacy. *Syst. Res.* **1977**, *22*, 237–245. [CrossRef]
23. Saaty, T.L. How to make a decision: The Analytic Hierarchy Process. *Eur. J. Oper. Res.* **1991**, *48*, 9–26. [CrossRef]
24. Wedley, W.C. Consistency prediction for incomplete AHP matrices. *Math. Comput. Model.* **1993**, *17*, 151–161. [CrossRef]
25. Saaty, T.L. *Mathematical Principles of Decision Making, Kindle Edition*; RWS: Pittsburgh, PA, USA, 2013.
26. Saaty, T.L.; Vargas, L.G. *The Logic of Priorities*, 1st ed.; Kluwer: Boston, MA, USA, 1982.

27. Forman, E.H. Random indices for incomplete pairwise matrices. *Eur. J. Oper. Res.* **1990**, *48*, 153–155. [CrossRef]
28. Donegan, H.A.; Dodd, F.J. A note on Saaty's random indexes. *Math. Comput. Model.* **1991**, *15*, 135–137. [CrossRef]
29. Pant, S.; Kumar, A.; Ram, M.; Klochkov, Y.; Sharma, H.K. Consistency indices in Analytic Hierarchy Process: A review. *Mathematics* **2022**, *10*, 1206. [CrossRef]
30. Saaty, T.L. *Decision Making with Dependence and Feedback: The Analytic Network Process*, 2nd ed.; RWS: Pittsburgh, PA, USA, 2001.
31. Aguaron, J.; Moreno-Jimenez, J. The geometric consistency index: Approximated threshold. *Eur. J. Oper. Res.* **2003**, *147*, 137–145. [CrossRef]
32. Mongeon, P.; Paul-Hus, A. The journal coverage of Web of Science and Scopus: A comparative analysis. *Scientometrics* **2016**, *106*, 213–228. [CrossRef]
33. Mazurek, J. Inconsistency reduction. In *Advances in Pairwise Comparisons. Multiple Criteria Decision Making*; Springer: Cham, Switzerland, 2023. [CrossRef]
34. Brunelli, M. Recent advances on inconsistency indices for pairwise comparisons—A commentary. *Fund. Inform.* **2016**, *144*, 321–332. [CrossRef]
35. Csato, L.; Ágoston, K.C.; Bozoki, S. On the coincidence of optimal completions for small pairwise comparison matrices with missing entries. *Ann. Oper. Res.* **2023**. [CrossRef]
36. Dong, Y.; Chen, X.; Li, C.; Hong, W.; Xu, Y. Consistency issues of interval pairwise comparison matrices. *Soft Comput.* **2015**, *19*, 2321–2335. [CrossRef]
37. Szybowski, J.; Kukalowski, K.; Prusak, A. New inconsistency indicators for incomplete pairwise comparisons matrices. *Math. Soc. Sci.* **2020**, *108*, 138–145. [CrossRef]

mathematics

MDPI

Article

Multi-Criteria Decision under Uncertainty as Applied to Resource Allocation and Its Computing Implementation

Petr Iakovlevitch Ekel [1,2], Matheus Pereira Libório [1,*], Laura Cozzi Ribeiro [1], Mateus Alberto Dorna de Oliveira Ferreira [1] and Joel Gomes Pereira Junior [2]

1 Programa de Pós-Graduação em Informática, Pontifícia Universidade Católica de Minas Gerais, Belo Horizonte 30535-901, MG, Brazil; petr.ekel2709@gmail.com (P.I.E.); laura.cozzi.ribeiro@gmail.com (L.C.R.); mateus.dorna.ferreira@gmail.com (M.A.D.d.O.F.)

2 Research Department, ASOTECH–Advanced System Optimization Technologies, Belo Horizonte 30535-630, MG, Brazil

* Correspondence: m4th32s@gmail.com

Abstract: This research addresses the problem of multi-objective resource allocation or resource deficits, offering robust answers to planning decisions that involve the elementary question: "How is it done?". The solution to the problem is realized using the general scheme of multi-criteria decision-making in uncertain conditions. The bases of the proposed scheme are associated with the possibilistic approach, which involves the generalization of fuzzy sets from the classical approach to process the uncertainty of information to produce robust (non-dominated) solutions in multi-criteria analysis. Applying this general scheme makes it possible to reduce regions of decision uncertainty through the maximum use of available quantitative information. In the case where quantitative information analysis is insufficient to obtain a unique solution, the proposed approach presupposes the appropriation of qualitative data extracted from experts, who express their opinions considering their knowledge, experience, and intuition. The information on the qualitative character can be represented in diverse preference formats processed by transformation functions to provide homogeneous information for decision procedures used at the final decision stage. The presented results have been implemented within the system of multi-criteria decision-making under uncertain conditions described in the paper. Its functioning is illustrated by solving the typical problem in investment planning activities.

Keywords: multiobjective allocation of resources; multi-criteria decision-making; uncertain conditions; non-dominated solutions

MSC: 03B52; 68U35; 68T37; 90C70

1. Introduction

By characterizing multi-criteria decision-making problems, it is necessary to distinguish two types of criteria: objectives and attributes. In such a manner, multi-criteria decision-making problems can be classified into multiobjective and multiattribute decision-making [1–3].

Multiobjective decision-making [4] is known as the continuous type of multi-criteria problem. Their main characteristic is the necessity to achieve multiple objectives. A multiobjective decision-making model includes a vector of decision variables (which can be continuous as well as discrete), objective functions (maximized or minimized) that describe the objectives, and constraints.

Multiattribute decision-making is associated with making preference decisions (which may be comparisons, choices, prioritizations, and/or ordering) over the alternatives [4]. The alternatives are characterized by multiple attributes. The main feature of multiattribute problems is that there are usually a limited number of considered alternatives. The decision is to be made on the basis of the attributes.

Thus, two classes of models, requiring the use of a multi-criteria approach, may be constructed: $< X, F >$ models (as multiobjective models) and $< X, R >$ models (as multiattribute models). The present paper describes these models and methods of their analysis, utilizing the modification of the Bellman–Zadeh [5] approach to decision-making in a fuzzy environment and on applying techniques of fuzzy preference modeling [6], respectively. In the case of using other preference formats for characterizing preferences among alternatives, the so-called transformation functions can be applied to convert all formats into a single one [7–9]. Other transformation functions exist to convert quantitative information into fuzzy preference relations [10,11]. The application of the transformation functions makes it possible to homogenize quantitative information and qualitative information (provided by experts) into different formats to solve decision-making problems, including group decision-making problems [12,13].

In general, our broad focus on the use of fuzzy set theory allows us to adequately take into account various types of uncertainty and combinations of different types of uncertainty.

The mutual construction and analysis of $< X, F >$ and $< X, R >$ models is founded from the substantial point of view. In any of our activities, for example, in planning (strategic, new business, innovation, etc.), two principal questions emerge: "What to do?" and "How to do it?". The answer to the first question can be developed by building and analyzing $< X, R >$ models. The construction and analysis of $< X, F >$ models permits one to answer the second question.

In addition, the consideration of $< X, F >$ models and $< X, R >$ models can serve as parts of a general scheme of multi-criteria decision in the conditions of uncertainty, which is the main subject of the present work. This scheme is associated with the generalization of the classic approach to dealing with information uncertainty [14,15] to multi-criteria problems. It is based on analyzing special aggregations of payoff matrices [16]. Its important characteristic is to apply existing quantitative information to the highest degree to reduce decision uncertainty regions. If the problem-resolving capacity related to quantitative information processing does not permit one to obtain unique solutions, the general scheme presumes the use of qualitative information based on knowledge, experience, and intuition of involved experts.

The motivation for developing a general scheme for multi-criteria decision-making under conditions of uncertainty is associated with the consequences of each action in terms of each criterion in multi-criteria models, which are generally based on deterministic assessments [17]. This deterministic approach subjects solutions and directions to a sensitivity level. Still, it is possible to recommend using the deterministic approach for situations where the primary source of complexity in the decision is not related to the uncertain nature of individual consequences but rather to the multi-criteria nature of the problem. Nevertheless, more formal uncertainty modeling is required when risks and uncertainties are as critical as issues of conflicting goals [17]. Considering this, it is necessary to stress three fundamental points.

First, it is impossible to talk about the future, and to plan the future only based on the tendencies of the past. Considering this, in the present work, we do not use the probabilistic approach. We use the possibilistic approach, which is based on aggregating information of a formal character (including a probabilistic one) and an informal character [18]. This aggregation opens the possibilities of obtaining representative combinations of initial data, states of nature or scenarios for uncertainty modeling.

Second, in uncertain conditions, optimal solutions do not exist: one solution may be optimal for one scenario and non-optimal for another scenario. Thus, what does the solution mean in uncertain conditions? So-called robust or non-dominated solutions can serve as such solutions. The robust solution is the solution that, to the highest degree, permits one to satisfy any scenario.

Third, all methods and strategies of operational research are based on the conception of existing optimal solutions and the "search for optimal solutions" that do not exist under uncertain conditions. Considering this, we do not search for the best solutions but for the worst ones, which are dominated by other solutions, applying any information or any type of preferences. Cutting out these solutions, we systematically reduce the decision uncertainty regions.

To deal with information uncertainty, it is possible to generalize the classical approach to considering the uncertainty factor [14,15] for analyzing multi-criteria models. In particular, this approach involves the construction and analysis of so-called payoff matrices, which reflect effects that can be obtained for different combinations of solution alternatives and representative combinations of initial data [4,6].

The generalization is based on applying fuzzy set theory and its combination with other branches of mathematics of uncertainty (in particular, game theory and interval analysis) [16]. This combination of branches of mathematics of uncertainty does not fit the general approaches discussed in Stewart [17], Durbach and Stewart [19], Eiself and Marianov [20], and Gaspars-Wieloch [21].

The merging of the generalization of the classical approach to considering the uncertainty factor with the analysis of ($< X, F >$) and ($< X, R >$) models permits us the development of the general scheme of multi-criteria decision-making under uncertain conditions, as is proposed in this research.

The results discussed below, as applied to the multi-criteria allocation of resources or their deficits under conditions of uncertainty, have been implemented within the computing system of multi-criteria decision-making in uncertain conditions, which is described as well.

Generally, the manuscript's results permit one to improve the adequateness of the constructed models and, consequently, the real efficiency of solutions obtained based on their analysis. From a practical point of view, these results are actively used to construct diverse portfolios in the projects developed for mining and energy companies. Furthermore, the research results represent an advancement concerning traditional decision-making methods [22,23]. Finally, the system of multi-criteria decision-making under uncertain conditions offers robust solutions to management problems in the most diverse fields, such as health [24,25], energy systems [26], information security [27], architecture [28], transport [29], floods [30], water allocation [31], education [32], project portfolio selection [33], business analytics [34], and sustainable development goals [35].

2. Problem Statement

The methods dedicated to the problem of resource allocation are based on three fundamental principles of allocation: proportional, optimal, and inverse priorities [36,37]. These methods present significant disadvantages that can be treated through the application of the $< XR >$ approach to maximize various positive consequences and minimize the negative consequences associated with the allocation of resources, considering or not considering the presence of deficiencies [6,38].

Two types of correlated multiobjective resource allocation problems were examined in this research. The fundamental difference between them is that the first allocation problem does not present resource limitations, while the second allocation problem considers limited resources, or rather, resource deficits. For the first type of problem, it is necessary to formulate specific decision objectives associated with investment planning activities to be evaluated by specialists. These objectives can be formulated using the following structure:

1. Predominant economic support for investments ensuring a greater quantity of supply of products overseas.
2. Predominant economic support for investments generating a greater profit percentage for every one million dollars invested.

The solution to achieve these objectives can be formulated by adding quantitative information. At the same time, the objective, which can be supported by qualitative information, for example, is the following:

3. Predominant economic support for investments generating a greater degree of innovation.

For the second type of problem, the objectives for investment planning activities can be formulated using the following structure:

4. Predominant economic limitation of investments ensuring a smaller quantity of supply of products overseas.
5. Predominant economic limitation of investments generating a worse profit percentage for every one million dollars invested.
6. Predominant economic limitation of investments generating a lower level of innovation.

Note that achieving the objectives formulated for the investment resource allocation problem involves maximizing or minimizing the presented objective functions [6]:

$$F_p(X) = \sum_{i=1}^{n} c_{pi}x_i, \setminus p = 1,2,\ldots,q \tag{1}$$

where x_i, $i = 1,2,\ldots,n$ are variables corresponding to intended resource amounts for the ith investment (e.g., new business project, strategic or tactic action, and expansion strategy); C_{pi}, $p = 1,2,\ldots,q$, $i = 1,2,\ldots,n$ are specific indicators corresponding to the pth specific objective for the ith investment.

Similarly, satisfying the objectives formulated for the second type of problem (allocation considering resource deficit) involves maximizing or minimizing the following objective functions [6]:

$$F_p(\Delta X) = \sum_{i=1}^{n} c_{pi}\Delta x_i, \setminus p = 1,2,\ldots,q \tag{2}$$

where Δx_i, $i = 1,2,\ldots,n$ are variables corresponding to the limitations of resource amounts intended for the i-th new business project, strategic or tactic action, or expansion strategy.

Other objective functions can be used to solve both resource allocation problems. In addition to functions (1) and (2), it is possible to formulate customized objective functions to reflect the core of specific objectives more consistently. Among the allocation models of available resources, resource deficits with unlimited cuts, and resource deficits with limited cuts [18], we selected the last one, which is of the most flexible character.

The demands D_I, $I = 1,2,\ldots,n$ and the minimally acceptable demands D_i^m, $I = 1,2,\ldots,$ n of the investment are assigned, and all resources are available; $R < \sum_{i=1}^{n} D_i$. In sequence, the deficit of resources for allocation is $A = \sum_{i=1}^{n} D_i - R$. Finally, the problem is resolved as follows [18]:

$$F_p(\Delta X) \Rightarrow \underset{X \in L}{max} \; or \; \underset{X \in L}{min}, \; p = 1,2, \ldots, q \tag{3}$$

taking into account the presence of the following restrictions

$$0 \leq \Delta x_i \leq A_i = D_i - D_i^m, \setminus i = 1,2, \ldots, n \tag{4}$$

and

$$\sum_{i=1}^{n} \Delta x_i = A \tag{5}$$

The models (3)–(5) allow for different problem assertions. Commonly, its investigation is noniterative. Nevertheless, the ultimate solution within the framework of $A_i = D_i - D_i^m$, $i = 1,2,\ldots,n$ may convert the investigation into an iteration when there is a need for negotiation.

3. Solution to the Problem in the Presence of Deterministic Information

The formal step in analyzing (3)–(5) is associated with defining the set of Ω's Pareto optimal solutions [2,39]. This phase is helpful but prevents one obtaining unique solutions. To solve this problem, it is possible to use the information provided by decision-makers and obtain a specific Pareto solution using three types of approaches: a priori, a posteriori, and adaptive [38]. When utilizing the last one, the enhancement of the quality of the solution is reached as modifications from $X^0_\alpha \in \Omega \subseteq L$ to $X^0_{\alpha+1} \in \Omega \subseteq L$ by considering information I_α of the decision-maker, presented in step α.

The elaboration of the multiobjective methods is realized in several directions [6,38]. Without detailing these directions, the quality of solutions in multiobjective models is a relevant point to be considered [40]. The concept of harmonious solutions is an effective and efficient way of checking the quality of solutions [41,42]. In short, solutions are considered harmonious when the objectives' satisfaction levels are equal or close to each other [43,44]. This concept can be expanded to situations in which the importance coefficients of the objective functions are not the same [6]. Considering this, the direction's validity and appropriateness concerning guaranteed results should be stressed [38]. Other forms may result in non-harmonious solutions in which satisfaction levels are high for certain objectives and low for others [6].

The complexity of solving multiobjective problems is methodological, derived from the lack of clarity regarding the concept of the "optimal solution." Within the decision-making approach in a fuzzy environment, the optimal solution is defined as the maximum degree of the implementation of objectives, which functions as an optimality criterion [43]. This corresponds to the principle of guaranteed results and allows for the generation of solutions in which the objectives' satisfaction levels are close or harmonious. The modification of Bellman and Zadeh's [5] approach to decision-making in a fuzzy environment allows one to generate $X^0 \in \Omega \subseteq L$ solutions based on applying computationally effective procedures [6]. Its application also permits dealing with the index, criterion, and restriction of qualitative characters.

Using the decision-making approach in a fuzzy environment [5], it is possible to replace the objective functions $F_p = (X),\ p = 1,2,\ldots,q$ with fuzzy sets $A_p = \{X,\ \mu_{A_p}(X)\}, X \in L$, $p = 1,2,\ldots q$. In this case, $\mu_{A_p}(X)$ is the membership function of A_p, with the fuzzy solution D defined as $D = \bigcap_{p=1}^{q} A_p$, with the following membership function:

$$\mu_D(X) = \min_{1\le p\le q} \mu_{A_p}(X),\ X \in L \tag{6}$$

Considering (6), it is possible to build the following problem:

$$max\ \mu_D(X) = \max_{X\in L}\ \min_{1\le p\le q} \mu_{A_p} \tag{7}$$

From (7), it is possible to obtain the following solution:

$$X^{\,0} = arg \max_{X\in L} \min_{1\le p\le q} \mu_{A_p} \tag{8}$$

This result can be reached using the nonlocal search algorithm [38], implemented within the system of multi-criteria decision-making in uncertain conditions.

4. Consideration of the Uncertainty of Information

The uncertainty of the initial data is treated through the transformation of the objective functions presented in (1) and (2). It is essential to highlight that these transformations must consider the corresponding description of the importance coefficients. Although aggregation procedures allow one to construct meaningful combinations of initial information, scenarios, or states of nature, considering information of a deterministic, interval, probabilistic, or fuzzy character, in the present work, for simplicity, we use the interval description, which has found diverse applications. Considering this, the objective function (1) can be represented as

$$F_p(X) = \sum_{i=1}^{n} \left[c'_{pi}, c''_{pi}\right] x_i, p = 1,2,\ldots,q \tag{9}$$

where c'_{pi} is the minimum value, and c''_{pi} is the maximum value, considering c_{pi}, $p = 1,2,\ldots,q,\ i = 1,2,\ldots,n$.

Similarly, the objective functions (2) can be reduced to

$$F_p(\Delta X) = \sum_{i=1}^{n} \left[c'_{pi}, c''_{pi}\right] \Delta x_i, p = 1, \ldots, q \tag{10}$$

The so-called LP_{τ}-sequences [45] can be applied to build representative combinations of initial data or scenarios. These sequences have greater properties of homogeneity among other non-homogeneous sequences, providing points $Q_s, s = 1,2,\ldots,S$ with coordinates q_{st}, $s = 1,2,\ldots,S$, $t = 1,2,\ldots,T$ in the relating element hypercube Q^T, where S gives the number of scenarios, and the number of coefficients of objective functions is given by T, for example, $T = qn$ in the analysis of problems involving objective functions (1) or (2).

In short, the initial constructed data are reduced to uniformly distributed sequence points in Q^T and their conversion to the hypercube C^T, which is formulated by the lower c'_t, $t = 1,2,\ldots,T$ and upper c''_t, $t= 1,2,\ldots,T$ boundaries of c'_{pi}, $p = 1,2,\ldots,q$, $i = 1,2,\ldots,n$ and c''_{pi}, $p = 1,2,\ldots,q$, $i = 1,2,\ldots,n$ for the objective functions (1) and (2), as follows [16]:

$$C_{st} = c'_t + (c''_t - c'_t)q_{st} = 1,2,\ldots,S, t = 1,2,\ldots,T \tag{11}$$

to form a uniformly distributed sequence in C^T.

Considering a given number S of the initial data, the coordinates of points calculated using (11) can be used to create S multiobjective problems, including deterministic coefficients. From these definitions and the application of (1), it is possible to represent the objective functions for each scenario Y_s, $s =1,\ldots,S$ as follows:

$$F_p(X,Y_s) = \sum_{i=1}^{n} c_{pis} x_i, p = 1,2,\ldots,q,\ s = 1,2,\ldots,S \tag{12}$$

Applying (6) to the resolution of problems (3)–(5) allows us to obtain the respective solutions S for each of the scenarios. From the set of solutions obtained for each of the S scenarios, a subset of $K \leq S$ different solutions (X_k, $K = 1,2,\ldots,K$) is selected to construct the payoff matrices.

Knowing the alternatives X_k, $k = 1,2,\ldots,K$ and the scenarios Y_s, $s = 1,2,\ldots,S$, it is possible to build the payoff matrix. This matrix is presented in Table 1 and reflects the effects (or consequences) of one or another solution X_k, $k = 1,2,\ldots,K$ for the related scenario Y_s, $s = 1,2,\ldots,S$.

Table 1. Payoff matrix for the pth objective function.

	Y_1	…	Y_s	…	Y_S
X_1	$F_p(X_1, Y_1)$	…	$F_p(X_1, Y_s)$	…	$F_p(X_1, Y_S)$
…	…	…	…	…	…
X_k	$F_p(X_k, Y_1)$	…	$F_p(X_k, Y_s)$	…	$F_p(X_k, Y_S)$
…	…	…	…	…	…
X_K	$F_p(X_K, Y_1)$	…	$F_p(X_K, Y_s)$	…	$F_p(X_K, Y_S)$

The assessment of payoff matrices and the selection of reasonable solution alternatives are based on characteristic estimates utilized within the selection criteria called Wald, Laplace, Savage, and Hurwicz [14,15].

The following characteristic estimates are used for each X_k, $k = 1,2,\ldots,K$ for the pth objective function:

- The minimum objective function level:

$$F_p^{min}(X_k) = \min_{1 \leq s \leq S} F_p(X_k, Y_s) \tag{13}$$

which is the most negative estimate for a maximizing objective function or the most positive estimate for a minimization objective function;

- The maximum objective function level:

$$F_p^{max}(X_k) = \max_{1 \le s \le S} F_p(X_k, Y_s) \tag{14}$$

which is the most negative estimate for a minimization objective function or the most positive estimate for a maximizing objective function;

- The mean objective function level:

$$\overline{F}_p(X_k) = \frac{1}{S}\sum_{s=1}^{S} F_p(X_k, Y_s) \tag{15}$$

- The maximum regret level:

$$R_p^{max}(X_k) = \max_{1 \le s \le S} R_p(X_k, Y_s) \tag{16}$$

where $R_P(X_k Y_s)$ is an overspending that occurs under a combination of the scenario Y_s and the selection of the X_k alternative rather than the local optimal solution Y_s. The estimates (13)–(16) serve to construct the corresponding matrix given in Table 2.

Table 2. Matrix with the characteristic estimations for the p-th objective function.

	$F_p^{max}(X_k)$	$F_p^{min}(X_k)$	$\overline{F}_p(X_k)$	$R_p^{max}(X_k)$
X_1	$F_p^{max}(X_1)$	$F_p^{min}(X_1)$	$\overline{F}_p(X_1)$	$R_p^{max}(X_1)$
...	...	...	...	...
X_k	$F_p^{max}(X_k)$	$F_p^{min}(X_k)$	$\overline{F}_p(X_k)$	$R_p^{max}(X_k)$
...	...	...	...	...
X_K	$F_p^{max}(X_K)$	$F_p^{min}(X_K)$	$\overline{F}_p(X_K)$	$R_p^{max}(X_K)$

The choice criteria [14,15] presented below are under the statement that the objective functions are to be minimized.

For the Wald criterion, the $F^{max}(X_k)$ estimate is used and enables one to choose the solution alternatives X^W, for which this estimate achieves the minimum:

$$\min_{1 \le k \le K} F^{max}(Y_s) = \min_{1 \le k \le K} \max_{1 \le s \le S} F(X_k, Y_s) \tag{17}$$

The criterion of Laplace is based on the estimate $\overline{F}(X_k)$ and is directed at the choice of the solution alternatives X^L. For this estimate, the minimum is considered:

$$\min_{1 \le k \le K} \overline{F}(X_k) = \min_{1 \le k \le K} \frac{1}{S}\sum_{s=1}^{S} F(X_k, Y_s) \tag{18}$$

To operationalize the Savage criterion, the estimate $R^{max}(X_k)$ is used. It permits to select the solution alternatives X^S, providing the minimum for this estimation:

$$\min_{1 \le k \le K} R^{max}(X_k) = \min_{1 \le k \le K} \max_{1 \le s \le S} R(X_k, Y_s) \tag{19}$$

Hurwicz's criterion combines convexly $F^{max}(X_k)$ and $F^{min}(X_k)$, allowing us to select the solution alternative X^H, where the resulting combination represents the minimum:

$$\min_{1 \le k \le K} [\alpha F^{max}(X_k) + (1-\alpha)F^{min}(X_k)] = \min_{1 \le k \le K} [\alpha \max_{1 \le s \le S} F(X_k, Y_s) + (1-\alpha) \min_{1 \le s \le S} F(X_k, Y_s)] \tag{20}$$

where $\alpha \in [0, 1]$ is the "pessimism–optimism" index specified by the expert.

Characterizing these criteria, it is possible to specify that the application of the criterion of Wald produces solution alternatives, providing the most unfavorable initial data combinations. The criterion of Wald guarantees that the level of the objective function is not greater than a certain value at future conditions. This is its merit. At the same time, the orientation on the most unfavorable initial data combinations is maximally cautious (conservative or pessimistic).

The application of the criterion of Laplace generates results corresponding to the principle of "insufficient reason,", i.e., to the situation that we have no basis for distinguishing initial data combinations. Thus, it is necessary to act since they are equally likely. This is its drawback. However, the average score is quite useful.

As in the case of applying the choice criterion of Wald, the use of the criterion of Savage is associated with the minimax principle. Hereby, the criterion of Savage can also be considered conservative. However, experience shows that the recommendations elaborated in its application are mismatched with the decisions obtained using the criterion of Wald. Operating with values of $R^{max}(X_k)$, we obtain a different evaluation of the situation, which may lead to more "daring" (less conservative) solutions.

If $\alpha = 1$ in (20), the choice criterion of Hurwicz becomes a criterion of Wald. If $\alpha = 0$, (20) becomes a criterion of "extreme optimism" (m/m) for which the most favorable combinations of initial data is considered. The author of [46] recommends one to choose α from 0.5 to 1.

The classical approach to dealing with the uncertainty of information is associated with analyzing the choice criteria of a particular objective function considering an environment with multiple scenarios Y_s, $s = 1, 2, \ldots, S$. Thus, considering that each choice criterion is associated with the estimates (13)–(16), one can consider these estimates as objective functions for the pth objective function:

$$F_p^W(X_k) = F_p^{max}(X_k) = \max_{1 \le s \le S} F_p(X_k, Y_s) \tag{21}$$

$$F_p^L(X_k) = \overline{F}_p(X_k) = \frac{1}{S} \sum_{s=1}^{S} F_p(X_k, Y_s) \tag{22}$$

$$F_p^S(X_k) = R_p^{max}(X_k) = \max_{1 \le s \le S} R_p(X_k, Y_s) \tag{23}$$

and

$$F_p^H(X_k) = \alpha F_p^{max}(X_k) + (1-\alpha) F_p^{min}(X_k) = \alpha \max_{1 \le s \le S} F_p(X_k, Y_s) + (1-\alpha) \min_{1 \le s \le S} F_p(X_k, Y_s) \tag{24}$$

The correlations (21)–(24) permit one to build $M \le 4$ problems (following the choice criteria of Wald, Laplace, Savage, and Hurwicz), as follows:

$$F_{r,p}(X) \to \underset{X \in L}{extr}, \ r = 1, 2, \ldots, M \le 4, \ p = 1, 2, \ldots, q \tag{25}$$

where $F_{1,p}(X) = F_p^W(X_k), F_{2,p}(X) = F_p^L(X_k), F_{3,p}(X) = F_p^S(X_k)$, and $F_{4,p}(X) = F_p^H(X_k)$.

From (25), it is possible to construct q matrices with the four choices of criteria. From the monobjective point of view, the matrix of Table 3 contains information for decision-making since it is possible to choose the alternative for a given pth objective function based on the alternatives, provided the minimum values of $F_p^W(X_k)$,$F_p^L(X_k)$, $F_p^S(X_k)$, and $F_p^H(X_k)$.

Table 3. Matrix of estimates of four selection criteria used in the research according to the first objective function.

	$F_p^W(X_k)$	$F_p^L(X_k)$	$F_p^S(X_k)$	$F_p^H(X_k)$
X_1	$F_p^W(X_1)$	$F_p^L(X_1)$	$F_p^S(X_1)$	$F_p^H(X_1)$
...	...	...	...	...
X_k	$F_p^W(X_k)$	$F_p^L(X_k)$	$F_p^S(X_k)$	$F_p^H(X_k)$
...	...	...	...	...
X_K	$F_p^W(X_K)$	$F_p^L(X_K)$	$F_p^S(X_K)$	$F_p^H(X_K)$

From a multiobjective point of view, it is possible to adapt the decision-making approach in a fuzzy environment defined by Bellman and Zadeh [5] to analyze multiobjective problems. This multiobjective analysis is carried out by normalizing the choice criteria estimates presented in Table 3. These normalization functions allow for the construction of membership functions for $F_{r,p}(X)$, $r = 1, 2, \ldots, M$. Next, these membership functions are used to obtain the fuzzy choice criteria levels for the pth objective function. Finally, the modified matrices are constructed with the two selection criteria presented in Table 4 from the matrices with the choice criteria levels.

Table 4. Modified matrix with the choice criteria estimations for the p-th objective function.

	$\mu^W_{A_p}(X_k)$	$\mu^L_{A_p}(X_k)$	$\mu^S_{A_p}(X_k)$	$\mu^H_{A_p}(X_k)$
X_1	$\mu^W_{A_p}(X_1)$	$\mu^L_{A_p}(X_1)$	$\mu^S_{A_p}(X_1)$	$\mu^H_{A_p}(X_1)$
X_k	$\mu^W_{A_p}(X_k)$	$\mu^L_{A_p}(X_k)$	$\mu^S_{A_p}(X_k)$	$\mu^H_{A_p}(X_k)$
X_K	$\mu^W_{A_p}(X_K)$	$\mu^L_{A_p}(X_K)$	$\mu^S_{A_p}(X_K)$	$\mu^H_{A_p}(X_K)$

Continuing analysis, it is possible to apply the min operator [47] or, generally, the ordered weighted averaging operator [48–51] for X_k, $K = 1, 2, \ldots, K$ to build a matrix containing the levels of two aggregated fuzzy selection criteria. Finally, applying (8) allows one to find non-dominated alternatives for each selection criterion, reflecting the estimates presented in Table 5.

Table 5. Matrix of criteria levels associated with aggregated fuzzy choice.

	$\mu^W_D(X_K)$	$\mu^L_D(X_k)$	$\mu^S_D(X_K)$	$\mu^H_D(X_k)$
X_1	$\mu^W_D(X_1)$	$\mu^L_D(X_1)$	$\mu^S_D(X_1)$	$\mu^H_D(X_1)$
...	...	...	...	...
X_k	$\mu^W_D(X_k)$	$\mu^L_D(X_k)$	$\mu^S_D(X_k)$	$\mu^H_D(X_k)$
...	...	...	...	...
X_K	$\mu^W_D(X_K)$	$\mu^L_D(X_K)$	$\mu^S_D(X_K)$	$\mu^H_D(X_K)$
	$\max\limits_{1\le k\le K} \mu^W_D(X_K)$	$\max\limits_{1\le k\le K} \mu^L_D(X_K)$	$\max\limits_{1\le k\le K} \mu^S_D(X_K)$	$\max\limits_{1\le k\le K} \mu^H_D(X_K)$

The estimates $\max\limits_{1\le k\le K} \mu^W_D(X_k)$, $\max\limits_{1\le k\le K} \mu^L_D(X_k)$, $\max\limits_{1\le k\le K} \mu^S_D(X_k)$, and $\max\limits_{1\le k\le K} \mu^H_D(X_k)$ lead to $X^W; X^L; X^S; X^H$. This evaluation approach is especially effective under conditions of uncertainty, as it ensures the selection of non-dominated alternatives when considering the Pareto optimality principle [4,16].

5. Construction and Analysis of $< X, R >$ Models

This section is dedicated to discussing and developing techniques that allow for modeling preferences in a fuzzy environment within the structure of $< X,\ R >$ models applicable to evaluating, comparing, choosing, prioritizing, and ordering alternatives [52]. Using these techniques provides an adequate and effective way to consider quantitative and qualitative character criteria.

Consider a set of alternatives X, for which q which can be quantitative or qualitative. Under these conditions, it is possible to establish the decision-making problem as a pair of $< X,\ R >$. In this case, $R = \{R_1, R_2 \ldots, R_p, \ldots, R_q\}$ corresponds to a fuzzy preference relation vector represented as [38]:

$$R_P[X \times X,\ \mu_{R_p}(X_k X_l)], p = 1, 2 \ldots, q,\ X_k X_l \in X \tag{26}$$

where $\mu_{R_p}(X_k X_l)$ is a membership function of the pth fuzzy preference relation.

The fuzzy preference relation (26) is also known as a non-strict fuzzy preference relation. The membership function $\mu_{R_p}(X_k, X_l)$ reflects how much X_k weakly dominates X_l, and thus, how much worse X_l is than X_k, taking into account criterion p.

The set X can be confined by applying data obtained from (26) to be constrained only by alternatives not dominated by other alternatives in X. On this matter, Ekel et al. [53] offer a coherent and satisfactory decision-making method for constructing R_p. In short, the method is associated with the conception of a membership function that allows for the treatment of preference relations in a generalized approach [54]. Specifically, the ease of use of linguistic or fuzzy estimates of alternatives $F_p(X_k)$, $p = 1, 2, \ldots, q$, $X_k \in X$ with the membership functions $\mu[F_P(X_k)]$, $p = 1, 2, \ldots,$ $X_k \in X$ enables one to construct $R_p, p, 1, 2, \ldots, q$, applying the following relationships:

$$\mu_{R_p}(X_k, X_l) = \sup_{\substack{X_k, X_l \in X \\ F_p(X_k) \leq F_p(X_l)}} \min\{\mu[F_p(X_k)], \mu[F_p(X_l)]\} \tag{27}$$

and

$$\mu_{R_p}(X_l, X_k) = \sup_{\substack{X_k, X_l \in X \\ F_p(X_l) \leq F_p(X_k)}} \min\{\mu[F_p(X_k)], \mu[F_p(X_l)]\} \tag{28}$$

if the p-th criterion is related to minimization. If the p-th criterion demands maximization, (27) and (28) must be written, respectively, for $F_p(X_k) \geq F_p(X_l)$ and $F_p(X_l) \geq F_p(X_k)$.

Eight formats can be applied to establish preferences among alternatives [55,56]. Naturally, their use requires a transformation of utilized formats to a distinctive one, enabling its processing and analysis. It is important to highlight that non-reciprocal diffuse preference relationships present important advantages for the objective of this research (see [10,12,57]. In this sense, it is possible to use so-called transformation functions to convert and homogenize the different evaluation formats into the non-reciprocal fuzzy preference relations format [58,59].

To obtain a strict fuzzy preference relation as follows, it is sufficient to process a single non-strict fuzzy preference relation R [54]:

$$R^S = R / R^{-1} \tag{29}$$

where R^{-1} is the inverse relation.

From (29), it is possible to obtain the following membership function:

$$\mu_R^S(X_k, X_l) = \max\{\mu_R(X_k, X_l) - \mu_R(X_l, X_k), 0\} \tag{30}$$

From this membership function, it is possible to obtain the evaluation of the level of non-dominance of each alternative X_k from the set of non-dominated alternatives through the following membership function:

$$\mu_R^{ND}(X_k) = \inf_{X_l \in X} [1 - \mu_R^S(X_l, X_k)] = 1 - \sup_{X_l \in X} \mu_R^S(X_l, X_k) \tag{31}$$

At this point, the process of choosing alternatives is carried out, observing the levels of non-dominance. Therefore, the choice of alternatives X^{ND} based on the highest levels of non-dominance is carried out as follows:

$$X^{ND} = \{X_k^{ND} | X_k^{ND} \in X, \mu_R^{ND}(X_k^{ND}) = \sup_{X_k \in X} \mu_R^{ND}(X_k)\} \tag{32}$$

Monocriteria problems involving the choice, evaluation, comparison, prioritization, and classification of alternatives can be solved by applying (30)–(32). These expressions also apply when R is a fuzzy preference relations vector in $< X, R >$ models, working as the first technique of multiattribute decision-making, as $R = \bigcap_{p=1}^{q} R_p$, i.e.,

$$\mu_R(X_k, X_l) = \min_{1 \leq p \leq q} \mu_{R_p}(X_k, X_l), X_k, X_l \in X \tag{33}$$

The advantage of applying (33) to the set X^{ND} is in fulfilment of the role of the Pareto set, which is processed using convolution [54]:

$$\mu_T(X_k, X_l) = \sum_{p=1}^{q} \lambda_p \mu_{R_p}(X_k, X_l), X_k, X_l \in X \tag{34}$$

It is possible to contract X^{ND}. In (34), $\lambda_p \geq 0$, $p = 1, 2, \ldots, q$ are important coefficients or weights for the corresponding criteria. Therefore, these weights must be normalized, respecting $\sum_{p=1}^{q} \lambda_p = 1$.

From $\mu_T(X_k, X_l)$, $X_k, X_l \in X$, it is possible to obtain the membership function $\mu_T^{ND}(X_k)$ of the non-dominated alternatives through the intersection of $\mu_R^{ND}(X_k)$ and $\mu_T^{ND}(X_k)$, formulated as

$$\mu^{ND}(X_k) = min\{\mu_R^{ND}(X_k), \mu_T^{ND}(X_k)\},\ X_k \in X \tag{35}$$

which generates X^{ND} following a correlation to (32).

The second technique applied in the research is lexicographic. In short, the second technique involves applying correlations (31) and (32) and the step-by-step inclusion of criteria. Using the second technique, it is possible to obtain $X^1, X^2, \ldots, X^q$, in such a way that $X \supseteq X^1 \supseteq X^2 \supseteq, \ldots, \supseteq X^q$ through the application of the following correlations:

$$\mu_{R_p}^{ND}(X_k) = \inf_{X_l \in X^{p-1}} [1 - \mu_{R_p}^{S}(X_l, X_k)] = 1 - \sup_{X_l \in X^{p-1}} \mu_{R_p}^{S}(X_l, X_k), p = 1, 2, \ldots, q \tag{36}$$

$$X^{\,p} = \{X_k^{ND,p} | X_k^{ND,p} \in X^{p-1}, \mu_{R_p}^{ND}(X_k^{ND,p}) = \sup_{X_l \in X^{p-1}} \mu_{R_p}^{ND}(X_k)\} \tag{37}$$

The expression (31) is written as

$$\mu_{R_p}^{ND}(X_k) = 1 - \sup_{X_l \in X} \mu_{R_p}^{S}(X_l, X_k),\ p = 1, 2, \ldots, q \tag{38}$$

This expression corresponds to a membership function of the set of non-dominated alternatives of fuzzy preference relations p.

These membership functions are the core of the third technique. They assume a central role and are used to replace the objective functions $F_p(X), p = 1, \ldots, q$. In this way, they are integrated into the Bellman and Zadeh [5] approach to decision-making in a confusing environment for solving ultra-objective problems, allowing us to obtain X^{ND} [60]:

$$\mu^{ND}(X_k) = \min_{1 \leq p \leq q} \mu_{R_p}^{ND}(X_k) \tag{39}$$

For problems in which preference relations have different weights, it is possible to reformulate (39) as follows:

$$\mu^{ND}(X_k) = \min_{1 \leq p \leq q} [\mu_{R_p}^{ND}(X_k)]^{\lambda_p} \tag{40}$$

Note that, to use (40), it is not necessary to perform the normalization of λ_p, $p = 1, 2, \ldots, q$.

Once presented with three techniques, it is possible to indicate their weaknesses and strengths. Applying the first technique does not guarantee the indication of a solution equal to that obtained in the second technique. Different solutions can also occur between the results of the first and third techniques, even though to a lesser extent, as they have the same fundamental basis. Although the combination of quantitative and qualitative information (preferences in different homogenized formats) reduces regions of uncertainty in solutions and offers an answer to situations in which the first and second techniques point to distinct decisions, the decision makers' preferences included in the third technique are, in itself, sources of uncertainty. In this sense, it is recommended that decision-makers consider the results of the third technique as a casting vote when the solutions of the first and second techniques are divergent.

However, it is important to highlight that the third technique is the most rigorous. For this reason, the alternatives indicated by the third technique are more likely to present a degree of non-dominance equal to one. This can occur even if the alternatives are not the best, considering all preference relationships. The third technique produces this result only when the alternatives present the best solutions in all fuzzy preferences.

Considering the above, it should be stressed that the possibility of obtaining different solutions based on distinct approaches is accepted, and the preference of the technique is a decision-maker's entitlement.

It is important to highlight that the three techniques presented in this research are elementary since they are related to the explicit ordering of the criteria, requiring the distinction of the approach that enables the representation of information on the importance of fuzzy preference relations in a fuzzy format [54]. The findings associated with analyzing alternatives using the conception of a fuzzy majority through the ordered weighted averaging operator and some of its modifications are presented in Ekel et al. [4]. Finally, PROMETHEE's fuzzy outranking approach was adopted to analyze alternatives [61,62].

It is also important to highlight that applying the three preference modeling techniques in a fuzzy environment directly applies to decision-making problems in a multi-criteria group [10]. This is a considerable advantage, as most decision-making problems involve groups of experts [63–65].

6. General Scheme of Multi-Criteria Decision-Making under Uncertain Conditions

The results presented above enable the construction of the general multi-criteria decision-making scheme under conditions of uncertainty, including the three stages shown in Figure 1.

The first stage (steps 1 and 2) consists of building S representative combinations of initial data, states of nature, or scenarios and solving S multi-criteria problems of allocating resources or their deficits. This stage is also associated with constructing q-payoff matrices. The number of matrices depends on the objective functions considered in the model for each of the solutions X_k, $k = 1, 2, \ldots, K \leq S$, as well as for each of the scenarios Y_s, $s = 1, 2, \ldots S$.

The second stage (steps 3 to 6) analyzes the constructed payoff matrices. The execution of this stage is performed using the generalization of the classic approach. This approach enables us to deal with the information uncertainty in monocriteria decision-making to multi-criteria problems discussed in the present paper. The execution of this stage provides robust (non-dominated) solutions.

The third stage (step 7) is related to constructing and analyzing $< X, R >$ models for contracting the regions of decision uncertainty, if necessary. As shown above, the multiattribute models permit the consideration of quantitative and qualitative character indices. In the case of using preference formats other than non-reciprocal fuzzy preference relations, transformation functions are applied to provide homogeneous information for applying decision procedures [7,8,12].

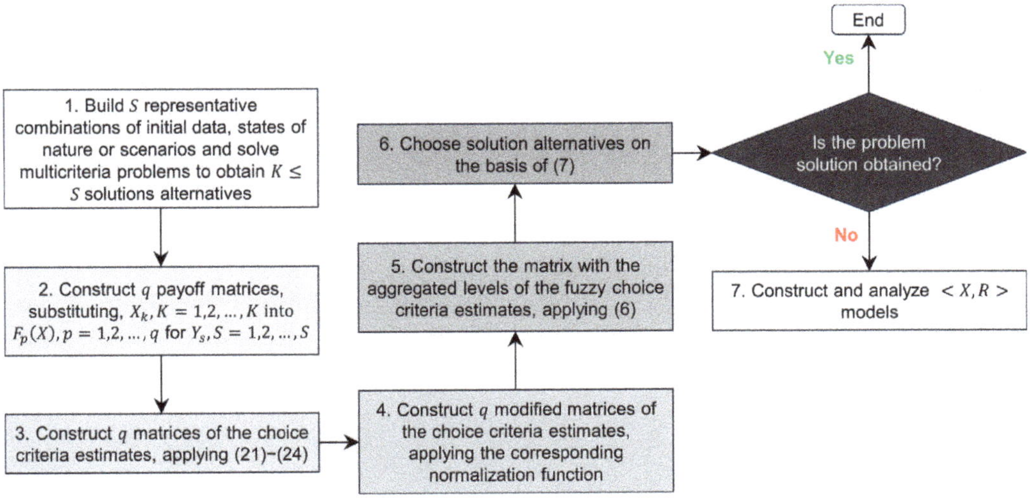

Figure 1. General scheme of multi-criteria decision-making under conditions of information uncertainty.

7. Computing System

The results presented above have been implemented within the system of multi-criteria decision-making in uncertain conditions. It was developed using the C#.NET framework and is executable in the Microsoft Windows® operational system. Although we discuss the use of the system of multi-criteria decision-making in uncertain conditions for solving the problem of the multi-criteria allocation of resources or their deficits, this software is of a universal character. The system helps resolve a wide range of planning problems.

The system of multi-criteria decision-making in uncertain conditions includes two subsystems. The first one corresponds to step 1 and is directed at solving problems of multi-criteria allocation of resources or their deficits for each scenario to obtain $K \leq S$ solution alternatives. The second subsystem corresponds to step 7 and is responsible for constructing and analyzing $< X, R >$ models. The subsystems have a complete and independent character and can be utilized for solving the corresponding problems. For instance, five preference formats can be considered and processed: ordering vector, utility vector, fuzzy estimates vector, multiplicative preference relations matrices, and fuzzy preference relations matrices [8,66].

The system of multi-criteria decision-making in uncertain conditions can solve allocation problems within three models of multiobjective decision-making, considering the uncertainty factor based on the possibilistic approach (see Figure 2).

The three models included in the system of multi-criteria decision-making in uncertain conditions are as follows:

- The allocation of available resources;
- The allocation of resource deficits with unlimited cuts;
- The allocation of resource deficits with limited cuts.

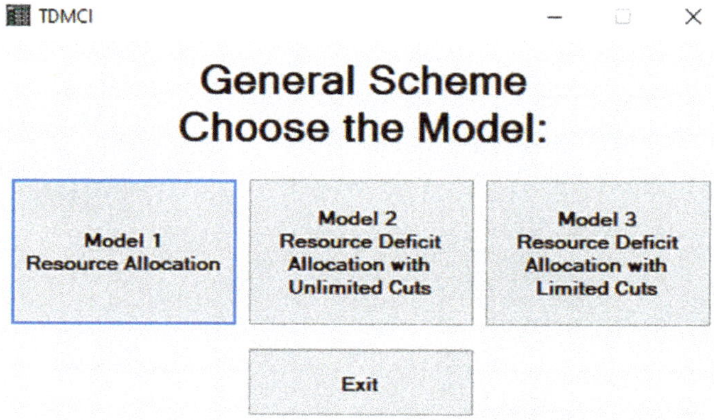

Figure 2. System of multi-criteria decision-making in uncertain conditions.

As seen in Figure 3, it is necessary to enter values relating to objective functions, available resources, and the number of scenarios (S) to be created to solve resource allocation problems in the system of multi-criteria decision-making in uncertain conditions.

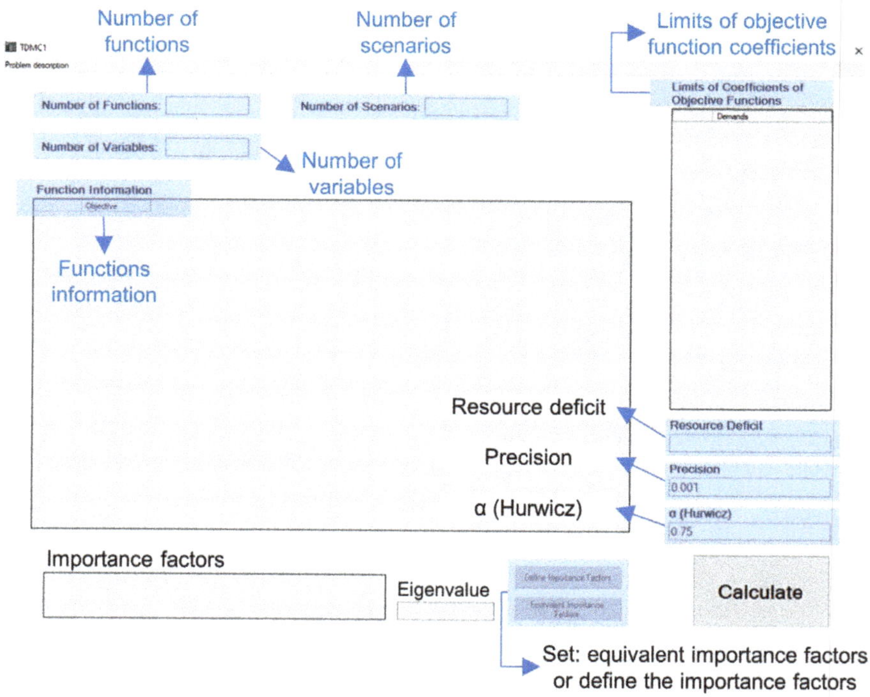

Figure 3. Screen for entering values related to objective functions, available resources, and scenarios for resolving resource allocation problems.

In addition to this mandatory data, the software allows for the insertion of importance coefficients into the objective functions by applying procedures described by Pedrycz et al. [38].

The availability of initial data permits one to execute steps 1–6, reflected by the flow chart in Figure 1. Suppose the execution of steps 1 to 6 produces two or more alternatives that cannot be convincingly distinguished. In that case, the software gives the user the option to start step 7, associated with constructing and analyzing $< X, R >$ models with the preliminary reduction in all the used preference formats.

The use of the multi-criteria decision-making system under conditions of uncertainty is illustrated by the analysis of multiobjective allocation, considering the scarcity of economic resources in an investment planning activity problem within the model of the allocation of resource deficits with a limited cuts framework. The problem is to be resolved for the fourth project by applying the following objectives:

1. Predominant economic limitations of investments ensure a smaller level of product trade overseas.
2. Predominant economic limitation of investments generating a lower percentage of profits for every one million dollars invested.

In the problems addressed in this research, fifty-two million dollars must be invested in the projects presented in Table 6.

Table 6. Initial data used in the investment allocation problem considering a deficit of economic resources.

Investment	D_i [M U$]	D_i^m [M U$]	A_i [M U$]
Project 1	12.53	10.02	2.51
Project 2	17.53	16.13	1.4
Project 3	9.74	7.77	1.98
Project 4	19.13	16.23	2.91

The analysis of this data allows us to obtain the following solution: $A = \sum_{i=1}^{n} D_i - R = 58.93 - 52.00 = 6.93$. Thus, we have to consider the following constraint:

$$\Delta x1 + \Delta x2 + \Delta x3 + \Delta x4 = 6927.00 \quad (41)$$

In addition, we have to take into account that

$$0 \le \Delta x1 \le 2512.00 \quad (42)$$

$$0 \le \Delta x2 \le 1398.00 \quad (43)$$

$$0 \le \Delta x3 \le 1976.00 \quad (44)$$

and

$$0 \le \Delta x4 \le 2910.00 \quad (45)$$

The initial data for constructing objective functions are associated with the use of $LP\tau$-sequences, and by considering the upper and lower limits of the importance coefficients of the objective functions. In this way, the seven scenarios S result in seven multiobjective problems. Then, the payoff matrices with characteristic estimates, matrices with the choice criteria estimates, and modified matrices with the choice criteria estimates for both objective functions were constructed by applying the results described above. In this way, the initial data to build the objective functions are presented as follows:

- Project 1: $c'_{11} = 20.5$; $c''_{11} = 31.5$; $c'_{21} = 2.21$; $c''_{21} = 2.85$.
- Project 2: $c'_{12} = 42.5$; $c''_{12} = 49.5$; $c'_{22} = 1.85$; $c''_{22} = 2.35$.
- Project 3: $c'_{13} = 25.5$; $c''_{13} = 33.5$; $c'_{23} = 1.70$; $c''_{23} = 1.95$.
- Project 4: $c'_{14} = 10.5$; $c''_{14} = 14.5$; $c'_{43} = 2.34$; $c''_{24} = 2.95$.

The matrix with the aggregated levels of the fuzzy choice criteria is obtained by applying the min operator. The results taken directly from the system of multi-criteria decision-making in uncertain conditions and that are presented in Figure 4 convincingly demonstrate that the problem's solution is ΔX2 = {Δx1 = 1971.689; Δx2 = 360.733; Δx3 = 1976.000; Δx4 = 2618.578}.

TDMC1

Problem description Step 1 Step 2 Step 3 Step 4 Step 5 Step 6

STEP1: Solve S multicriteria problems to obtain K ≤ S alternative solutions:

Solution Alternatives found:

	Δx_01	Δx_02	Δx_03	Δx_04
1	1 681,214	642,887	1 976,000	2 626,899
2	1 971,689	360,733	1 976,000	2 618,578
3	2 083,692	340,168	1 976,000	2 527,139
4	2 221,119	69,114	1 976,000	2 660,768
5	2 244,887	185,166	1 976,000	2 520,947
6	2 440,874	59,381	1 976,000	2 450,745
7	2 512,000	183,686	1 976,000	2 255,314

TDMC1

Problem description Step 1 Step 2 Step 3 Step 4 Step 5 Step 6

STEP5: Construct the matrix with the aggregated levels of the fuzzy choice criteria estimates:

Aggregated Payoff Matrix of estimates:

	F_W	F_L	F_S	F_H
ΔX1	0,000	0,000	0,000	0,000
ΔX2	0,446	0,463	0,478	0,469
ΔX3	0,314	0,395	0,456	0,423
ΔX4	0,051	0,000	0,165	0,047
ΔX5	0,221	0,263	0,354	0,220
ΔX6	0,000	0,080	0,000	0,000
ΔX7	0,075	0,349	0,171	0,179

Figure 4. Results taken directly from the system of multi-criteria decision-making in uncertain conditions.

However, there are situations where the aggregate levels of the fuzzy choice criteria point to more than one solution. In these situations, criteria of a qualitative nature can be inserted in the final stage of the decision.

8. Conclusions and Future Development

In this paper, the general scheme of multi-criteria decision-making under uncertain conditions within the framework of the possibilistic approach was considered. Since optimal solutions do not exist in uncertain conditions, the scheme is associated with constructing robust (non-dominated) solutions. The general scheme is based on generalizing the classical approach to dealing with information uncertainty in monocriteria decision-making for multi-criteria problems. This generalization permits one to simultaneously consider the characteristic estimates applied within the classic approach's choice criteria as objective functions in the multiobjective model framework. The general scheme includes three stages. The first consists of building representative combinations of initial data, states of nature, or scenarios and solving the corresponding multiobjective problems, formalized within the framework of $< X, F >$ models, for each. The second stage is constructing and analyzing payoff matrices to form robust (non-dominated) solutions. The third stage is associated with constructing and analyzing $< X, R >$ models and can be carried out when the solutions obtained in the second stage are not unique. The general scheme of multi-criteria decision-making under uncertain conditions helps us to reduce regions of decision uncertainty by maximizing the use of available quantitative information. However, the scheme presumes to apply qualitative information if convincing solutions are not achieved. In these situations, experts are consulted, and qualitative information is obtained based on their knowledge, experience, and intuition. At this point, it is worth noting that the proposed scheme and its computational implementation allow experts to express their preferences through the different formats, with transformation functions being applied to homogenize the information provided and use it in the decision process.

The general scheme of multi-criteria decision-making under uncertainty was applied to solving a significant real-world problem of the multiobjective allocation of resources (or their deficits), answering the fundamental question "How is this done?", which arises in diverse planning activities.

Finally, it is necessary to point out that our search for literature sources permits us to conclude that the existing literature does not contain the following topics, which are the main research objects of the present work:

- Ways of constructing robust multi-criteria solutions.
- The solution of decision-making problems that require the simultaneous analysis of multiobjective models ($< X, F >$ models) as well as multiattribute models ($< X, R >$ models).
- The generation of harmonious solutions in analyzing $< X, F >$ models.
- Structuration problems of the multicriteia allocation of resources or their deficits.

The results presented have been implemented within the system of multi-criteria decision-making in uncertain conditions. Its functioning is illustrated by solving the typical problem in investment planning activities, as well as urban logistic planning problems [67], future elective courses in undergraduate programs [68], the control of production processes [69], environment, social, corporate governance, and economic efficiency [70], the recovery of historic bridges [71], teaching–learning in higher education [72], processes of construction supply chains [73], the installation of low-carbon energy technologies [74], and governmental strategic planning [75].

The future development of the results of the present work is associated with producing robust, non-dominated solutions. These solutions are based on representative combinations of initial data, states of nature, or scenarios built by applying heterogeneous quantitative and qualitative information. The possibility of processing heterogeneous qualitative information offers specialists the possibility of choosing the evaluation format of their preference, increasing their psychological comfort and reducing the cognitive efforts of the evaluation process. This possibility is made possible by applying transformation functions that homogenize qualitative information, allowing for its combination with quantitative information, providing information fusion mechanisms within two multiobjective models, applied to the multiobjective allocation of resources or their deficits. In addition, procedures are to be developed to achieve dialog at any stage of the decision process, applying qualitative information. Finally, to improve the efficiency of the use of qualitative information, procedures of consensus construction are to be developed and implemented.

Author Contributions: Conceptualization, J.G.P.J. and P.I.E.; data curation, L.C.R.; formal analysis, L.C.R. and M.A.D.d.O.F.; funding acquisition, P.I.E.; investigation, L.C.R. and M.A.D.d.O.F.; methodology, J.G.P.J. and P.I.E.; project administration, P.I.E.; software, M.A.D.d.O.F. and J.G.P.J.; supervision, M.P.L.; validation, M.A.D.d.O.F.; visualization, M.P.L.; writing—original draft, L.C.R.; writing—review and editing, M.P.L.and P.I.E. All authors have read and agreed to the published version of the manuscript.

Funding: This study was financed in part by Vale S.A. (within the Research, Development, and Innovation Partnership Agreement), and the National Council for Scientific and Technological Development of Brazil (CNPq), Productivity grants 311922/2021-0 (P.I.E.), and Postdoctoral fellowship 151518/2022-0 (M.P.L.).

Data Availability Statement: The authors confirm that the data supporting the findings of this study are available within the article.

Conflicts of Interest: The authors declare no conflict of interest.

References

1. Xu, Z. *Uncertain Multiple Attribute Decision Making: Methods and Applications*; CRC Press: Boca Raton, FL, USA, 2014.
2. Hwang, C.L.; Masud, A.C. *Multiple Objective Decision Making: Methods and Applications*; Springer Science & Business Media: Berlin/Heidelberg, Germany, 1979.
3. Hwang, C.L.; Yoon, K. *Multiple Attribute Decision Making: Methods and Applications—A State-of-the-Art Survey*; Springer Science & Business Media: Berlin/Heidelberg, Germany, 1981.
4. Ekel, P.; Kokshenev, I.; Parreiras, R.; Pedrycz, W.; Pereira, J., Jr. Multiobjective and multiattribute decision making in a fuzzy environment and their power engineering applications. *Inf. Sci.* **2016**, *361*, 100–119. [CrossRef]
5. Bellman, R.E.; Zadeh, L.A. Decision-making in a fuzzy environment. *Manag. Sci.* **1970**, *17*, 141–164. [CrossRef]
6. Ekel, P.; Pedrycz, W.; Pereira, J., Jr. *Multicriteria Decision-Making under Conditions of Uncertainty: A Fuzzy Set Perspective*; John Wiley & Sons: Hoboken, NJ, USA, 2020.
7. Büyüközkan, G.; Tüfekçi, G. A decision-making framework for evaluating appropriate business blockchain platforms using multiple preference formats and VIKOR. *Inf. Sci.* **2021**, *571*, 337–357. [CrossRef]
8. Chao, X.; Kou, G.; Peng, Y.; Viedma, E.H. Large-scale group decision-making with non-cooperative behaviors and heterogeneous preferences: An application in inclusion. *Eur. J. Oper. Res.* **2021**, *288*, 271–293. [CrossRef]
9. Wu, Z.; Liao, H. A consensus reaching process for large-scale group decision making with heterogeneous preference information. *Int. J. Intell. Syst.* **2021**, *36*, 4560–4591. [CrossRef]
10. Herrera-Viedma, E.; Herrera, F.; Chiclana, F. A consensus model for multiperson decision making with different preferences structures. *IEEE Trans. Syst. Man Cybern.* **2002**, *A-32*, 394–402. [CrossRef]
11. Zhang, Q.; Xu, Z.; Zhou, B. Group consensus measurement in MADM with multiple preference formats. *J. Decis. Syst.* **2015**, *24*, 146–158. [CrossRef]
12. Chiclana, F.; Herrera, F.; Herrera-Viedma, E. Integrating three representation models in fuzzy multipurpose decision making based on fuzzy preference relations. *Fuzzy Sets Syst. Financ.* **1998**, *97*, 33–48. [CrossRef]
13. Ervural, B.; Kabak, Ö. A cumulative belief degree approach for group decision-making problems with heterogeneous information. *Expert Syst.* **2019**, *36*, e12458. [CrossRef]
14. Luce, R.D.; Raiffa, H. *Games and Decisions*; John Wiley & Sons: Hoboken, NJ, USA, 1957.
15. Raiffa, H. *Decision Analysis*; Addison-Wesley: Boston, MA, USA, 1968.
16. Pereira, J.G., Jr.; Ekel, P.Y.; Palhares, R.M.; Parreiras, R.O. On multicriteria decision making under conditions of uncertainty. *Inf. Sci.* **2015**, *324*, 44–59. [CrossRef]
17. Stewart, T. Dealing with uncertainties in MCDA. In *Multiple Criteria Decision Analysis—State of the Art Annotated Surveys, International Series in Operations Research and Management Science*; Springer: New York, NY, USA, 2005; pp. 445–470.
18. Ramalho, F.D.; Ekel, P.; Pedrycz, W.; Pereira, J.G., Jr.; Soares, G.L. Multicriteria decision making under conditions of uncertainty in application to multiobjective allocation of resources. *Inf. Fusion* **2019**, *48*, 249–261. [CrossRef]
19. Durbach, I.N.; Stewart, T.J. Modeling uncertainty in multi-criteria decision analysis. *Eur. J. Oper. Res.* **2012**, *223*, 1–14. [CrossRef]
20. Eiself, H.A.; Marianov, V. Multicriteria decision making under uncertainty: A visual approach. *Int. Trans. Oper. Res.* **2014**, *21*, 525–540. [CrossRef]
21. Gaspars-Wieloch, H. The use of a modification of the Hurwicz's decision rule in multicriteria decision making under complete uncertainty. *Bus. Manag. Econ. Eng.* **2014**, *12*, 283–302. [CrossRef]
22. Alvarez, P.A.; Ishizaka, A.; Martínez, L. Multiple-criteria decision-making sorting methods: A survey. *Expert Syst. Appl.* **2021**, *183*, 115368. [CrossRef]
23. Sahoo, S.K.; Goswami, S.S. A comprehensive review of multiple criteria decision-making (MCDM) Methods: Advancements, applications, and future directions. *Decis. Mak. Adv.* **2023**, *1*, 25–48. [CrossRef]
24. Khan, I.; Pintelon, L.; Martin, H. The application of multicriteria decision analysis methods in health care: A literature review. *Med. Decis. Mak.* **2022**, *42*, 262–274. [CrossRef] [PubMed]
25. Chakraborty, S.; Raut, R.D.; Rofin, T.M.; Chakraborty, S. A comprehensive and systematic review of multi-criteria decision-making methods and applications in healthcare. *Healthc. Anal.* **2023**, *4*, 100232. [CrossRef]
26. Bohra, S.S.; Anvari-Moghaddam, A. A comprehensive review on applications of multicriteria decision-making methods in power and energy systems. *Int. J. Energy Res.* **2022**, *46*, 4088–4118. [CrossRef]
27. Maček, D.; Magdalenić, I.;Ređep, N.B. A systematic literature review on the application of multicriteria decision making methods for information security risk assessment. *Int. J. Saf. Secur. Eng.* **2020**, *10*, 161–174. [CrossRef]
28. Nadkarni, R.R.; Puthuvayi, B. A comprehensive literature review of Multi-Criteria Decision Making methods in heritage buildings. *J. Build. Eng.* **2020**, *32*, 101814. [CrossRef]
29. Yannis, G.; Kopsacheili, A.; Dragomanovits, A.; Petraki, V. State-of-the-art review on multi-criteria decision-making in the transport sector. *J. Traffic Transp. Eng. (Engl. Ed.)* **2020**, *7*, 413–431. [CrossRef]
30. Abdulrahman, S.; Bwambale, J. A review on flood risk assessment using multicriteria decision making technique. *World Water Policy* **2021**, *7*, 209–221. [CrossRef]

31. Gebre, S.L.; Cattrysse, D.; Van Orshoven, J. Multi-criteria decision-making methods to address water allocation problems: A systematic review. *Water* **2021**, *13*, 125. [CrossRef]
32. Hii, P.K.; Goh, C.F.; Rasli, A.; Tan, O.K. A review of two decades of research on the application of multicriteria decision making techniques to evaluate e-learning's effectiveness. *Knowl. Manag. E-Learn.* **2022**, *14*, 466.
33. Kandakoglu, M.; Walther, G.; Ben Amor, S. The use of multi-criteria decision-making methods in project portfolio selection: A literature review and future research directions. *Ann. Oper. Res.* **2023**, *332*, 807–830. [CrossRef]
34. Yalcin, A.S.; Kilic, H.S.; Delen, D. The use of multi-criteria decision-making methods in business analytics: A comprehensive literature review. *Technol. Forecast. Soc. Chang.* **2022**, *174*, 121193. [CrossRef]
35. Sousa, M.; Almeida, M.F.; Calili, R. Multiple criteria decision making for the achievement of the UN sustainable development goals: A systematic literature review and a research agenda. *Sustainability* **2021**, *13*, 4129. [CrossRef]
36. Seleznev, P.S.; Naumov, V.N.; Zorin, V.Y.; Zelenov, V.I.; Tsyplenkov, D.S.; Vasiliev, V.G. Research and Development of a Unified Methodology for Assessing the Resource Efficiency of International Digital Platform Promotion for E-Learning. *Symmetry* **2022**, *14*, 497. [CrossRef]
37. Chupina, Z.; Chursin, A.; Boginsky, A.; Kvasov, I. Sustainable Economic Development of Enterprises: A Methodology Based on the Toolkit. *Sustainability* **2023**, *15*, 12682. [CrossRef]
38. Pedrycz, W.; Ekel, P.; Parreiras, R. *Fuzzy Multicriteria Decision-Making: Models, Methods, and Applications*; John Wiley & Sons: Hoboken, NJ, USA, 2011.
39. Ehrgott, M. *Multicriteria Optimization*; Springer: Berlin/Heidelberg, Germany, 2005.
40. Jin, X.; He, T.; Lin, Y. Multi-objective model selection algorithm for online sequential ultimate learning machine. *EURASIP J. Wirel. Commun. Netw.* **2019**, *2019*, 156. [CrossRef]
41. Zhang, W.; Li, F.; Tolbert, L.M. Review of reactive power planning: Objectives, constraints, and algorithms. *IEEE Trans. Power Syst.* **2007**, *22*, 2177–2186. [CrossRef]
42. Neto, B.M.; Araújo, G.H.; Guimarães, F.G.; Mesquita, R.C.; Ekel, P.Y. A fuzzy genetic algorithm for automatic orthogonal graph drawing. *Appl. Soft Comput.* **2012**, *12*, 1379–1389. [CrossRef]
43. Ekel, P.Y. Fuzzy sets and models of decision making. *Comput. Math. Appl.* **2002**, *44*, 863–875. [CrossRef]
44. Singh, D.; Singh, D.; Verma, K. Multiobjective optimization for DG planning with load models. *IEEE Trans. Power Syst.* **2009**, *24*, 427–436. [CrossRef]
45. Sobol, I. On the systematic search in a hypercube. *SIAM J. Numer. Anal.* **1979**, *16*, 790–793. [CrossRef]
46. Belyaev, L.S. *A Practical Approach to Choosing Alternative Solutions to Complex Optimization Problems under Uncertainty*; IIASA: Laxenburg, Austria, 1977.
47. Shu, Z.; Carrasco, R.A.; García-Miguel, J.P.; Sánchez-Montañés, M. Multiple Scenarios of Quality of Life Index Using Fuzzy Linguistic Quantifiers: The Case of 85 Countries in Numbeo. *Mathematics* **2022**, *10*, 2091. [CrossRef]
48. Yager, R.R. On ordered weighted averaging aggregation operators in multi-criteria decision making. *IEEE Trans. Syst. Man Cybern.* **1988**, *18*, 183–190. [CrossRef]
49. Yager, R.R.; Kacprzyk, J.; Beliakov, G. (Eds.) *Recent Developments in the Ordered Weighted Averaging Operators: Theory and Practice*; Springer: Berlin/Heidelberg, Germany, 2011.
50. Yager, R.R. OWA aggregation with an uncertainty over the arguments. *Inf. Fusion* **2019**, *52*, 206–212. [CrossRef]
51. Li, W.; Yi, P.; Zhang, D. Quantile-induced vector-based heavy OWA operator and the application in dynamic decision making. *Int. J. Intell. Syst.* **2020**, *35*, 250–266. [CrossRef]
52. Fodor, J.; Roubens, M. *Fuzzy Preference Modelling and Multicriteria Decision Support*; Springer Science & Business Media: Dordrecht, The Netherlands, 1994.
53. Ekel, P.; Pedrycz, W.; Schinzinger, R. A general approach to solving a wide class of fuzzy optimization problems. *Fuzzy Sets Syst.* **1998**, *97*, 49–66. [CrossRef]
54. Orlovsky, S.A. *Problems of Decision Making with Fuzzy Information*; IIASA: Laxenburg, Austria, 1981. (In Russian)
55. Zhang, Q.; Chena, J.C.H.; Chong, P.P. Decision consolidation: Criteria weight determination using multiple preference formats. *Decis. Support Syst.* **2004**, *38*, 247–258. [CrossRef]
56. Zhang, Q.; Wang, Y.; Yang, Y. Fuzzy multiple attribute decision making with eight types of preference information. In Proceedings of the 2007 IEEE Symposium on Computational Intelligence in Multicriteria Decision Making, Honolulu, HI, USA, 1–5 April 2007; IEEE: Piscataway, NY, USA, 2007; pp. 288–293.
57. Chiclana, F.; Herrera, F.; Herrera-Viedma, E. A note on the internal consistency of various preference representations. *Fuzzy Sets Syst.* **2002**, *131*, 75–78. [CrossRef]
58. Büyüközkan, G.; Cifci, G. An extended quality function deployment incorporating fuzzy logic and GDM under different preference structures. *Int. J. Comput. Intell. Syst.* **2015**, *8*, 438–454. [CrossRef]
59. Büyüközkan, G.; Feyzioğlu, O.; Ruan, D. Fuzzy group decision-making to multiple preference formats in quality function deployment. *Comput. Ind.* **2007**, *58*, 392–402. [CrossRef]
60. Ekel, P.Y.; Neto, F.H.S. Algorithms of discrete optimization and their application to problems with fuzzy coefficients. *Inf. Sci.* **2006**, *176*, 2846–2868. [CrossRef]
61. Brans, J.P.; Vincke, P. A preference ranking organisation method: The PROMETHEE method for multiple criteria decision making. *Manag. Sci.* **1985**, *31*, 647–656. [CrossRef]

62. Wei-xiang, L.; Bang-yi, L. An extension of the Promethee II method based on generalized fuzzy numbers. *Expert Syst. Appl.* **2010**, *37*, 5314–5319.
63. Wang, G.; Wu, L.; Liu, Y.; Ye, X. A review on fuzzy preference modeling methods for group decision-making. *J. Intell. Fuzzy Syst.* **2021**, *40*, 10645–10660. [CrossRef]
64. Gupta, M.; Mohanty, B.K. An algorithmic approach to group decision making problems under fuzzy and dynamic environment. *Expert Syst. Appl.* **2016**, *55*, 118–132. [CrossRef]
65. Kokshenev, I.; Parreiras, R.; Ekel, P.; Alves, G.; Menicucci, S. A web-based decision support center for electrical energy companies. *IEEE Trans. Fuzzy Syst.* **2015**, *23*, 16–28. [CrossRef]
66. Libório, M.P.; Ekel, P.I.; Bernardes, P.; Gomes, L.F.A.M.; Vieira, D.A.G. Specialists' knowledge and cognitive stress in making pairwise comparisons. *Opsearch* **2023**, *61*, 51–70. [CrossRef]
67. Kovač, M.; Tadić, S.; Krstić, M.; Veljović, M. A methodology for planning city logistics concepts based on city-dry port micro-consolidation centres. *Mathematics* **2023**, *11*, 3347. [CrossRef]
68. Hashemkhani Zolfani, S.; Nemati, A.; Reyes-Norambuena, P.J.; Monardes-Concha, C.A. A Novel MCDM Approach Based on OPA-WINGS for Policy Making in Undergraduate Elective Courses. *Mathematics* **2022**, *10*, 4211. [CrossRef]
69. Fattoruso, G.; Scognamiglio, S.; Violi, A. A New Dynamic and Perspective Parsimonious AHP Model for Improving Industrial Frameworks. *Mathematics* **2022**, *10*, 3138. [CrossRef]
70. Chang, S.C.; Lu, M.T.; Chen, M.J.; Huang, L.H. Evaluating the Application of CSR in the High-Tech Industry during the COVID-19 Pandemic. *Mathematics* **2021**, *9*, 1715. [CrossRef]
71. Rogulj, K.; Kilić Pamuković, J.; Ivić, M. Hybrid MCDM based on VIKOR and cross entropy under rough neutrosophic set theory. *Mathematics* **2021**, *9*, 1334. [CrossRef]
72. Laguna-Sánchez, P.; Palomo, J.; de la Fuente-Cabrero, C.; de Castro-Pardo, M. A Multiple Criteria Decision Making Approach to Designing Teaching Plans in Higher Education Institutions. *Mathematics* **2020**, *9*, 9. [CrossRef]
73. Keshavarz-Ghorabaee, M.; Amiri, M.; Hashemi-Tabatabaei, M.; Zavadskas, E.K.; Kaklauskas, A. A new decision-making approach based on Fermatean fuzzy sets and WASPAS for green construction supplier evaluation. *Mathematics* **2020**, *8*, 2202. [CrossRef]
74. Kokkinos, K.; Karayannis, V. Supportiveness of low-carbon energy technology policy using fuzzy multicriteria decision-making methodologies. *Mathematics* **2020**, *8*, 1178. [CrossRef]
75. Majidi, A.; Mirzapour Al-e-Hashem, S.M.; Hashemkhani Zolfani, S. Sustainability ranking of the Iranian major ports by using MCDM methods. *Mathematics* **2021**, *9*, 2451. [CrossRef]

mathematics

MDPI

Article

Are Brazilian Higher Education Institutions Efficient in Their Graduate Activities? A Two-Stage Dynamic Data-Envelopment-Analysis Cooperative Approach

Lívia Mariana Lopes de Souza Torres [1,*] and Francisco S. Ramos [1,2]

1 Department of Production Engineering, Graduate Program in Management Engineering, Federal University of Pernambuco (UFPE), Architecture Avenue, s/n, Recife 50740-550, Brazil
2 Department of Economics, Laboratory of Risk Management, Governance and Compliance—LabGRC, Federal University of Pernambuco (UFPE), Professor Moraes Rego, s/n, Recife 50670-901, Brazil; francisco.ramos@ufpe.br
* Correspondence: livia.mariana@ufpe.br; Tel.: +55-84996708424

Abstract: Higher education evaluation presents itself as a worldwide trend. It aims to improve performance due to its importance for economic and personal growth. Graduate activities are essential for Brazilian research and innovation systems. However, previous studies have disregarded the importance of this educational level and have evaluated efficiency by jointly considering teaching and research or only undergraduate courses. Therefore, this study contributes to Brazilian reality by proving a national graduate activities efficiency evaluation that considers them as a two-stage system (formative and scientific production stages). The study provides three main methodological contributions by presenting a new centralized two-stage dynamic network data envelopment analysis (DNDEA) model with shared resources. Besides measuring efficiency, an efficiency decomposition based on a leader–follower assumption shows managers how much efficiency can alter when one of the stages needs to be prioritized. Finally, a new framework based on modified virtual inputs and outputs provides a bi-dimensional representation of the efficiency frontier. Results indicate the usefulness of the approach for ranking universities, and the need to improve scientific production, highlighting the negative impacts of COVID-19 on the formative process efficiency and showing no significant regional discrepancies regarding performance.

Keywords: universities; shared resources; dynamic network data envelopment analysis; game theory; bi-dimensional representation; efficiency decomposition

MSC: 90C90; 90C05; 91A12; 91A35; 91A80

1. Introduction

Universities represent a driving force of science and knowledge in countries, and they are crucial for providing a skilled and expert workforce in the job market [1]. Evaluating their results is a complex process due to the existence of different indicators to obtain an overview of system performance [2]. Understanding how to increase the universities' performance is challenging for governments, leading operators, and funders [3,4]. The last decades have shown a worldwide trend for implementing exercises about evaluation and for a comparison of various estimation methods [5].

In Brazil, it is possible to verify the same trend. Law 10.861/2004 instituted the National Higher Education Assessment System (SINAES) in 2004. SINAES aims to improve the results of the Brazilian higher education system and consists of three main components: the performance evaluation of institutions, courses, and students. Higher education censuses occur annually to collect data about the three dimensions to obtain indicators used in SINAES to assess and accredit courses and institutions.

Brazilian graduate courses are responsible for a large part of the research and innovation generated in Brazil [6]. Regardless of its importance, the report on investments in research and development in the world carried out by the United Nations Educational, Scientific, and Cultural Organization (UNESCO) considering 2014 to 2018, shows that the budget reduction of the Ministry of Science, Technology and Innovation (MCTI) in the same period was around 50% [7]. Considering 2012 to 2021, the reduction corresponds to 84% (from BRL 11.5 billion to BRL 1.8 billion, in inflation-adjusted values). Despite the reduction, the report indicates the continuous growth in scientific production.

Governmental assessment of Brazilian HEIs' graduate activities is crucial. However, the General Index of Courses (IGC) is the sole indicator that accounts for graduate accomplishments. Despite its importance, there are several criticisms regarding this indicator.

Efficiency estimates in complex systems with multiple inputs and outputs, such as education, are achievable using data envelopment analysis (DEA) [8]. DEA is an instructive tool in the educational context, and the ability to deal with multiples inputs and multiple outputs represents one of the reasons for the range of applications [8]. Surveys regarding educational applications and types of evaluations within the DEA field [8–11] show that analysis regarding cost efficiency, technical efficiency, research performance, administrative services evaluation, university rankings, assessing academics on teaching and research activities, and student performance were developed. Despite the extensive volume of research already carried out, discussions about the differential emphasis given to education or to research in each institution are still scarce [5].

In this paper, we propose an innovative way to measure the efficiency of graduate activities to provide a national view of this level of education in Brazilian HEIs. We developed a new centralized two-stage dynamic DEA model with shared resources and a bi-dimensional representation of its results. We contemplate the formative and scientific production processes. A leader–follower framework is also used to investigate the impacts on efficiency decomposition when one of the stages needs to be prioritized.

The choice of a network structure lies in the fact that simple black-box structures cannot accurately reflect the complex production process in real life, making it easy to overlook important information in production activities [12]. Although relevant, most investigations visualize universities as "black boxes" and do not consider internal processes. Several studies contemplated teaching or research activities, disregarding the existence of both processes at the graduate level. Following this reasoning, the use of multi-stage models is required in order to adequately portray such features.

Besides being multi-product and multi-process organizations, the educational process usually takes several years, and investigating productivity changes across time is necessary to comprehend whether universities have improved, stagnated, or regressed in their performance [13]. Due to this multi-period feature, suitable models are required to portray the situation adequately. Dynamic DEA (DDEA), and dynamic DEA models with network structure (DNDEA) represent DEA alternatives available to incorporate temporal aspects into efficiency measures. It is possible to identify works that have addressed university evaluation, considering DDEA [13–15] or DNDEA [16–20].

Clarity and simplifications can be very valuable in a world where data are increasingly abundant [20]. In Brazil, partial evaluations for 2021 indicate the existence of 27,711, 1054, 829, and 37 graduate programs in federal, state, private, and municipal institutions, respectively. Therefore, analyzing such dimensions requires significant effort from the committees and evaluation teams. It is also noteworthy that both in the Brazilian case and international assessments, the commissions are multidisciplinary, and not all members are always familiar with mathematical programming models.

Considering such particularities, we develop a bi-dimensional representation to visually display the efficiency frontier and the DMUs' positions concerning the frontier. Since DNDEA models provide several efficiency levels, modified virtual inputs and outputs constitute the selected tool used to represent all the different efficiency scores obtained with the DNDEA model. The bi-dimensional representation summarizes the information in a simple and straightforward way. This tool allows direct efforts, helps persuade managers and policymakers about the validity of the results, and translates recommendations into actions [21].

This work presents four major contributions to the literature and governmental management actions. We develop a new DNDEA framework to reflect a vital but minimally explored level of education in Brazil, graduate activities. To the authors' knowledge, we are the first to investigate the efficiency of the graduate level considering its internal processes (formative and scientific production) in a dynamic manner.

The proposed models open up a way to enrich DNDEA studies by considering the shared inputs among the stages of the network. To our knowledge, only few studies considered shared resources in the DNDEA framework. In addition to proposing a new model, we also discuss the impact on the efficiency decomposition of the stages when there is a need to prioritize one of them for managerial reasons. Although different DNDEA models have been used to evaluate university performance, optimizing resource allocation is most important for the Brazilian case, since approximately half of the Brazilian graduate courses are developed in public universities and Brazilian public research agencies finance scholarships for master's and doctoral students in these institutions. Therefore, this analysis supports the best use of public resources.

We provide comprehensive efficiency analysis of teaching and research activities at graduate level. The decomposition of the overall efficiency for the entire horizon can better aid managers in finding the process that requires more attention to prioritize resources. Therefore, relevant insights are provided for the government and HEIs on improving performance. Lastly, we also developed a simple but effective way to present the results for managers and policymakers, aiding in better comprehension and reducing cognitive efforts in the decision-making process.

The following section details a literature review on the main topics relating to the work: the Brazilian evaluation system, DNDEA models and visual representation in the DEA context. Section 3 presents methodology and data. Section 4 presents the results, and Section 5 concludes the review.

2. Literature Review

2.1. The Brazilian Evaluation System

In Brazil, there are four groups of higher education institutions (HEIs): universities, university centers, faculties, and federal institutes. We can classify HEIs into four administrative categories: federal, state, municipal, and private. These institutions are evaluated annually by SINAES with the aid of micro data collected by the Anísio Teixeira National Institute for Educational Research and Studies (INEP) in the higher education census.

According to Normative Ordinance n. 550 [22], SINAES is composed of six quality metrics: Institutional Evaluation (AVALIES), Course Evaluation (ACG), General Index of Courses (IGC), Preliminary Concept of Courses (CPC), Indicator of the Difference between Observed and Expected Performances (IDD) and the National Student Performance Examination (ENADE). The last four converge in their results but they do not communicate much with the first two, and only these four have their results released annually: ENADE since 2004, CPC and IGC since 2007, and IDD since 2014.

The evaluation processes are coordinated and supervised by the National Commission for the Evaluation of Higher Education (CONAES), while the operation is the responsibility of INEP [23]. On the other hand, the evaluation of graduate programs is performed by the Coordination for the Improvement of Higher Education Personnel (CAPES), which gives a score ranging from one to seven.

Considering all the above, it is possible to affirm that SINAES is a complex process involving different time periods and multiple tools, and it also enables the production, dissemination and management of indicators and information for Brazilian HEIs [24].

The General Course Index (IGC) is the quality indicator used to rank and guide HEIs' evaluation. It considers metrics of the quality of all undergraduate, master, and doctoral courses at an HEI, aggregating them all into one indicator. Despite the importance of this indicator for HEIs, it is possible to find several criticisms of its construction in the literature.

First, as shown in Figure 1, SINAES indicators directly impact each other. This implies that several composite indicators are applied in the construction of the IGC. Therefore, problems in these indicators can impact the final result of the IGC. It is also possible to verify problems regarding the weighting of the considered criteria.

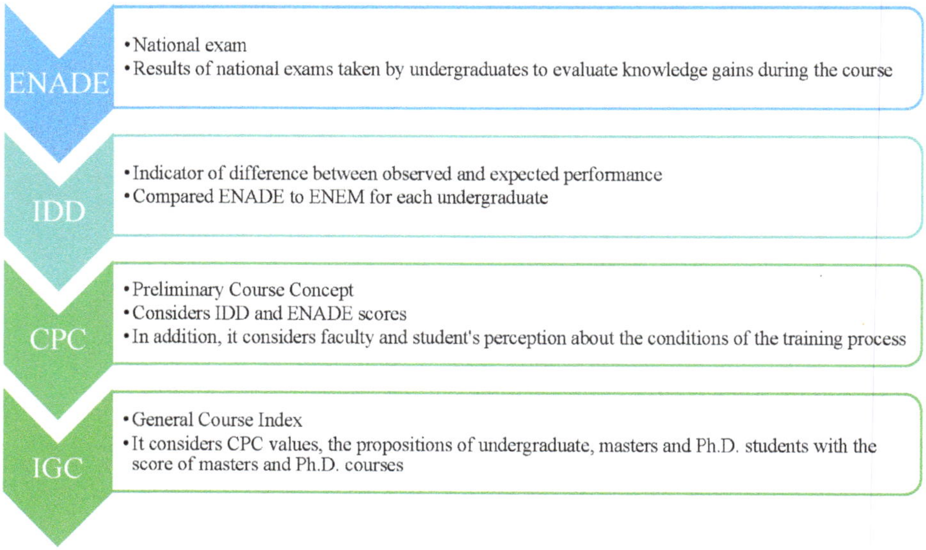

Figure 1. Description of Brazilian quality indicators.

Technical notes issued by the government do not justify the choice of weighting for each criteria, and minor weight variations can significantly change the results [25,26]. Another criticism relates to using the same criteria for courses in different areas, in different types of institutions, and for different regions of the country [27].

In addition to these issues, another point deserves attention. There are individual and in-depth assessments for undergraduate and graduate courses. However, the indicator relating to higher education institutions (IGC) aggregates information from all undergraduate and graduate courses at these HEIs without any distinction between these levels of education. This aggregation does not allow for providing targets or projections of how each educational level should improve to enhance the institution's performance.

Due to the mentioned problems and to the fact that there is no indicator to aggregate and show a global overview of graduate activities, the current work proposes a method to limit this gap.

The implementation of graduate studies in Brazil took place with the creation of CAPES in 1951 and through the standards defined by Report CFE 977/65 of 1965 [28]. National discussions are taking place to reformulate the evaluation process of graduate programs in Brazil. CAPES evaluates graduate programs concerning the National Graduate Plan (PNPG) guidelines. Currently, the seventh PNPG is in effect, but we are using data for the period (2019–2020) contemplated by the sixth plan. Thus, our results can help in this discussion and foster the evaluation for the seventh plan, which is still ongoing.

For the sixth plan, we had political and economic crises (2011–2020). After 2015 and the impeachment of then-president Dilma Roussef, there was a reduction in federal government transfers to higher education, with budget cuts in science and technology. The scenario becomes even worse after 2019, with the contingency of part of the budget directed to discretionary spending by federal universities, including payment of academic grants and research inputs.

The scenario of scarce resources and high public investment in the sector has motivated research to measure efficiency in this field. Ref. [29] investigated the impact of information asymmetry on organizational efficiency using data about Brazilian undergraduate courses. Ref. [30] used DEA and SFA to investigate the efficiency of undergraduate business administration courses. Ref. [31] applied ordinary least squares and SFA to investigate differences in private and public Brazilian universities' performance. Ref. [32] addresses the efficiency of public expenditure in federal universities, and their results indicate most federal universities analyzed are still inefficient in allocating public expenditures. Ref. [33] focused on Brazil's Federal Institute of Education, Science and Technology. They created several efficiency measures based on DEA and TOPSIS and evaluated the correlation of such scores with performance indicators applied by the Ministry of Education between 2014 and 2017. The authors verified the fact that HEIs did not improve significantly in the considered time frame. Ref. [34] is the only study that considered graduate activities in their investigations. A multistage network DEA model is applied to investigate HEIs regarding their financial, undergraduate and graduate performance.

Although Ref. [34] consider graduate aspects, they focus on allocating public resources among undergraduate and graduate activities. Their discussions also disregard time effects and can present bias, since they used quality indicators as outputs. The referred indicators contemplate the same inputs used in their evaluation, therefore being redundant. It is also important to mention that none focused exclusively on graduate activities. Therefore, the present study is the first to propose a dynamic evaluation of graduate activities and consider their internal structure. Considering the network structure will facilitate the identification of aspects that need reinforcing to foster improvement and guide managers to use public resources better. In order to achieve such goals, we developed a new dynamic DEA model with network structure, and the following section discusses the characteristics of these models.

2.2. *Dynamic DEA with Network Structure*

The results of the survey in [35] indicate that network and dynamic models must be highlighted among the main research fronts in the DEA literature. This statement is corroborated by the research developed by [36–39], which states several areas of applications and the development of distinct propositions for both modelings.

When it comes to efficiency measurement, there is the quantification of the conversion of inputs into outputs of the unit in focus. In the literature, static models are predominant in which there is an assumption of consumption and production in the same period tempo [36]. Dynamic models measure the efficiency of several periods in an aggregated perspective, where a link variable interconnects the periods [37].

The distinction between dynamic models (DDEA) and classical DEA models is the existence of variables, called carry-over, to link two consecutive periods. Ref. [40] proposes categorizing carry-overs into four groups: (1) desirable (good), (2) undesirable (bad), (3) discretionary (free) and (4) discretionary (fixed). This inter-period temporal interdependence can be attributed to a combination of five factors associated with the dynamic aspects of production: (1) production delays; (2) inventories; (3) capital or quasi-fixed factors; (4) cost adjustments; and (5) incremental improvements and learning models [36].

Refs. [41–43] can be considered the first to address the interdependence between periods for efficiency measurement, while Refs. [40,42,43] served as a basis for other dynamic DEA formulations [37]. From there, theoretical models [44] and applied ones were proposed, among which were the areas of agriculture [45], education [46], energy [40,47,48] and forests [49].

On the other hand, network models (NDEA) consider that the overall system efficiency consists of combining the DMUs' subdivision performance, whose pioneering spirit can be attributed to [50]. Considering the DMU internal structure is necessary to avoid misleading results, such as deeming systems efficient when they are not [51], and for identifying cases in which all processes of a DMU have lower performance than other DMUs. However, the overall system efficiency indicates superior results when compared to others [52].

It is also important to note that a network can be arranged in different ways. There are two basic structures, series and parallel [53]. Complex organizations formed by the combination of the basic ones are also found in the literature and can also portray specific situations such as shared inputs [54], shared inputs and outputs [55], and also shared intermediate measures [56].

These situations are analyzed by [57] in their consideration of multi-stage models and by [39], who make a unified classification of two-stage models. The last authors suggested four classes: (i) two independent stages; (ii) two connected stages considering the interaction between them; (iii) relational models, and (iv) game-theoretic models, considering cooperation and non-cooperation.

Regarding the DEA specifications, several models are also found in the literature, such as the slack-based model (SBM) [48], additive propositions [58], the inefficiency SBM measure [59], the relational model [53], models that combined relational and SBM aspects [60], and models that simultaneously consider multi-stage and multi-level aspects [61–64].

It is also important that the proposition of distinct network and dynamic models allows the investigation of different situations and areas of applications. This diversity can be verified in the broad range of applications of both models.

Most organizational structures can be characterized by processes structured in networks and related through multiple inputs and outputs over time. Under this scenario, multiple dynamic stages connected by network structure links in each period analysis are necessary to represent reality properly [65].

The combination of dynamic and network models enables the observation of the DMUs' overall efficiency over the entire observed period, and also to conduct further analysis; that is, observing the dynamic change in the period efficiency and dynamic change in the DMUs' divisional efficiency [66]. This framework enables considerations about the heterogeneous organizations of DMUs, in which the divisions are mutually connected by link-type variables and by the internal exchange of intermediate products [67,68]. In order to assess this broad range of analysis, the dynamic model with a network structure (DNDEA) considers a structure that consists of a finite number of static models' interaction [69].

Despite recent development, distinct mathematical developments have been made to propose new DNDEA models. It is possible to find in the literature approaches to deal with input uncertainties [70], with non-homogeneous DMUs [71], super-efficiency models [72], and the use of common weights to measure efficiency [73], and they have been used to investigate distinct areas of application, such as energy [66,74], transportation [75,76], supply chain [77], banks [78], and insurance companies [79].

Regarding higher education, it is possible to identify some investigations using DNDEA models to measure efficiency. Ref. [19] investigated the efficiency of the knowledge production process for nanobiotechnology research in US universities. Ref. [20] considered the financial and academic divisions to measure the efficiency of Vietnamese public colleges. Ref. [17] extended the discussions of [20] to investigate the impacts of financial and academic divisions on overall efficiency with the aid of DNDEA. The authors also applied a regression analysis to verify the effects of contextual factors on the efficiency of the financial division. Ref. [18] focused on the Australian vocational education and its subprocesses (teaching and industry responsiveness) to measure the efficiency of the teaching–industry linkage. Ref. [16] applied the DNDEA model to investigate Chile's higher education system. They aimed to compare the results of the three-stage system proposed (teaching, research and grant application) with the current one used to rank and accredit HEIs.

It is imperative at this point to differentiate dynamic network models from multi-level multi-stage models. In this last type of model, similar to network models, several internal stages to the network are considered. However, for same cases, instead of having multiple production stages and multiple levels, where the DMUs operating also exist, these could be geographical divisions (e.g., a DMU operating within a region, which in turn is part of a whole country) or functional (e.g., sub-units of an organization, divisions and subdivisions). In these cases, a hierarchical modeling seems appropriate [64]. Therefore, this type of modeling allows for observing the DMUs as a part of a larger system.

We highlight the works of [61–64] in this context, since they provided methodological advances and approached a situation that is somehow related to the university context, the innovation systems. Refs. [63,64] developed a multi-level multi-stage approach with a soft hierarchy to investigate the knowledge production process (KPP) and knowledge commercialization process (KCP). Refs. [61,62] also approach multi-level multi-stage models. Differently from Refs. [63,64], refs. [61,62] addressed the topics under the microeconomic theory. The new studies dealt with more stages [62] or the application of the Spence distortion principle to the hierarchy of a system [61].

By observing the previous studies, it is possible to verify that there are no propositions directed toward the investigations of graduate activities. In addition to the application, it is noteworthy that all applications used the model of [66] in their investigations. Therefore, in addition to proposing a new discussion, the current study develops a new model that allows simultaneously the consideration of shared resources to discuss resource allocation and the observation of the efficient decomposition of the stages.

Because of the relevance of DNDEA models and the large amount of information generated, the current study proposes a bi-dimensional representation of the DNDEA model proposed in the following section. The following section presents a brief overview of frontier representation alternatives and how our approach diverges from them.

2.3. Visual Representation in the DEA Field

Because of the relevance of DNDEA models and the large amount of information generated, the current study proposes a bi-dimensional representation of the DNDEA model proposed in the following section. The following section presents a brief overview of frontier representation alternatives and how our approach diverges from them. The original idea behind DEA was to provide a methodology whereby, within a set of comparable DMUs, those exhibiting best practice could be identified, and would form an efficient frontier, with this frontier allowing for the identification of benchmarks against which such inefficient units can be compared [57].

A significant part of the theoretical foundation of DEA comes from the proposition of [80]. Since this initial foundation, the graphic representation of the efficiency frontier has been of significant concern, because visual representation is a powerful tool for decision-makers, allowing them to ascertain how far the DMUs are from the efficient frontier or to look for concentrations of DMUs in some areas on the graph [81].

Ref. [80] presents different isoquants to discuss the efficient frontier when the production function is known and to estimate an efficient production function from observations of the inputs and outputs for some firms. In their seminal paper, ref. [82] considered two inputs and one output, transforming the input/unit of the output, plotting this information in a bi-dimensional graph. The same idea applies to the case of one input and two outputs. Each axis corresponds to one ratio of input/output or vice versa. However, this structure becomes unfeasible for cases with multiple variables. Since this, some discussions can be found in the literature to propose alternatives to represent the frontier.

Ref. [83] proposed an interactive visual DEA (VIDEA) consisting of an extension of the multiple criteria analysis model developed previously by the same authors. They employed a multiple-criteria hierarchical model to adapt the DEA model into an aggregate measure of input and output used to plot a two-dimensional graph. Ref. [84] proposed a set of two-dimensional charts to make the presentation to the managerial community more quickly. The authors compared efficiencies with individual factors, the impacts of virtual outputs, and the use of reference units to understand inefficient DMUs' performance better.

Ref. [85] developed a combination of DEA and Sammon Mapping to visualize the efficiency and the reference relations. They highlighted the fact that several questions could be answered directly from observations of the two-dimensional images. For example, which DMUs are efficient and which are not, which DMUs exhibit influence on the efficiency scores of other DMUs, and how strong the influence of a specific reference unit on an inefficient DMU is.

Ref. [86] used Co-Plot, with the ratio of outputs to inputs rather than the actual DEA results, stating that efficient DMUs are around the ring sector. However, in their proposition, there must exist an efficient frontier. Ref. [87] proposed a bi-dimensional representation using one input and four outputs and considered normalization to adapt the CCR results and a defined efficiency frontier.

In the literature about software designed to represent DEA results visually, ref. [21] introduced the interactive data envelopment analysis laboratory (IDEAL) as a tool to plot 3D frontiers. Although this type of graph helps see the results, the software is limited to three variables, and visualization becomes more challenging with the increase in DMUs. Ref. [88] proposed the SmartDEA, combining DEA and data mining to develop a general decision support system (DSS) framework to analyze the results of basic DEA models.

Ref. [81] proposed a more general approach to a bi-dimensional representation of CCR and BCC models. The authors used weight normalization based on the development of [83] to obtain the modified virtual inputs and outputs. These metrics are then plotted on a graph for each DMU with an efficient frontier. The main advantage of this proposition lies in its simplicity: it does not require modifications to the original model, the frontier is defined and easily obtained, the distance of the DMUs is obtainable, and visualization is easy even with a large number of DMUs.

Ref. [89] developed an extension of [81] focused on the dynamic approach of [49]. They used virtual outputs and inputs to represent divisional efficiency and applied an average of virtual inputs and outputs to represent the global efficiency of DMUs in a two-dimensional approach. Ref. [90] extended the approach of [81] to the network DEA models of [53,91]. The authors developed modified virtual inputs and outputs to represent the overall efficiency and sub-process efficiency. The model can handle multiple inputs, outputs, and intermediate measures, but is limited to two stages.

The current study is related to the propositions of [81,89]. Differently from [89], we use different types of modified virtual outputs and inputs to represent each efficiency level provided by the DNDEA model. DNDEA models provide different levels of information, ranging from global efficiency to divisional efficiency, by period. Evidently, as the number of stages or periods increases, the volume of information increases significantly, making it challenging to understand the results. Thus, the proposition of a visual tool to understand all levels of the results provided is of paramount importance, since the similar nomenclature for the different types of efficiency can represent an obstacle for decision-makers to understand the results. This feature is another significant contribution of the current study.

The bi-dimensional representation summarizes this information in a simple and straightforward way. This can help decision-makers who need to make faster and more accurate decisions. To the authors' knowledge, we are the first to propose a bi-dimensional representation of the frontier for DNDEA models. We are also the first to deepen the discussion of bi-dimensional representation for all efficiency types measured by this type of modeling, and this is particularly important in the educational field because national assessments contemplate voluminous amounts of information and reducing the cognitive effort in these processes helps significantly in the decision-making process.

3. Materials and Methods

The application of DEA in the educational context goes back to the beginning of applied studies using the technique. The discussions employing DEA includes analysis at distinct education levels and for distinct types of investigations. Ref. [92] indicates that there are two main paths when analyzing DEA development in the education field:

higher education and basic education. In the context of higher education, ref. [10] details a broad range of topics covered with DEA studies, such as university efficiency, the efficiency of individual academic departments, programs within an institution, and the central administration or services across universities. Ref. [11] also highlights using student ratings to assess performance in tertiary education, while ref. [8] details a new range of investigations such as cost efficiency, technical efficiency, research performance, rankings, and personal and teaching evaluations in higher education with DEA. In this section, we present the developed approach in Sections 3.1–3.3, while the Brazilian context is presented along with its data in Section 3.4.

3.1. *Two-Stage Dynamic DEA with Shared Inputs: A Centralized Approach*

The developed model aims to investigate resource sharing in a two-stage network model and to measure efficiency in a dynamic manner. The framework considered to develop our model is displayed in Figure 2. We considered the presence of shared and specific inputs. However specific inputs are present only in the first stage. The following models are designed to deal with shared inputs among the two stages. Therefore, the shared input p is divided into parcel α_{pj} which is consumed by the first division, and parcel $(1-\alpha_{pj})$, which is used by the second division. It is important to highlight that α is a decision variable, and it will be determined by the model. However, we proposed the use of upper and lower bounds for α because the stages share the resource and a parcel must be allocated to both of them.

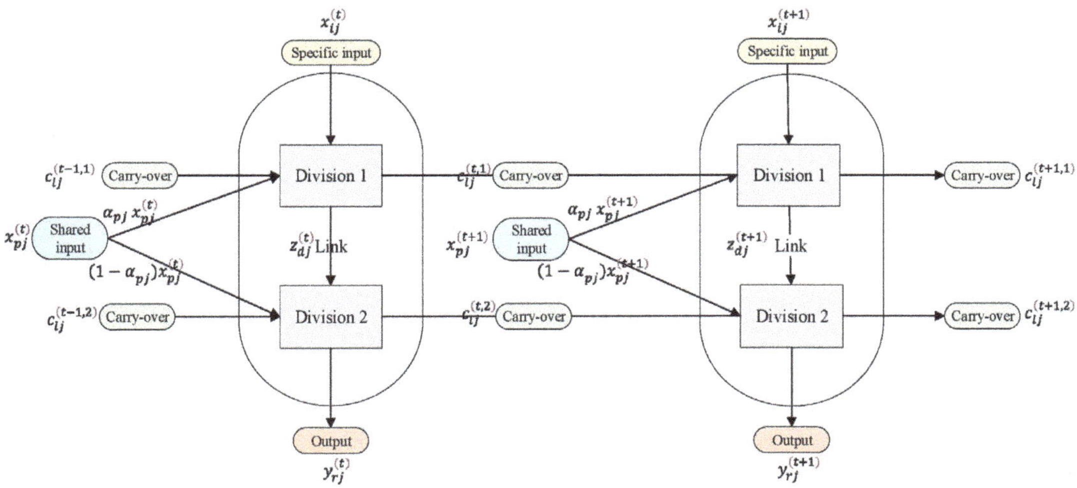

Figure 2. Two-stage dynamic DEA model with shared inputs.

No exogenous inputs are entering the second division. It is also considered that all intermediate measures produced by the first stage are consumed by the second. With these assumptions in mind, we proposed two distinct frameworks to investigate the referred context, a cooperative and a non-cooperative one.

The notations, summarized in Table 1, present the indexes, parameters, and variables considered in a centralized relational DNDEA model and a leader–follower form of the DNDEA model.

Table 1. Indexes, parameters and variables of the model.

	Indexes
$j = 1, \ldots, n$	Index for jth DMU;
$t = 1, \ldots, T$	Index for tth period;
$k = 1, \ldots, K$	Index for kth division;
$i = 1, \ldots, m$	Index for ith specific input;
$p = 1, \ldots, P$	Index for pth input shared between the divisions;
$r = 1, \ldots, s$	Index for rth output;
$d = 1, \ldots, D$	Index for dth link;
$l = 1, \ldots, L$	Index for lth carry-over.
	Parameters
$x_{ij}^{(t)}$	ith specific input of DMU j in division 1 in period t;
$x_{pj}^{(t)}$	pth shared input of DMU j between divisions 1 and 2 at period t;
$y_{rj}^{(t)}$	rth output of DMU j at division 2 at period t;
$z_{dj}^{(t)}$	dth link of DMU j leaving division 1 to division 2 at period t;
$c_{lj}^{(t,k)}$	lth carry-over at DMU j in division k that connects period t to the next one; $(l = 1, \ldots, l_k, \ldots L; j = 1, \ldots, n; t = 1, \ldots, T-1, k = 1, \ldots, K)$.
	Variables
$\propto_{pj}$	The proportion of the shared input of DMU j that will be used by division 1;
v_i^*, v_p^*, u_r^* w_l^*, f_d^*	The optimal weights attached to specific inputs, shared inputs, outputs, carry-overs and links, respectively.

For the two-stage system illustrated in Figure 2, the divisions of an observed DMU_0 can be evaluated considering constant returns to scale by Model (1) and (2) in each period. In Model (1) and (2), the objective function portrays the efficiency of the DMU under evaluation, while the restrictions ensure that the efficiency scores do not exceed one.

$$\begin{gathered}
E_j^{(t,1)} = max \frac{\sum_{l \in l^1} f_l c_{lo}^{(t,1)} + \sum_{d=1}^{D} w_d z_{do}^{(t)}}{\sum_{l \in l^1} f_l c_{lo}^{(t-1,1)} + \sum_{p=1}^{P} \alpha_{pj} v_p x_{po}^{(t)} + \sum_{i=1}^{m} v_i x_{io}^{(t)}} \\
\text{s.t.} \ \frac{\sum_{l \in l^1} f_l c_{lj}^{(t,1)} + \sum_{d=1}^{D} w_d z_{dj}^{(t)}}{\sum_{l \in l^1} f_l c_{lj}^{(t-1,1)} + \sum_{p=1}^{P} \alpha_{pj} v_p x_{pj}^{(t)} + \sum_{i=1}^{m} v_i x_{ij}^{(t)}} \leq 1 \ (j = 1, \ldots, n) \\
L_{pj}^1 \leq \alpha_{pj} \leq L_{pj}^2 \\
v_i, w_l, f_d, v_p, \geq \varepsilon; \ i = 1, \ldots, m; l = 1, \ldots, L; d = 1, \ldots, D; \ p = 1, \ldots, P
\end{gathered} \tag{1}$$

$$\begin{gathered}
E_j^{(t,2)} = max \frac{\sum_{r=1}^{s} u_r y_{ro}^{(t)} + \sum_{l \in l^2} f_l c_{lo}^{(t,2)}}{\sum_{l \in l^2} f_l c_{lo}^{(t-1,2)} + \sum_{p=1}^{P} (1-\alpha_{pj}) v_p x_{po}^{(t)} + \sum_{d=1}^{D} w_d z_{do}^{(t)}} \\
\text{s.t} \ \frac{\sum_{r=1}^{s} u_r y_{rj}^{(t)} + \sum_{l \in l^2} f_l c_{lj}^{(t,2)}}{\sum_{l \in l^2} f_l c_{lj}^{(t-1,2)} + \sum_{p=1}^{P} (1-\alpha_{pj}) v_p x_{pj}^{(t)} + \sum_{d=1}^{D} w_d z_{dj}^{(t)}} \leq 1 \ (j = 1, \ldots, n) \\
L_{pj}^1 \leq \alpha_{pj} \leq L_{pj}^2 \\
v_i, u_r, w_l, f_d, v_p \geq \varepsilon; \ i = 1, \ldots, m; r = 1, \ldots, s; l = 1, \ldots, L; d = 1, \ldots, D; \ p = 1, \ldots, P
\end{gathered} \tag{2}$$

Therefore, similar to the [51] assumption of the centralized model, we considered the same weights for the variables in all periods. We proposed a weighted average of stages 1 and 2 for each period, as displayed in (3).

$$E_j^{(t,sys)} = w_1^t E_j^{(t,1)} + w_2^t E_j^{(t,2)} \quad (3)$$

In order to define w_1^t and w_2^t, the consideration of [54] was selected. The authors argued that the proportion of total resources devoted to each stage presents one reasonable choice of weight to reflect the relative size of a stage. It is important to note that in dynamic models with network structures, carry-overs, and links play a dual role. Carry-overs represent both the output of one period and an input of the following one, while links consist of outputs from the first stage and inputs from the second. Therefore, we define w_1^t and w_2^t in (4) and (5).

$$w_1^t = \frac{\sum_{i=1}^m v_i\, x_{ij}^{(t)} + \sum_{p=1}^P \alpha_{pj} v_p x_{pj}^{(t)} + \sum_{l=1}^{l_1} f_l c_{lj}^{(t-1,1)}}{\sum_{i=1}^m v_i\, x_{ij}^{(t)} + \sum_{p=1}^P v_p x_{pj}^{(t)} + \sum_{d=1}^D w_d z_{dj}^{(t)} + \sum_{k=1}^K \sum_{l=1}^L f_l c_{lj}^{(t-1,k)}} \quad (4)$$

$$w_2^t = \frac{\sum_{p=1}^P (1-\alpha_{pj}) v_p x_{pj}^{(t)} + \sum_{d=1}^D w_d z_{dj}^{(t)} + \sum_{l=1}^{l_2} f_l c_{lj}^{(t-1,2)}}{\sum_{i=1}^m v_i\, x_{ij}^{(t)} + \sum_{p=1}^P v_p x_{pj}^{(t)} + \sum_{d=1}^D w_d z_{dj}^{(t)} + \sum_{k=1}^K \sum_{l=1}^L f_l c_{o}^{(t-1,k)}} \quad (5)$$

In (4) and (5), $\sum_{i=1}^m v_i x_{ij}^{(t)} + \sum_{p=1}^P v_p x_{pj}^{(t)} + \sum_{d=1}^D w_d z_{dj}^{(t)} + \sum_{k=1}^K \sum_{l=1}^L f_l c_o^{(t-1,k)}$ represents the total amount of resources (inputs) used by the stages in a period t. On the other hand, $\sum_{i=1}^m v_i x_{ij}^{(t)} + \sum_{p=1}^P \alpha_{pj} v_p x_{pj}^{(t)} + \sum_{l=1}^{l_1} f_l c_{lj}^{(t-1,1)}$ and $\sum_{p=1}^P (1-\alpha_{pj}) v_p x_{pj}^{(t)} + \sum_{d=1}^D w_d z_{dj}^{(t)} + \sum_{l=1}^{l_2} f_l c_{lj}^{(t-1,2)}$ indicates the resource size of stage 1 and 2, respectively. Therefore, the system efficiency in each period is detailed in (6).

$$E_j^{(t,\,sys)} = \frac{\sum_{r=1}^s u_r y_{rj}^{(t)} + \sum_{d=1}^D w_d z_{dj}^{(t)} + \sum_{k=1}^K \sum_{l=1}^L f_l c_{lj}^{(t,k)}}{\sum_{i=1}^m v_i\, x_{ij}^{(t)} + \sum_{p=1}^P v_p x_{pj}^{(t)} + \sum_{d=1}^D w_d z_{dj}^{(t)} + \sum_{k=1}^K \sum_{l=1}^L f_l c_{lj}^{(t-1,k)}} \quad (6)$$

We also considered that the overall efficiency is a weighted average of the system efficiency in each period. The proportion of total resources devoted to each period presents the choice to reflect the relative size of the period. Therefore, we define w^t in (7).

$$w^t = \frac{\sum_{i=1}^m v_i x_{ij}^{(t)} + \sum_{p=1}^P v_p x_{pj}^{(t)} + \sum_{k=1}^K \sum_{l=1}^L f_l c_{lj}^{(t-1,k)} + \sum_{d=1}^D w_d z_{dj}^{(t)}}{\sum_{t=1}^T \sum_{i=1}^m v_i x_{ij}^{(t)} + \sum_{t=1}^T \sum_{p=1}^P v_p x_{pj}^{(t)} + \sum_{t=1}^T \sum_{k=1}^K \sum_{l=1}^L f_l c_{lj}^{(t-1,k)} + \sum_{t=1}^T \sum_{d=1}^D w_d z_{dj}^{(t)}} \quad (7)$$

In (7), $\sum_{t=1}^T \sum_{i=1}^m v_i x_{ij}^{(t)} + \sum_{t=1}^T \sum_{p=1}^P v_p x_{pj}^{(t)} + \sum_{t=1}^T \sum_{k=1}^K \sum_{l=1}^L f_l c_{lj}^{(t-1,k)} + \sum_{t=1}^T \sum_{d=1}^D w_d z_{dj}^{(t)}$ represents the total amount of resources (inputs) used in all time frames considered. On the other hand, it indicates the resource size of each period t. Therefore, the overall system efficiency is detailed in (8).

$$E_j^{(sys)} = \frac{\sum_{t=1}^T \sum_{r=1}^s u_r y_{rj}^{(t)} + \sum_{t=1}^T \sum_{d=1}^D w_d z_{dj}^{(t)} + \sum_{t=1}^T \sum_{k=1}^K \sum_{l=1}^L f_l c_{lj}^{(t,k)}}{\sum_{t=1}^T \sum_{i=1}^m v_i x_{ij}^{(t)} + \sum_{t=1}^T \sum_{p=1}^P v_p x_{pj}^{(t)} + \sum_{t=1}^T \sum_{k=1}^K \sum_{l=1}^L f_l c_{lj}^{(t-1,k)} + \sum_{t=1}^T \sum_{d=1}^D w_d z_{dj}^{(t)}} \quad (8)$$

Thus, under CRS, the overall efficiency score can be evaluated by solving the following fractional program as presented in Model (9). In Model (9), the objective function corresponds to the overall system efficiency. The first constraint relates to the system efficiency in each period, the second one relates to the first stage, the third one relates to the second stage, and the fourth limits α_{pj} between the upper and lower bounds. The last ensures that the weights do not assume negative values.

$$
\begin{aligned}
&\theta_o^* = Max \frac{\sum_{t=1}^{T}\sum_{r=1}^{s} u_r y_{ro}^{(t)} + \sum_{t=1}^{T}\sum_{d=1}^{D} w_d z_{do}^{(t)} + \sum_{t=1}^{T}\sum_{k=1}^{K}\sum_{l=1}^{L} f_l c_{lo}^{(t,k)}}{\sum_{t=1}^{T}\sum_{i=1}^{m} v_i x_{io}^{(t)} + \sum_{t=1}^{T}\sum_{p=1}^{P} v_p x_{po}^{(t)} + \sum_{t=1}^{T}\sum_{k=1}^{K}\sum_{l=1}^{L} f_l c_{lo}^{(t-1,k)} + \sum_{t=1}^{T}\sum_{d=1}^{D} w_d z_{do}^{(t)}} \\
&\text{s.t } \frac{\sum_{r=1}^{s} u_r y_{rj}^{(t)} + \sum_{d=1}^{D} w_d z_{dj}^{(t)} + \sum_{k=1}^{K}\sum_{l=1}^{L} f_l c_{lj}^{(t,k)}}{\sum_{i=1}^{m} v_i x_{ij}^{(t)} + \sum_{p=1}^{P} v_p x_{pj}^{(t)} + \sum_{d=1}^{D} w_d z_{dj}^{(t)} + \sum_{k=1}^{K}\sum_{l=1}^{L} f_l c_{lj}^{(t-1,k)}} \le 1 \; (j = 1, \ldots, n; t = 1, \ldots, T) \\
&\frac{\sum_{l\in l^1} f_l c_{lj}^{(t,1)} + \sum_{d=1}^{D} w_d z_{dj}^{(t)}}{\sum_{l\in l^1} f_l c_{lj}^{(t-1,1)} + \sum_{p=1}^{P} \alpha_{pj} v_p x_{pj}^{(t)} + \sum_{i=1}^{m} v_i x_{ij}^{(t)}} \le 1 \; (j = 1, \ldots, n; t = 1, \ldots, T) \\
&\frac{\sum_{r=1}^{s} u_r y_{rj}^{(t)} + \sum_{l\in l^2} f_l c_{lj}^{(t,2)}}{\sum_{l\in l^2} f_l c_{lj}^{(t-1,2)} + \sum_{p=1}^{P} (1-\alpha_{pj}) v_p x_{pj}^{(t)} + \sum_{d=1}^{D} w_d z_{dj}^{(t)}} \le 1 \; (j = 1, \ldots, n; t = 1, \ldots, T) \\
&L_{pj}^1 \le \alpha_{pj} \le L_{pj}^2 \\
&v_i, u_r, w_l, f_d, v_p \ge \varepsilon; \; i = 1, \ldots, m; r = 1, \ldots, s; l = 1, \ldots, L; d = 1, \ldots, D; \; p = 1, \ldots, P
\end{aligned} \tag{9}
$$

With the aid of the Charnes–Cooper transformation, the fractional program proposed in Model (9) can be converted into Model (10).

$$
\begin{aligned}
&\theta_o^* = max \sum_{t=1}^{T}\sum_{r=1}^{s} \mu_r y_{ro}^{(t)} + \sum_{t=1}^{T}\sum_{k=1}^{K}\sum_{l=1}^{L} \gamma_l c_{lo}^{(t,k)} + \sum_{t=1}^{T}\sum_{d=1}^{D} \mu_d z_{do}^{(t)} \\
&\sum_{t=1}^{T}\sum_{i=1}^{m} v_i x_{io}^{(t)} + \sum_{t=1}^{T}\sum_{p=1}^{P} v_p x_{po}^{(t)} + \sum_{t=1}^{T}\sum_{k=1}^{K}\sum_{l=1}^{L} \gamma_l c_{lo}^{(t-1,k)} + \sum_{t=1}^{T}\sum_{d=1}^{D} \mu_d z_{do}^{(t)} = 1 \\
&\sum_{r=1}^{s} \mu_r y_{rj}^{(t)} + \sum_{d=1}^{D} \mu_d z_{dj}^{(t)} + \sum_{k=1}^{K}\sum_{l=1}^{L} \gamma_l c_{lj}^{(t,k)} - \sum_{i=1}^{m} v_i x_{ij}^{(t)} - \sum_{p=1}^{P} v_p x_{pj}^{(t)} - \sum_{d=1}^{D} \mu_d z_{dj}^{(t)} - \sum_{k=1}^{K}\sum_{l=1}^{L} \gamma_l c_{lj}^{(t-1,k)} \\
&\le 0 \;\; (j = 1, \ldots, n; t = 1, \ldots, T) \\
&\sum_{l\in l^1} \gamma_l c_{lj}^{(t,1)} + \sum_{d=1}^{D} \mu_d z_{dj}^{(t)} - \sum_{i=1}^{m} v_i x_{ij}^{(t)} - \sum_{p=1}^{P} \alpha_{pj} v_p x_{pj}^{(t)} - \sum_{l\in l^1} \gamma_l c_{lj}^{(t-1,1)} \le 0 \;\; (j = 1, \ldots, n; t = 1, \ldots, T) \\
&\sum_{r=1}^{s} \mu_r y_{rj}^{(t)} + \sum_{l\in l^2} \gamma_l c_{lj}^{(t,2)} - \sum_{l\in l^2} \gamma_l c_{lj}^{(t-1,2)} - \sum_{d=1}^{D} \mu_d z_{dj}^{(t)} - \sum_{p=1}^{P} (1-\alpha_{pj}) v_p x_{pj}^{(t)} \le 0 \;\; (j = 1, \ldots, n; t = 1, \ldots, T) \\
&L_{pj}^1 \le \alpha_{pj} \le L_{pj}^2 \\
&v_i, v_p, \mu_r, \gamma_l, \mu_d \ge \varepsilon; \; i = 1, \ldots, m; r = 1, \ldots, s; l = 1, \ldots, L; d = 1, \ldots, D; \; p = 1, \ldots, P
\end{aligned} \tag{10}
$$

Model (10) is non-linear since $\alpha_{pj} v_p$ is present in the constraints related to stage efficiency. It is possible to obtain a linear model considering that $\beta_{pj} = \alpha_{pj} v_p (p = 1, \ldots, P, j = 1, \ldots, n)$. After this substitution, Model (10) can be converted into Model (11).

$$
\begin{aligned}
&\theta_o^* = max \sum_{t=1}^{T}\sum_{r=1}^{s} \mu_r y_{ro}^{(t)} + \sum_{t=1}^{T}\sum_{k=1}^{K}\sum_{l=1}^{L} \gamma_l c_{lo}^{(t,k)} + \sum_{t=1}^{T}\sum_{d=1}^{D} \mu_d z_{do}^{(t)} \\
&\sum_{t=1}^{T}\sum_{i=1}^{m} v_i x_{io}^{(t)} + \sum_{t=1}^{T}\sum_{p=1}^{P} v_p x_{po}^{(t)} + \sum_{t=1}^{T}\sum_{k=1}^{K}\sum_{l=1}^{L} \gamma_l c_{lo}^{(t-1,k)} + \sum_{t=1}^{T}\sum_{d=1}^{D} \mu_d z_{do}^{(t)} = 1 \\
&\sum_{r=1}^{s} \mu_r y_{rj}^{(t)} + \sum_{d=1}^{D} \mu_d z_{dj}^{(t)} + \sum_{k=1}^{K}\sum_{l=1}^{L} \gamma_l c_{lj}^{(t,k)} - \sum_{i=1}^{m} v_i x_{ij}^{(t)} - \sum_{p=1}^{P} v_p x_{pj}^{(t)} - \sum_{d=1}^{D} \mu_d z_{dj}^{(t)} - \sum_{k=1}^{K}\sum_{l=1}^{L} \gamma_l c_{lj}^{(t-1,k)} \\
&\le 0 \;\; (j = 1, \ldots, n; t = 1, \ldots, T) \\
&\sum_{l\in l^1} \gamma_l c_{lj}^{(t,1)} + \sum_{d=1}^{D} \mu_d z_{dj}^{(t)} - \sum_{i=1}^{m} v_i x_{ij}^{(t)} - \sum_{p=1}^{P} \beta_p x_{pj}^{(t)} - \sum_{l\in l^1} \gamma_l c_{lj}^{(t-1,1)} \le 0 \;\; (j = 1, \ldots, n; t = 1, \ldots, T) \\
&\sum_{r=1}^{s} \mu_r y_{rj}^{(t)} + \sum_{l\in l^2} \gamma_l c_{lj}^{(t,2)} - \sum_{l\in l^2} \gamma_l c_{lj}^{(t-1,2)} - \sum_{d=1}^{D} \mu_d z_{dj}^{(t)} - \sum_{p=1}^{P} (v_p - \beta_{pj}) x_{pj}^{(t)} \le 0 \;\; (j = 1, \ldots, n; t = 1, \ldots, T) \\
&v_p L_{pj}^1 \le \beta_{pj} \le v_p L_{pj}^2 \\
&v_i, v_p, \mu_r, \gamma_l, \mu_d \ge \varepsilon; \; i = 1, \ldots, m; r = 1, \ldots, s; l = 1, \ldots, L; d = 1, \ldots, D; \; p = 1, \ldots, P
\end{aligned} \tag{11}
$$

3.2. Efficiency Decomposition

After solving Model (11), it is possible to obtain all efficiency scores discussed previously, namely, process efficiency, system efficiency and overall efficiency. Still, it is possible for Model (11) to present alternative optimal solutions. This multiplicity implies that the efficiency decomposition may not be unique. To investigate this, we adopted a leader–

follower approach. This type of analysis has been employed in several DEA studies, such as [51,93,94].

We employed a similar framework to [51,54] in which the first division has its efficiency maximized while the overall and system efficiency is maintained at the level identified with the aid of Model (11). Let $\upsilon_i^*, \upsilon_{p,}^*, \mu_r^*, \gamma_l^*, \mu_d^*$ be the optimal weights, while θ_o^*, $\theta_o^{(t,sys)*}$, $\theta_o^{(t,1)*}$ and $\theta_o^{(2,sys)*}$ represents the optimal overall, the optimal system efficiency by period, and the division 1 and division 2 at period t optimal efficiency θ_o^* of an observed $\mathrm{DMU_o}$. Suppose we focus on the maximization of the first stage: while maintaining the system by period and overall score, we have:

$$
\begin{aligned}
&\theta_o^{(t,1)} = max \frac{\sum_{l\in l^1} f_l c_{lo}^{(t,1)} + \sum_{d=1}^{D} w_d z_{do}^{(t)}}{\sum_{i=1}^{m} v_i x_{io}^{(t)} + \sum_{l\in l^1} f_l c_{lo}^{(t-1,1)} + \sum_{p=1}^{P} \alpha_{pj} v_p x_{po}^{(t)}} \\
&\text{s.t. } \frac{\sum_{l\in l^1} f_l c_{lj}^{(t,1)} + \sum_{d=1}^{D} w_d z_{dj}^{(t)}}{\sum_{i=1}^{m} v_i x_{ij}^{(t)} + \sum_{p=1}^{P} \alpha_{pj} v_p x_{pj}^{(t)} + \sum_{l\in l^1} f_l c_{lj}^{(t-1,1)}} \le 1\ (j = 1, \ldots, n) \\
&\frac{\sum_{r=1}^{s} u_r y_{rj}^{(t)} + \sum_{l\in l^2} f_l c_{lj}^{(t,2)}}{\sum_{l\in l^2} f_l c_{lj}^{(t-1,2)} + \sum_{p=1}^{P} (1-\alpha_{pj}) v_p x_{pj}^{(t)} + \sum_{d=1}^{D} w_d z_{dj}^{(t)}} \le 1\ (j = 1, \ldots, n) \\
&\frac{\sum_{r=1}^{s} u_r y_{ro}^{(t)} + \sum_{d=1}^{D} w_d z_{do}^{(t)} + \sum_{k=1}^{K} \sum_{l=1}^{L} f_l c_{lo}^{(t,k)}}{\sum_{i=1}^{m} v_i x_{io}^{(t)} + \sum_{p=1}^{P} v_p x_{po}^{(t)} + \sum_{d=1}^{D} w_d z_{do}^{(t)} + \sum_{k=1}^{K} \sum_{l=1}^{L} f_l c_{lo}^{(t-1,k)}} = \theta_o^{(t,sys)*} \\
&\frac{\sum_{t=1}^{T} \sum_{r=1}^{s} u_r y_{ro}^{(t)} + \sum_{t=1}^{T} \sum_{d=1}^{D} w_d z_{do}^{(t)} + \sum_{t=1}^{T} \sum_{k=1}^{K} \sum_{l=1}^{L} f_l c_{lo}^{(t,k)}}{\sum_{t=1}^{T} \sum_{i=1}^{m} v_i x_{io}^{(t)} + \sum_{t=1}^{T} \sum_{p=1}^{P} v_p x_{po}^{(t)} + \sum_{t=1}^{T} \sum_{k=1}^{K} \sum_{l=1}^{L} f_l c_{lo}^{(t-1,k)} + \sum_{t=1}^{T} \sum_{d=1}^{D} w_d z_{do}^{(t)}} = \theta_o^* \\
&w_1^{t*} * \frac{\sum_{l\in l^1} f_l c_{lo}^{(t,1)} + \sum_{d=1}^{D} w_d z_{do}^{(t)}}{\sum_{l\in l^1} f_l c_{lo}^{(t-1,1)} + \sum_{p=1}^{P} \alpha_{pj} v_p x_{po}^{(t)} + \sum_{i=1}^{m} v_i x_{io}^{(t)}} \le \theta_o^{(t,sys)*} \\
&L_{pj}^1 \le \alpha_{pj} \le L_{pj}^2 \\
&v_i, u_r, w_l, f_d, v_p \ge \varepsilon;\ i = 1, \ldots, m; r = 1, \ldots, s; l = 1, \ldots, L; d = 1, \ldots, D;\ p = 1, \ldots, P
\end{aligned} \tag{12}
$$

Model (12) can be converted into linear programming, as displayed in Model (13).

$$
\begin{aligned}
&\theta_o^{(t,1)*} = max \sum_{l\in l^1} \gamma_l c_{lj}^{(t,1)} + \sum_{d=1}^{D} \mu_d z_{dj}^{(t)} \\
&\sum_{i=1}^{m} \nu_i x_{io}^{(t)} + \sum_{p=1}^{P} \beta_p x_{po}^{(t)} + \sum_{l\in l^1} \gamma_l c_{lo}^{(t-1,1)} = 1 \\
&\sum_{l\in l^1} \gamma_l c_{lj}^{(t,1)} + \sum_{d=1}^{D} \mu_d z_{dj}^{(t)} - \sum_{i=1}^{m} \nu_i x_{io}^{(t)} - \sum_{p=1}^{P} \beta_p x_{pj}^{(t)} - \sum_{l\in l^1} \gamma_l c_{lj}^{(t-1,1)} \le 0\ (j = 1, \ldots, n) \\
&\sum_{r=1}^{s} \mu_r y_{rj}^{(t)} + \sum_{l\in l^2} \gamma_l c_{lj}^{(t,2)} - \sum_{l\in l^2} \gamma_l c_{lj}^{(t-1,2)} - \sum_{p=1}^{P} (\nu_p - \beta_{pj}) \nu_p x_{pj}^{(t)} - \sum_{d=1}^{D} \mu_d z_{dj}^{(t)} \le 0\ (j = 1, \ldots, n) \\
&\sum_{r=1}^{s} \mu_r y_{ro}^{(t)} + \sum_{d=1}^{D} \mu_d z_{do}^{(t)} + \sum_{k=1}^{K} \sum_{l=1}^{L} \gamma_l c_{lo}^{(t,k)} - \theta_o^{(t,sys)*} \left(\sum_{i=1}^{m} \nu_i x_{io}^{(t)} + \sum_{p=1}^{P} \nu_p x_{po}^{(t)} + \sum_{d=1}^{D} \mu_d z_{do}^{(t)} + \sum_{k=1}^{K} \sum_{l=1}^{L} \gamma_l c_{lo}^{(t-1,k)} \right) \\
&\quad \le 0 \\
&\sum_{t=1}^{T} \sum_{r=1}^{s} \mu_r y_{ro}^{(t)} + \sum_{t=1}^{T} \sum_{k=1}^{K} \sum_{l=1}^{L} \gamma_l c_{lo}^{(t,k)} + \sum_{t=1}^{T} \sum_{d=1}^{D} \mu_d z_{do}^{(t)} \\
&\quad -\theta_o^* \left(\sum_{t=1}^{T} \sum_{i=1}^{m} \nu_i x_{io}^{(t)} + \sum_{t=1}^{T} \sum_{p=1}^{P} \nu_p x_{po}^{(t)} + \sum_{t=1}^{T} \sum_{k=1}^{K} \sum_{l=1}^{L} \gamma_l c_{lo}^{(t-1,k)} + \sum_{t=1}^{T} \sum_{d=1}^{D} \mu_d z_{do}^{(t)} \right) \le 0 \\
&w_1^{t*} * \left(\sum_{l\in l^1} \gamma_l c_{lj}^{(t,1)} + \sum_{d=1}^{D} \mu_d z_{dj}^{(t)} \right) \le \theta_o^{(t,sys)*} \\
&\nu_p L_{pj}^1 \le \beta_{pj} \le \nu_p L_{pj}^2 \\
&\nu_i, \nu_p, \mu_r, \gamma_l, \mu_d \ge \varepsilon;\ i = 1, \ldots, m; r = 1, \ldots, s; l = 1, \ldots, L; d = 1, \ldots, D;\ p = 1, \ldots, P
\end{aligned} \tag{13}
$$

As previously discussed, the system efficiency is a weighted average of the stages; therefore, is possible to obtain the efficiency of the second stage as $\theta_o^{(t,2)} = \frac{\theta_o^{(t,sys)*} - w_1^{t*} \theta_o^{(t,1)*}}{w_2^{t*}}$. It is important to highlight that $\theta_o^{(t,sys)*}$, w_1^{t*} and w_2^{t*} are obtained with the optimal solution of

Model (11), and $\theta_o^{(t,1)*}$ indicates that the efficiency of Stage 1 was prioritized and optimized first. Based on this assumption, we maintained overall, system efficiency in each period and proportion of total resources devoted to each stage in each period unchanged. Therefore, it possible to proceed to the efficiency decomposition. The same hypotheses can be used to investigate Stage 2 efficiency, as shown in Model (14).

It is possible to obtain the efficiency of the first stage as $\theta_o^{(t,1)} = \frac{\theta_o^{(t,sys)*} - w_2^{t*}\theta_o^{(t,2)*}}{w_1^{t*}}$. It is important to mention that the proposed models and evaluation must be used for each period t under analysis. If $\theta_o^{(t,1)} = \theta_o^{(t,1)*}$ or, $\theta_o^{(t,2)} = \theta_o^{(t,2)*}$, there is a unique decomposition.

$$
\begin{aligned}
&\theta_o^{(t,2)*} = max \sum_{r=1}^{s} \mu_r y_{ro}^{(t)} + \sum_{l \in l^2} \gamma_l c_{lo}^{(t,2)} \\
&\sum_{l \in l^2} \gamma_l c_{lo}^{(t-1,2)} - \sum_{d=1}^{D} \mu_d z_{do}^{(t)} - \sum_{p=1}^{P} (v_p - \beta_{po}) v_p x_{po}^{(t)} = 1 \\
&\sum_{l \in l^1} \gamma_l c_{lj}^{(t,1)} + \sum_{d=1}^{D} \mu_d z_{dj}^{(t)} - \sum_{i=1}^{m} v_i x_{io}^{(t)} - \sum_{p=1}^{P} \beta_p x_{pj}^{(t)} - \sum_{l \in l^1} \gamma_l c_{lj}^{(t-1,1)} \le 0 \ (j = 1, \ldots, n) \\
&\sum_{r=1}^{s} \mu_r y_{rj}^{(t)} + \sum_{l \in l^2} \gamma_l c_{lj}^{(t,2)} - \sum_{l \in l^2} \gamma_l c_{lj}^{(t-1,2)} - \sum_{p=1}^{P} (v_p - \beta_{pj}) v_p x_{pj}^{(t)} - \sum_{d=1}^{D} \mu_d z_{dj}^{(t)} \le 0 \ (j = 1, \ldots, n) \\
&\sum_{r=1}^{s} \mu_r y_{ro}^{(t)} + \sum_{d=1}^{D} \mu_d z_{do}^{(t)} + \sum_{k=1}^{K} \sum_{l=1}^{L} \gamma_l c_{lo}^{(t,k)} \\
&\qquad -\theta_o^{(t,sys)*} \left(\sum_{i=1}^{m} v_i x_{io}^{(t)} + \sum_{p=1}^{P} v_p x_{po}^{(t)} + \sum_{d=1}^{D} \mu_d z_{do}^{(t)} + \sum_{k=1}^{K} \sum_{l=1}^{L} \gamma_l c_{lo}^{(t-1,k)} \right) \le 0 \\
&\sum_{t=1}^{T} \sum_{r=1}^{s} \mu_r y_{ro}^{(t)} + \sum_{t=1}^{T} \sum_{k=1}^{K} \sum_{l=1}^{L} \gamma_l c_{lo}^{(t,k)} + \sum_{t=1}^{T} \sum_{d=1}^{D} \mu_d z_{dj}^{(t)} \\
&\qquad -\theta_o^{*} \left(\sum_{t=1}^{T} \sum_{i=1}^{m} v_i x_{io}^{(t)} + \sum_{t=1}^{T} \sum_{p=1}^{P} v_p x_{po}^{(t)} + \sum_{t=1}^{T} \sum_{k=1}^{K} \sum_{l=1}^{L} \gamma_l c_{lo}^{(t-1,k)} + \sum_{t=1}^{T} \sum_{d=1}^{D} \mu_d z_{do}^{(t)} \right) \le 0 \\
&w_2^{t*} * \left(\sum_{r=1}^{s} \mu_r y_{ro}^{(t)} + \sum_{l \in l^2} \gamma_l c_{lo}^{(t,2)} \right) \le \theta_o^{(t,sys)*} \\
&v_p L_{pj}^1 \le \beta_{pj} \le v_p L_{pj}^2 \\
&\qquad v_i, v_p, \mu_r, \gamma_l, \mu_d \ge \varepsilon; \ i = 1, \ldots, m; r = 1, \ldots, s; l = 1, \ldots, L; d = 1, \ldots, D; \ p = 1, \ldots, P
\end{aligned}
\tag{14}
$$

3.3. Bi-Dimensional Representation

We use virtual inputs and outputs to obtain the bi-dimensional representation. The main issue is the constraint that states that the virtual input or virtual output equals 1 (added to linearize the mathematical model). So, in a virtual-input or virtual-output plot, all DMUs would be located on the same vertical straight line, and such a graphical representation would be meaningless [81]. In the case of dynamic models, it is necessary to add a parcel related to the other variables that also play the role of system input. The authors introduced a constraint that limits the sum of input weights to be equal to 1. In order to bypass this limitation, we follow the proposition of [81].

In the case of DNDEA models, it is necessary to add parcels related to all variables presented in the constraint that is equal to one in Model (11) because they also represent the system's input in input-oriented cases. Then, we must consider the total sum of the weights for all the variables in the constraint referred to, which is equal to 1.

A new model is developed by adding this constraint. However, with a simple mathematical operation, it is possible to apply the results of the Model (11) by dividing the resulting weights by the total sum of the weights of the DMU under observation.

Let S_j be the total sum of the shared inputs, specific inputs, carry-overs and link weights of DMU j:

$$S_j = \sum_{i=1}^{m} v_{ij} + \sum_{p=1}^{P} v_{pj} + \sum_{l=1}^{L} \gamma_{lj} + \sum_{d=1}^{D} \mu_{dj} \tag{15}$$

v_{ij} is the weight of the specific input i of DMU j, v_{pj} is the weight of the shared input p of DMU j, γ_{lj} is the weight of carry-over l in DMU j and μ_{dj} is the weight of the link d in DMU j. To obtain the representation with virtual variables, let v'_{ij}, $(v'_{pj}; \beta'_{pj})$, γ'_{lj}, μ'_{dj} and μ'_{rj} be the modified weights of the specific input i, shared input p, carry-over l, link d and output r of DMU j, respectively:

$$v'_{ij} = \frac{v_{ij}}{S_j}; v'_{pj} = \frac{v_{pj}}{S_j}; \beta'_{pj} = \frac{\beta_{pj}}{S_j} \gamma'_{lj} = \frac{\gamma_{lj}}{S_j}; \mu'_{dj} = \frac{\mu_{dj}}{S_j}; \mu'_{rj} = \frac{\mu_{rj}}{S_j} \quad (16)$$

When DNDEA models are used, different efficiency results are obtained. We proposed a distinct set of modified virtual inputs and outputs to represent visually all levels of results. We start with overall system efficiency, following the system's efficiency in each period, and to conclude, we present the process efficiency in each period. Appendix B presents proof that the efficiency values obtained with modified virtual inputs and outputs do not change the scores provided by the original DNDEA model.

3.3.1. Overall System Efficiency

Let $x_{ij} = \sum_{t=1}^{T} x_{ij}^{(t)}$, $x_{pj} = \sum_{t=1}^{T} x_{pj}^{(t)}$, $z_{dj} = \sum_{t=1}^{T} z_{dj}^{(t)}$, and $y_{rj} = \sum_{t=1}^{T} y_{rj}^{(t)}$. We shall consider $I_j'^{(sys)}$ the system's virtual input and $O_j'^{(sys)}$ the system's virtual output of DMU j.

$$I_j'^{(sys)} = \sum_{i=1}^{m} v'_{ij} x_{ij} + \sum_{p=1}^{P} v'_{pj} x_{pj} + \sum_{d=1}^{D} \mu'_{dj} z_{dj} + \sum_{t=1}^{T} \sum_{k=1}^{K} \sum_{l=1}^{L} \gamma'_{lj} c_{lj}^{(t-1,k)} \quad (17)$$

$$O_j'^{(sys)} = \sum_{r=1}^{s} \mu'_{rj} y_{rj} + \sum_{d=1}^{D} \mu'_{dj} z_{dj} + \sum_{t=1}^{T} \sum_{k=1}^{K} \sum_{l=1}^{L} \gamma'_{lj} c_{lj}^{(t,k)} \quad (18)$$

We obtain the efficiency of DMU_0, $E_o^{(sys)}$, by dividing the virtual output by the virtual input. Hereafter, we refer to (17) as the system virtual input and we do the same for (18) in the case of outputs.

3.3.2. System Efficiency in Each Period

We shall also consider $I'^{(t,sys)}_j$ the virtual input of the system in period t in DMU j and $O'^{(t,sys)}_j$ the virtual output of the system in period t in DMU j. However, since there are differences in the role of the variables for the stage, two distinct virtual inputs and outputs are required.

$$I_j'^{(t,sys)} = \sum_{i=1}^{m} v'_{ij} x_{ij}^{(t)} + \sum_{p=1}^{P} v'_{pj} x_{pj}^{(t)} + \sum_{d=1}^{D} \mu'_{dj} z_{dj}^{(t)} + \sum_{k=1}^{K} \sum_{l=1}^{L} \gamma'_{lj} c_{lj}^{(t-1,k)} \quad (19)$$

$$O_j'^{(t,sys)} = \sum_{r=1}^{s} \mu'_{rj} y_{rj}^{(t)} + \sum_{d=1}^{D} \mu'_{dj} z_{dj}^{(t)} + \sum_{k=1}^{K} \sum_{l=1}^{L} \gamma'_{lj} c_{lj}^{(t,k)} \quad (20)$$

Hereafter, we refer to (19) and (20) as the virtual inputs of the modified system by period and the virtual outputs of the modified system by period, respectively. We obtain $E_O^{(t,sys)}$ by dividing the virtual output by the virtual input.

3.3.3. Process Efficiency in Each Period

Let $I'^{(t,k)}_j$ be the virtual input of division k in period t and $O'^{(t,k)}_j$ the virtual output of division k in period t:

$$I_j'^{(t,1)} = \sum_{i \in i^1} v'_{ij} x_{ij}^{(t)} + \sum_{p=1}^{P} \beta'_{pj} x_{pj}^{(t)} + \sum_{l \in l^1} \gamma'_{lj} c_{lj}^{(t-1,1)} \quad (21)$$

$$I\prime_j^{(t,2)} = \sum_{l\in l^2} \gamma'_{lj} c_{lj}^{(t-1,2)} + \sum_{d=1}^{D} \mu'_{dj} z_{dj}^{(t)} + \sum_{p=1}^{P} \beta'_{pj} x_{pj}^{(t)} - \sum_{p=1}^{P} v'_{pj} x_{pj}^{(t)} \tag{22}$$

$$O_j^{\prime(t,1)} = \sum_{l\in l^1} \gamma'_{lj} c_{lj}^{(t,1)} + \sum_{d=1}^{D} \mu'_{dj} z_{dj}^{(t)} \tag{23}$$

$$O_j^{\prime(t,2)} = \sum_{r=1}^{s} \mu'_{rj} y_{rj}^{(t)} + \sum_{l\in l^2} \gamma'_{lj} c_{lj}^{(t,2)} \tag{24}$$

Summarizing our approach in a step-by-step procedure, as in [81]:

1. Run the input-oriented DNDEA Model (11) for each DMU j;
2. Calculate S_j for each DMU j;
3. Calculate the modified variable weights $v'_{ij}; v'_{pj}; \beta'_{pj}; \gamma'_{lj}; \mu'_{dj}; \mu'_{rj}$ according to Equation (16);
4. Calculate the modified virtual input- and virtual output overall system using Equations (17) and (18) for each DMU j;
5. Calculate the virtual input and output of the modified system in each period using Equations (19) and (20) for each DMU j;
6. Calculate the process-modified virtual input and output using Equations (21)–(24) for each DMU j;
7. Use the modified virtual input $I\prime_j^{(sys)}$ in the x-axis and the modified virtual output $O\prime_j^{(sys)}$ in the y-axis in a bi-dimensional graph for each DMU j for overall efficiency;
8. Use the modified virtual input $I\prime_j^{(t,sys)}$ in the x-axis and the modified virtual output $O\prime_j^{(t,sys)}$ in the y-axis in a bi-dimensional graph for each DMU j for system efficiency for each period;
9. Use the modified virtual input $I\prime_j^{(t,k)}$ in the x-axis and the modified virtual output $O\prime_j^{(t,k)}$ in the y-axis in a bi-dimensional graph for each DMU j for process efficiency in each period;
10. Draw the 45° line representing the efficient frontier where efficient DMU presents $I\prime_j^{(t,k)} = O\prime_j^{(t,k)}$, $I\prime_j^{(t,sys)} = O\prime_j^{(t,sys)}$, and $I\prime_j^{(sys)} = O\prime_j^{(sys)}$ for process efficiency, system efficiency in each period, and overall system efficiency.

3.4. Data

In the current discussion, we aim to evaluate the graduate activities of Brazilian HEIs with the aid of a DNDEA model. As previously mentioned, universities present a multi-activity framework, and contemplating the productivity changes of these institutions is of high importance. Our DNDEA model considers two stages: the formative process and the scientific production process. The proposed framework is displayed in Figure 3.

In the first one, a parcel of faculty and enrolled student workload represents the inputs. It is important to clarify that when considering these two inputs, we are not allocating part of the students and faculty to the formative process and another to the scientific production process. We consider that all students and all faculty divide their workload between these activities. The number of programs available in a university represents the carry-over variable, while master's dissertations and Ph.D. theses correspond to the intermediate factor linking the stages. Variable dropout represents an undesirable output and reflects the reality that some students do not finish their master's or Ph.D. training. Since it consists of an undesirable variable, it requires treatment to be properly used in the DEA framework. We subtracted values from a large number, ensuring the results were isotonic, as discussed by [95].

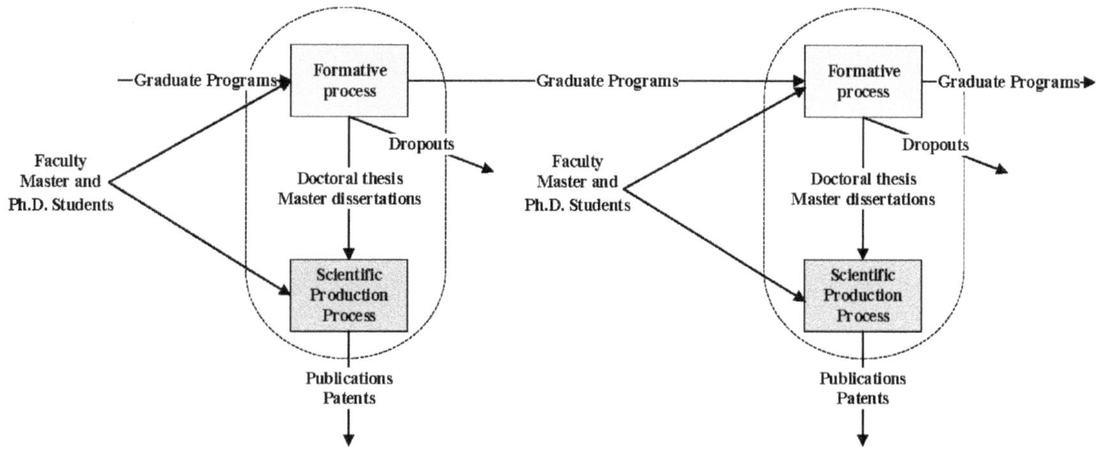

Figure 3. Two-stage dynamic DEA model with shared inputs for graduate activities.

The second stage (the scientific production process) converts the other portion of faculty and enrolled students' workload and the desirable products of the formative process into research products: dissertations and theses correspond to the research developed, representing the basis for generating papers and patents. The publications considered are in the SCOPUS database.

As previously mentioned, faculty and students divide their workloads between both processes. Therefore, they correspond to the shared inputs, and entirely allocating these inputs to the first stage would be inappropriate and penalize its efficiency. The analysis of these resource allocations responds to the question of whether they are being efficiently used or not, and this information can benefit HEIs' performance.

These variables were selected due to their relevance to the national reality; they are already used for the individual evaluation of programs, and most of them are also used in the international literature and university rankings. We would also like to highlight the fact that for the Brazilian context, master's dissertations should be considered as inputs to the scientific production process. Firstly, in Brazil, there are extremely rare cases in which students enter directly onto a doctorate. For the most part, students enroll and complete their master's degree before applying for a doctorate position, with the master's degree being a prerequisite for most universities in the Ph.D. application process.

Furthermore, due to changes that have occurred in recent years in the process of monitoring postgraduate courses, many programs have been applying publication requirements for students to obtain a master's degree. Therefore, master's dissertations have contributed to the Brazilian scientific production process. However, for other countries, this variable could be considered an output of the first stage, with the model being easily modified to adapt to such a situation.

We emphasize that the choice to use DEA aims to mitigate one of the main criticisms verified among the government's already-used indicators. Brazil is a country with very different regions in socio-economic and demographic terms. This national characteristic is reflected in the universities' very different missions and objectives. Therefore, the flexibility of the weights for weighing the criteria is essential, so that each university has the autonomy to reflect these characteristics and so that the final result is not questioned, with the claim that the weighting of the criteria benefited some to the detriment of others.

In this study, we focus on the graduate activities in federal universities, because (i) they are responsible for more than half of the country's master's and doctoral courses and students, and produce most of the national science [6]; (ii) they represent a set of more homogeneous institutions; and (iii) they use public funds to finance their activities.

This analysis is vital, given federal government spending. Approximately half of the Brazilian graduate courses are developed in public universities. Data from 2020 indicate that the federal government spent 23 billion in federal universities to finance personnel and charges in the same year. In addition, it is worth mentioning that Brazilian research agencies such as Coordination for the Improvement of Higher Education Personnel (CAPES) and the National Council for Scientific and Technological Development (CNPq) finance scholarships for master's and doctoral students in these institutions. Therefore, this analysis helps with the best use of public resources.

According to data released by the 2020 Higher Education Census, there are 68 federal universities in Brazil. Reports generated by CAPES correspond to the data source used, since CAPES is responsible for evaluating and consolidating information regarding individual graduate activities in Brazil. The reduction in the number of universities analyzed was due to a lack of data on one or more variables, mainly in patents and Ph.D. theses. Consequently, our sample contains 32 universities, with data from 2019 to 2020.

The selected time frame aims to evaluate the most-recent available data and obtain a glimpse of the COVID-19 pandemic's impact on graduate activities. The descriptive statistics of the sample are presented in Table 2.

Table 2. Descriptive statistics of data.

Variable	Category	Average	SD	Minimum	Maximum
Formative process					
Faculty (number)	Input	1033.42	700.22	235	2913
Enrollments (number)	Input	2833.43	2289.97	321	9163
Programs (number)	Carry-over	100.94	71.71	9	321
Dropouts (number)	Output	40.35	23.49	7	89
Ph.D. theses (number)	Link	41.95	23.53	9	91
Master's dissertations (number)	Link	219.78	218.70	8	954
Scientific production process		584.60	408.58	84	1786
Publications (number)	Output	3019.22	2228.38	326	10,400
Patents (number)	Output	52.32	51.55	1	210

4. Discussion

The results are divided into three sections. First, we present the DNDEA efficiencies of the 32 federal universities. Second, with the aid of the bidimensional representation, we deepen the performance discussion. Then, the efficiency decomposition under the leader–follower assumption with the procedure detailed in Section 3.2 is presented.

4.1. DNDEA Efficiency Results

The framework and variables in the proposition for investigation of graduate activities are displayed in Figure 3. We applied the developed DNDEA model discussed in Section 3 to investigate graduate activities in Brazilian federal universities.

First, we applied Model (11), considering 0.40 and 0.70 as lower and upper bounds for both shared inputs, and Table 3 displays the descriptive statistics for all the efficiency results. Table 3 shows, in the second column, the overall efficiency. Columns three and four report the system efficiency, while five to eight present the process efficiencies for 2019 and 2020.

Table 3. Descriptive results of the efficiencies.

	$E^{(sys)}$	$E^{(1,sys)}$	$E^{(2,sys)}$	$E^{(1,1)}$	$E^{(1,2)}$	$E^{(2,1)}$	$E^{(2,2)}$
Mean	80.97%	79.59%	82.75%	88.63%	65.55%	81.87%	82.14%
S.D	5.07%	7.13%	5.89%	7.74%	11.23%	7.68%	10.97%
Max	89.57%	92.31%	94.47%	100%	89.29%	97.91%	100%
Min	68.94%	66.40%	66.77%	71.62%	46.42%	64.45%	58.66%

The average overall efficiency of the considered period is 80.97%. When observing the periods, 2019 obtained an average result of 79.59%, while 2020 returned 82.75%. When analyzing the average division values, it is possible to verify that 2020 returned higher efficiency scores, and the increase in performance in scientific production can explain such results. The training process showed an efficiency decline of 6.76% (88.63% in 2019 to 81.87% in 2020). A total of 26 of the 32 DMUs showed reduced efficiency when comparing the periods. On the other hand, there was an increase of 16.59% in efficiency (65.55% in 2019 to 82.14% in 2020) in the scientific production process. A total of 30 of the 32 DMUs displayed increased performance.

Considering DNDEA scores, federal universities could increase their efficiency in a network structure of the formative process and scientific production by approximately 19.03%. The scores in Table 3 indicate that, on average, the training process had better results than the scientific production process before the COVID-19 pandemic. However, in 2020, the average values are closer (81.87% and 82.14%), but with better results for the scientific production process.

The number of publications explains the better performance of the scientific production process in 2020. When comparing 2019 with 2020, there is a reduction in thesis and dissertation numbers for more than 90% of the DMUs. However, the number of publications grew for all DMUs, and approximately 60% of DMUs also saw increased patent numbers.

The performance fluctuations in 2020 may also be related to the COVID-19 pandemic. Teaching activities were suspended for several periods in Brazilian HEIs, which corresponded to most of the year. During this interval, research activities and, consequently, publications derived from this research continued remotely. In addition, the significant impacts of the pandemic on the most diverse areas of knowledge and the need for quick responses stimulated the development of a high amount of research, as seen in special COVID-19 specific discussion sections at scientific events and special issues in various journals.

However, the verified impact on teaching activities was negative. Learning in remote teaching requires a learning curve for both students and teachers. It is also worth noting that, unfortunately, access to the internet with the minimum conditions necessary to participate in activities was a problem for some of the students, with classes being one of the activities most affected by these issues, directly impacting teaching and learning.

The scientific process plays an indispensable role in disseminating the research produced in the university to the academic community and society. It is important to note that in the period before the pandemic, the performance of this stage was significantly lower than the training process. These results indicate that the investigation of more recent data is necessary to verify whether the increase in performance remains or if the difficulties verified in 2019 persist, indicating a significant difficulty in disseminating the produced knowledge beyond the university.

The investigation of more recent data is indispensable because funds directed to graduate activities in Brazil have been reduced drastically over the last decade. As pointed out by the UNESCO report, the increase in publications over recent years indicates that Brazilian research is resilient. However, resilience also has its limits. Therefore, it is relevant to understand whether the lower performance in the scientific production verified in 2019 can be related to difficulties in research funding. This topic becomes even more critical in the context of the migration of several journals to the open-access format, consequently increasing publishing costs. The increasing costs in a scenario of successive cuts in public funds can negatively impact the number of publications in Brazilian public universities.

The correlation between formative process and scientific production is negative. This result supports the previous discussion of the possible difficulty of transforming knowledge into products. Given that the theses and dissertations consist of second-stage inputs, there must be an effort to increase them in order to obtain better results for this process. However, although most universities increased their performance in 2020 in this process, there is still room for improvement.

As previously mentioned, the lower limits of α_1 and α_2 were defined a priori. A sensitivity analysis was performed to verify the impact of this choice on efficiency values. We performed two types of investigations. First only one parameter was altered, and then we altered both, simultaneously.

Figure 4 displays the overall and system efficiency values when only α_1 or α_2 were changed. The graphs present the efficiency values for all evaluated DMUs. We performed a similar analysis for all the efficiency levels. It is possible to verify changes in some DMU scores.

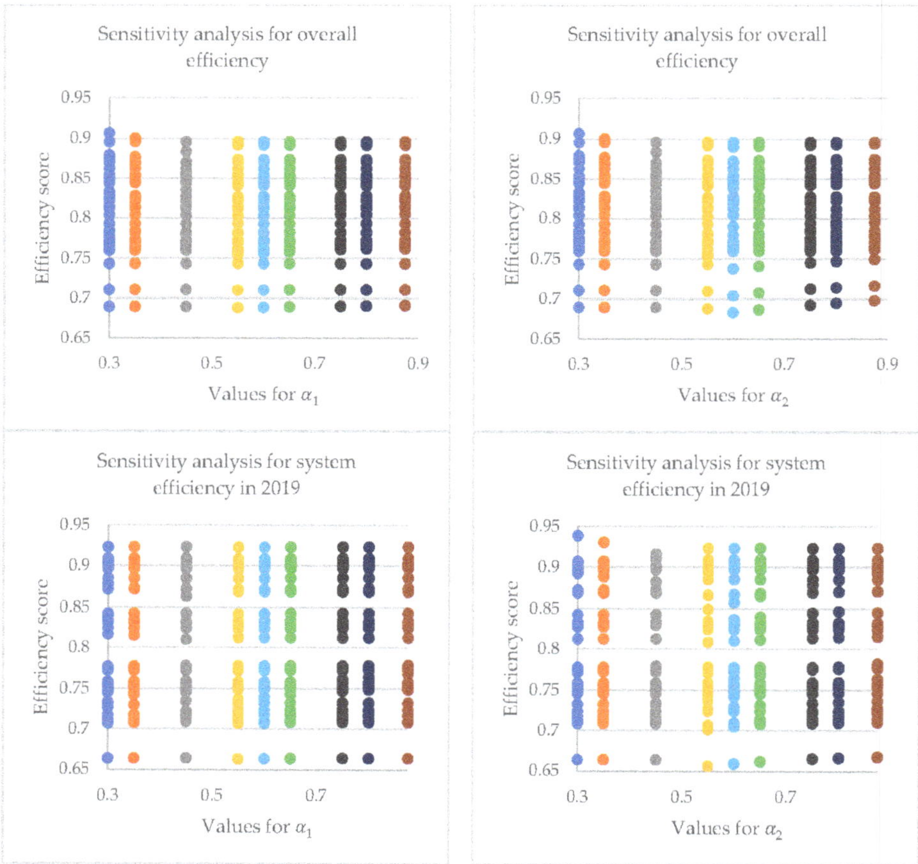

Figure 4. *Cont.*

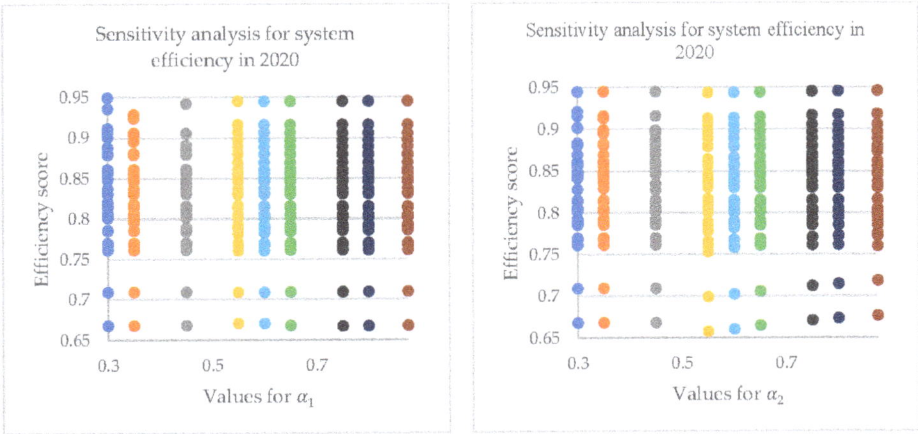

Figure 4. Sensitivity analysis for α values.

However, these changes were not very significant. Although small, it is possible to notice that greater alterations occurred in α_2 values. It is also observed that for α_1 and α_2, the greater alterations occurred for the smallest observed values, that is, for reductions in the lower limits.

We also observed that for α_2 values, some variations related to the reductions in the upper limit, that is, for values between 0.5 and 0.7. These variations can be particularly observed for system efficiency in 2020. These observations initially indicate that a greater variation in the allocation of students' workload would have a greater impact on efficiency than a variation in the allocation of teachers' workload.

Appendix C presents additional graphs relating to sensitivity analyses for process efficiencies and for cases where the two alpha values are changed. The results of these tests converge to ensure that the model is robust. The variations observed in all levels of efficiency are minimal, occurring in many cases only in the fourth decimal place.

4.2. Bi-Dimensional Representation

Following the procedure described in Section 3.3, the first step requires running the DNDEA model. In this subsection, we present the graphs and the empirical findings. Appendix B details the mathematical proof that the efficiency values are maintained with the bi-dimensional representation.

Figure 5 displays the frontier for the system efficiency in 2019 and 2020. The green line leaving the origin (0, 0) corresponds to the efficiency frontier in our bi-dimensional representation. The different colors in the graphs relate to the five Brazilian macro-regions. The choice to highlight the macro-regions relates to the significant social and economic discrepancies among them. Also, previous literature findings indicate that the DMU location can impact on the efficiency score.

The graphics in Figure 5 indicate that for the system efficiencies per year, there are no significant discrepancies among the Brazilian macro-regions. It is also noteworthy that no DMU obtained maximum performance in 2019 and 2020. This fact is also true for overall efficiency values. The results presented so far show the greater power of discrimination of the proposed DNDEA model.

We further examine the bi-dimensional representation of process efficiencies. The results in Figure 6 show no significant discrepancies among the Brazilian macro-regions for both processes. The Kruskal–Wallis non-parametric test was used to investigate whether the differences among the macro-regions are significant. Table 4 present the test results for the overall efficiency and system's efficiency for all years, while Table 5 presents the test results for process efficiency. At a 5% and 10% significance level, it is possible to infer that there are no differences in the median of the Brazilian macro-regions for all efficiency levels.

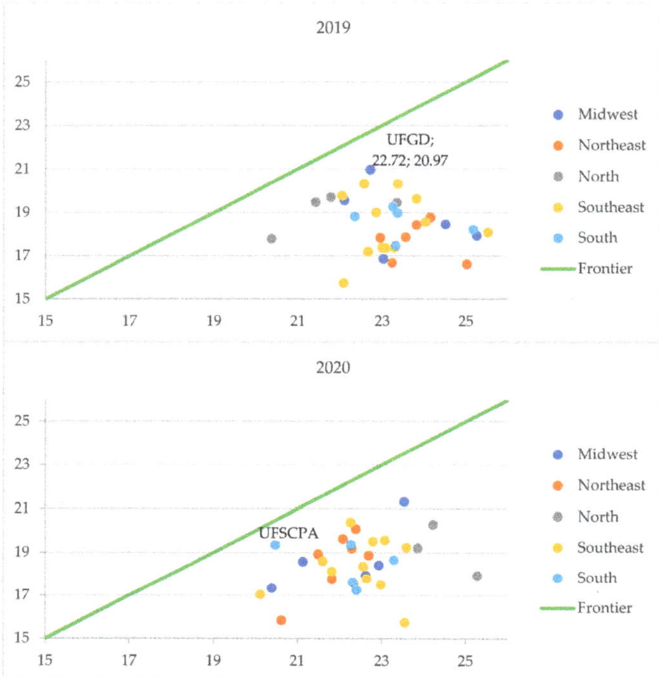

Figure 5. System efficiency efficiencies 2019–2020.

Table 4. Kruskal–Wallis test results for overall and system efficiencies.

Efficiency	Degrees of Freedom	Chi-Square	*p*-Value
Overall	4	3.7004	0.4481
2019	4	7.2126	0.1251
2020	4	2.3201	0.6771

Table 5. Kruskal–Wallis test results for process efficiencies.

Efficiency	Degrees of Freedom	Chi-Square	*p*-Value
Formative Process in 2019	4	3.5366	0.4723
Formative Process in 2020	4	3.8443	0.4275
Scientific Production in 2019	4	4.7731	0.3114
Scientific Production in 2020	4	6.3686	0.1734

However, unlike overall and system efficiencies for which no DMU obtained maximum performance in 2019 and 2020, four DMUs were considered efficient in the training process in 2019, and four were considered efficient in the scientific process in 2020. It is also important to mention the findings that the average performance of the formative process is superior to that of scientific production in 2019 and that the pattern reversed in 2020 is easily verifiable when observing that most DMUs are further away from the efficiency frontier in the respective graphics of Figure 6. The results presented so far show the greater power of discrimination of the proposed DNDEA model.

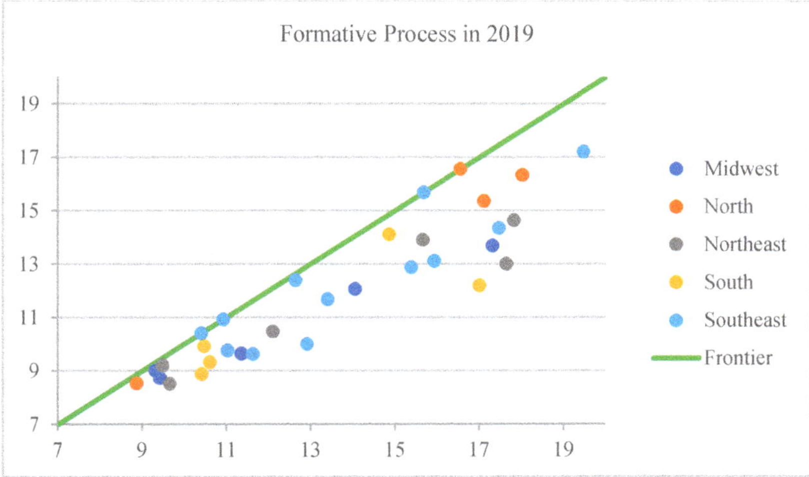

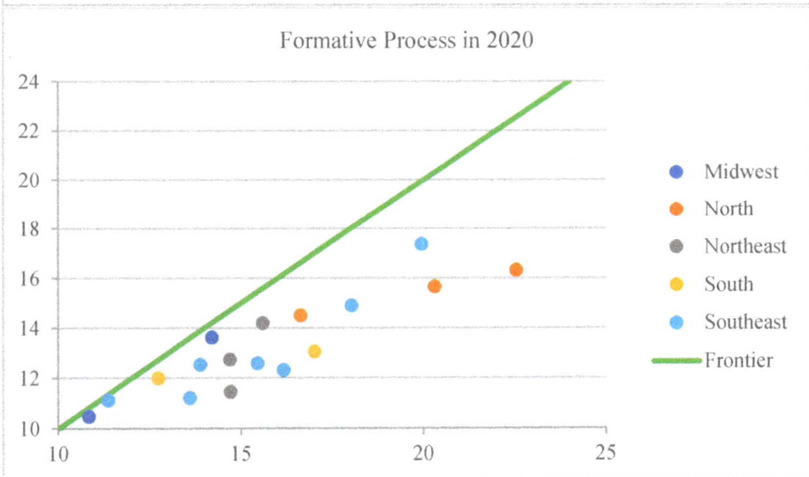

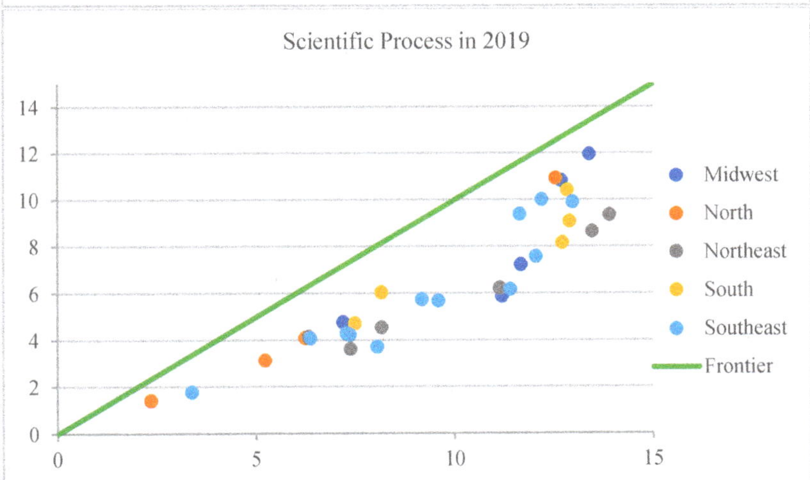

Figure 6. *Cont.*

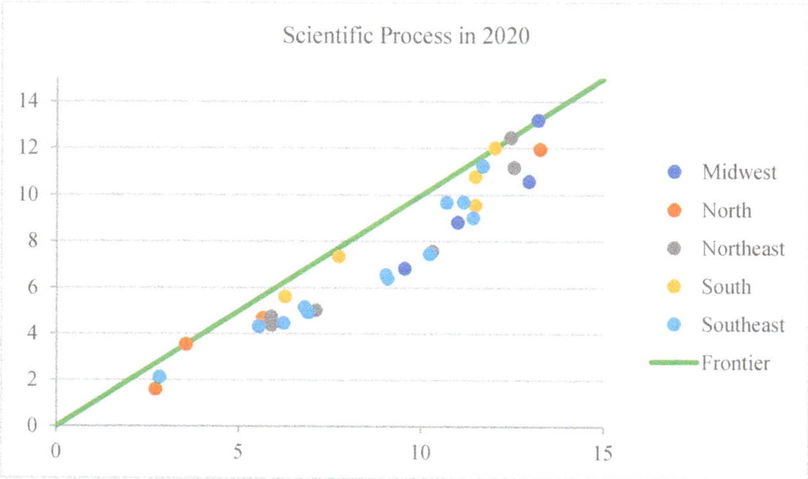

Figure 6. Process efficiencies 2019–2020.

Table 2 details the same results that are presented in Figures 5 and 6. However, it is simpler to identify patterns and obtain a quicker understanding of the results. It is also noteworthy that both in the Brazilian case and international assessments, the commissions responsible for evaluations are multidisciplinary, and not all the members involved are always familiar with mathematical programming models.

The visual analysis will also aid in faster identification of DMU performance patterns and faster identification of the best-performing units. It can also help to verify the existence of performance discrepancies between geographical regions, and these checks are faster than analyzing large data tables. Although it does not represent the main objective of the analysis, the value of the DMU's efficiency is easily obtained by observing the graph. We can quickly obtain the value of the DMU's efficiency by observing Figure 5: for the highlighted DMU (UFGD), its virtual output corresponds to 22.72, whereas the virtual input corresponds to 20.97, and the efficiency corresponds to 0.9230 (20.97/22.72).

Table 6 also shows that the proposed method makes ranking universities based on efficiency values possible. In addition to these values, the results related to the proportion of the allocation of resources shared between the stages and the importance of the stages reflected by the proportion of inputs are presented.

Table 6. Results of the centralized model.

University	θ^*	$\theta^{(1,sys)*}$	$\theta^{(2,sys)*}$	w^1	w^2	α_1	α_2	Formative Process $\theta^{(1,1)*}$	Scientific Production $\theta^{(1,2)*}$	w_1^1	w_2^1	Formative Process $\theta^{(2,1)*}$	Scientific Production $\theta^{(2,2)*}$	w_1^2	w_2^2
UFSCPA	0.823	0.724	0.945	0.551	0.449	0.40	0.70	0.716	0.739	0.676	0.324	0.941	0.950	0.622	0.378
UFMS	0.896	0.885	0.906	0.484	0.516	0.70	0.40	0.928	0.853	0.426	0.574	0.785	1.000	0.439	0.561
UFRR	0.782	0.872	0.709	0.446	0.554	0.70	0.70	0.906	0.610	0.885	0.115	0.724	0.587	0.892	0.108
UFS	0.826	0.777	0.881	0.529	0.471	0.40	0.59	0.821	0.655	0.738	0.262	0.909	0.805	0.726	0.274
UNIPAMPA	0.821	0.842	0.801	0.490	0.510	0.70	0.70	0.949	0.631	0.665	0.335	0.766	0.896	0.731	0.269
UFPI	0.846	0.833	0.860	0.512	0.488	0.40	0.70	0.897	0.658	0.733	0.267	0.871	0.828	0.746	0.254
UNB	0.772	0.710	0.850	0.553	0.447	0.40	0.70	0.858	0.525	0.557	0.443	0.967	0.716	0.532	0.468
UFBA	0.711	0.664	0.767	0.548	0.452	0.69	0.70	0.736	0.491	0.705	0.295	0.778	0.741	0.714	0.286
UFGD	0.863	0.923	0.803	0.498	0.502	0.70	0.40	0.966	0.893	0.410	0.590	0.784	0.818	0.435	0.565
UFPB	0.793	0.773	0.814	0.522	0.478	0.43	0.70	0.887	0.555	0.657	0.343	0.866	0.705	0.674	0.326
UFAL	0.804	0.777	0.831	0.503	0.497	0.70	0.40	0.969	0.641	0.413	0.587	0.757	0.891	0.448	0.552
UNIFAL-MG	0.844	0.831	0.856	0.501	0.499	0.70	0.55	0.884	0.528	0.852	0.148	0.871	0.754	0.876	0.124
UFCG	0.821	0.758	0.889	0.516	0.484	0.40	0.40	0.881	0.672	0.410	0.590	0.746	1.000	0.437	0.563
UFG	0.762	0.732	0.792	0.504	0.496	0.40	0.70	0.849	0.619	0.493	0.507	0.782	0.803	0.514	0.486
UNIFEI	0.765	0.745	0.785	0.510	0.490	0.70	0.70	0.823	0.577	0.684	0.316	0.815	0.718	0.692	0.308
UFJF	0.769	0.708	0.846	0.559	0.441	0.40	0.70	0.821	0.464	0.684	0.316	0.903	0.718	0.690	0.310
UFLA	0.874	0.900	0.848	0.495	0.505	0.40	0.70	1.000	0.807	0.484	0.516	0.798	0.906	0.537	0.463
UFMT	0.811	0.753	0.879	0.537	0.463	0.70	0.40	0.790	0.664	0.706	0.294	0.959	0.716	0.672	0.328
UFMG	0.814	0.772	0.861	0.527	0.473	0.70	0.70	0.981	0.541	0.526	0.474	0.979	0.730	0.526	0.474
UFOP	0.760	0.758	0.761	0.497	0.503	0.70	0.70	0.837	0.592	0.679	0.321	0.762	0.759	0.704	0.296
UFPEL	0.862	0.829	0.897	0.510	0.490	0.40	0.40	0.852	0.811	0.448	0.552	0.778	1.000	0.464	0.536
UFPE	0.743	0.718	0.770	0.509	0.491	0.50	0.70	0.865	0.557	0.521	0.479	0.800	0.735	0.540	0.460
UNIR	0.852	0.904	0.805	0.477	0.523	0.40	0.40	1.000	0.602	0.760	0.240	0.771	1.000	0.851	0.149
UFSC	0.769	0.749	0.790	0.511	0.489	0.40	0.70	0.877	0.642	0.455	0.545	0.744	0.833	0.486	0.514
UFSM	0.840	0.813	0.870	0.512	0.488	0.40	0.70	0.946	0.704	0.448	0.552	0.796	0.939	0.485	0.515
UFSCAR	0.782	0.751	0.813	0.506	0.494	0.40	0.70	0.885	0.629	0.478	0.522	0.836	0.790	0.494	0.506
UFSJ	0.827	0.824	0.830	0.522	0.478	0.70	0.40	0.827	0.821	0.488	0.512	0.789	0.870	0.489	0.511
UNIFESP	0.855	0.897	0.815	0.483	0.517	0.70	0.58	1.000	0.643	0.711	0.289	0.826	0.781	0.765	0.235
UFU	0.770	0.755	0.786	0.504	0.496	0.62	0.70	0.871	0.593	0.583	0.417	0.825	0.727	0.601	0.399
UFV	0.892	0.869	0.916	0.512	0.488	0.40	0.70	1.000	0.764	0.445	0.555	0.864	0.962	0.476	0.524
UFABC	0.689	0.713	0.668	0.484	0.516	0.70	0.70	0.774	0.626	0.585	0.415	0.644	0.705	0.614	0.386
UFAC	0.871	0.909	0.837	0.469	0.531	0.70	0.40	0.963	0.871	0.414	0.586	0.759	0.902	0.453	0.547

4.3. Efficiency Decomposition

Tables 7 and 8 present the efficiency decomposition results. The first one portrays the case when the first stage is prioritized, while the second views Stage 2 as a leader. Besides efficiency values, these tables also present the optimal proportions of each shared input for all years under investigation.

Table 7. Results with Stage 1 as leader.

University	2019				2020			
	α_1	α_2	Formative Process	Scientific Production	α_1	α_2	Formative Process	Scientific Production
UFSCPA	0.4	0.7	0.7166	0.7764	0.4	0.7	0.9977	0.8574
UFMS	0.5568	0.7	0.9306	0.8258	0.7	0.7	0.7998	0.9886
UFRR	0.7	0.7	0.9094	0.6362	0.7	0.7	0.7246	0.5801
UFS	0.4	0.7	0.8657	0.6422	0.4	0.7	0.9577	0.6762
UNIPAMPA	0.7	0.7	0.9499	0.5514	0.4	0.7	0.7910	0.8292
UFPI	0.7	0.7	0.9198	0.5797	0.7	0.7	0.8995	0.7453
UNB	0.4	0.7	0.8717	0.5572	0.4	0.7	0.9974	0.6817
UFBA	0.4	0.7	0.7460	0.5461	0.6336	0.7	0.8010	0.6830
UFGD	0.7	0.7	0.9944	0.8355	0.4	0.7	0.8032	0.8027
UFPB	0.7	0.7	0.9228	0.5548	0.7	0.7	0.9023	0.6311
UFAL	0.6656	0.7	0.9706	0.5901	0.4	0.7	0.7573	0.8913
UNIFAL-MG	0.4	0.7	0.9480	0.1355	0.7	0.7	0.8724	0.7413
UFCG	0.7	0.7	0.8816	0.6751	0.7	0.7	0.7834	0.9708
UFG	0.4	0.7	0.8554	0.6269	0.7	0.7	0.7910	0.7938
UNIFEI	0.7	0.7	0.8366	0.5961	0.7	0.7	0.8161	0.7159
UFJF	0.4	0.7	0.9264	0.4152	0.4	0.7	0.9408	0.6337
UFLA	0.7	0.7	1.0000	0.7904	0.4	0.7	0.8531	0.8424
UFMT	0.7	0.7	0.8043	0.7364	0.4	0.7	1.0000	0.6308
UFMG	0.7	0.7	1.0000	0.5257	0.7	0.7	1.0000	0.7070
UFOP	0.6509	0.7	0.8807	0.5704	0.7	0.7	0.7628	0.7570
UFPEL	0.4	0.7	0.8794	0.7999	0.7	0.7	0.8049	0.9767
UFPE	0.5274	0.7	0.9008	0.5106	0.5134	0.7	0.8330	0.6958
UNIR	0.7	0.7	1.0000	0.4534	0.4	0.7	0.7727	0.9889
UFSC	0.4870	0.7	0.8994	0.6225	0.7	0.7	0.7860	0.7930
UFSM	0.4	0.7	0.9463	0.7150	0.7	0.7	0.8406	0.8968
UFSCAR	0.4	0.7	0.9112	0.6035	0.5090	0.7	0.8438	0.7832
UFSJ	0.7	0.7	0.8750	0.6844	0.4	0.7	0.8847	0.7779
UNIFESP	0.4333	0.7	1.0000	0.6712	0.7	0.7	0.8443	0.7208
UFU	0.7	0.7	0.9061	0.5626	0.7	0.7	0.8471	0.6934
UFV	0.4	0.7	1.0000	0.7490	0.7	0.7	0.9008	0.9294
UFABC	0.5236	0.7	0.7777	0.5634	0.7	0.7	0.6450	0.7038
UFAC	0.4	0.7	0.9816	0.8712	0.7	0.7	0.7862	0.8795

When analyzing the efficiency decomposition of the processes, it is possible to verify that the decomposition was unique only when one of the stages was considered efficient. This is the case of UFLA, UNIR, UNIFESP, and UFV in the formative process in 2019. This situation was also observed in UFMS, UFCG, UFPEL, and UNIR in 2020 for the scientific production process.

In Table 7, it is possible to identify that the number of efficient DMUs remains the same in 2019. However, in 2020, two DMUs became efficient when the first stage was the leader. In contrast, nine and seven are deemed efficient in 2019 and 2020, respectively, when the second stage becomes the leader, as displayed in Table 8.

Table 8. Results with Stage 2 as leader.

University	2019				2020			
	α_1	α_2	Formative Process	Scientific Production	α_1	α_2	Formative Process	Scientific Production
UFSCPA	0.59	0.70	0.5912	1.0000	0.4	0.7	0.9176	0.9893
UFMS	0.70	0.40	0.7304	1.0000	0.4	0.7	0.7853	1.0000
UFRR	0.70	0.70	0.8977	0.6770	0.7	0.7	0.7202	0.6163
UFS	0.40	0.70	0.7322	0.9045	0.4	0.4	0.8908	0.8533
UNIPAMPA	0.40	0.65	0.8310	0.8645	0.7	0.7	0.7282	1.0000
UFPI	0.40	0.48	0.7981	0.9292	0.4	0.7	0.8402	0.9195
UNB	0.40	0.70	0.6281	0.8130	0.4	0.7	0.8682	0.8284
UFBA	0.68	0.70	0.6388	0.7242	0.7	0.7	0.7522	0.8049
UFGD	0.40	0.40	0.8125	1.0000	0.4	0.4	0.7197	0.8671
UFPB	0.40	0.70	0.7222	0.8717	0.4	0.7	0.8048	0.8324
UFAL	0.58	0.70	0.7489	0.7960	0.7	0.4	0.7122	0.9278
UNIFAL-MG	0.40	0.40	0.8322	0.8273	0.4	0.4	0.8652	0.7917
UFCG	0.40	0.70	0.4090	1.0000	0.7	0.4	0.7457	1.0000
UFG	0.46	0.70	0.6540	0.8084	0.4	0.7	0.7579	0.8288
UNIFEI	0.70	0.70	0.7453	0.7454	0.7	0.7	0.7653	0.8300
UFJF	0.40	0.41	0.6915	0.7451	0.4	0.7	0.8770	0.7760
UFLA	0.40	0.70	0.7940	1.0000	0.4	0.7	0.7926	0.9125
UFMT	0.40	0.70	0.7087	0.8601	0.4	0.4	0.9526	0.7281
UFMG	0.40	0.70	0.6752	0.8794	0.4	0.7	0.8633	0.8591
UFOP	0.56	0.40	0.7271	0.8240	0.4	0.4475	0.7266	0.8430
UFPEL	0.40	0.70	0.6185	1.0000	0.4	0.7	0.7780	1.0000
UFPE	0.51	0.70	0.6346	0.8076	0.4	0.7	0.7370	0.8086
UNIR	0.40	0.70	0.8743	1.0000	0.7	0.7	0.7708	1.0000
UFSC	0.48	0.70	0.6164	0.8595	0.4	0.7	0.6870	0.8866
UFSM	0.45	0.70	0.6562	0.9396	0.4	0.7	0.7501	0.9821
UFSCAR	0.46	0.70	0.7065	0.7923	0.4	0.4	0.8137	0.8126
UFSJ	0.70	0.40	0.6396	1.0000	0.7	0.4	0.7095	0.9457
UNIFESP	0.69	0.70	0.9070	0.8719	0.4	0.7	0.8199	0.8000
UFU	0.67	0.70	0.7159	0.8100	0.4	0.4	0.7775	0.7981
UFV	0.40	0.70	0.7055	1.0000	0.4	0.7	0.8229	1.0000
UFABC	0.70	0.70	0.6485	0.8026	0.7	0.7	0.6338	0.7217
UFAC	0.70	0.70	0.7803	1.0000	0.7	0.4	0.6407	1.0000

It is also relevant to observe that when the first stage is prioritized, the allocation of students is maintained or even enlarged in 2020 for the majority of the DMUs. However, the same pattern is not verified for professors. On the other hand, the pattern verified for the second stage is similar for both years. The majority of DMUs are inclined to maintain or reduce both students' and professors' workloads when compared to the initial DNDEA results.

Efficiency decomposition analysis allows universities to evaluate different scenarios and consider the impact of prioritizing the performance of one process over another. In addition, the model used provides individual answers for each university, as well as the proportion of resource allocation for the investigated cases.

5. Conclusions

Universities are essential for social and economic development. Public funds used in these institutions have stimulated the development of proposals for evaluation. DEA has stood out in the field of efficiency measurements in education, with the application of models in distinct areas, such as primary education, secondary schools, teachers, students, research, and teaching.

Educational processes usually span several consecutive periods. Therefore, it is adequate to use models considering the temporal effects on efficiency. We also consider that there is a network structure when analyzing the processes of graduate activities. Thus, in this paper, there is a proposition of a two-stage dynamic network model that considers shared inputs among the stages. First, we propose a centralized approach that maximizes the efficiency of the system, considering all periods and stages under investigation. The overall efficiency is obtained with a weighted sum of the period and process efficiency. In this initial view, the approach considers that all stages cooperate and act in unity to obtain the best possible results, considering the entire time frame evaluated.

Considering resource sharing between the stages makes it possible to represent the context of graduate activities more accurately. Nevertheless, the proposed DNDEA use is broader than the educational context and can be applied to others where the stages share common resources. Also, Appendix A points out that our approach can easily be adapted to cases without shared inputs, and considers exogenous inputs in the second division of the DMU. After this initial analysis, we investigated the efficiency uniqueness of the centralized DNDEA with a decomposition based on a leader–follower approach. In this framework, we investigated the cases where the first stage takes priority and the situations where the second stage is the leader.

This paper also presents a new framework for a bi-dimensional representation of the DNDEA efficiency frontier and the location of DMUs regarding the frontier when multiple inputs and outputs are present. Then, we present the step-by-step procedure developed to generate the graphs to present all the distinct levels of DNDEA efficiency results. Through a linearization of weights, we obtain a new set of weights—the modified weights- to obtain modified virtual inputs and outputs, allowing a two-dimensional representation.

The bi-dimensional framework provides intelligible graphs. The proposed approach's advantages are related to the simplicity of the method. Also, dynamic models provide a more comprehensive range of information when compared with classical models. In this sense, graphical representation offers a critical and effective way to deliver all the information to the decision-maker.

Combining the models and their bi-dimensional representation indicates that the DNDEA model is more suitable for analyzing universities. We verify an increase in system efficiency from 2019 to 2020. Results indicate that the COVID pandemic impacted the formative and scientific production processes differently. We also evaluated whether there were significant performance differences when considering the five Brazilian macro-regions. No significant disparities were found when analyzing the bi-dimensional representation and the statistical tests.

The formative and scientific production process results inversed the patterns in 2019 and 2020. Before the pandemic, the formative process performed better, but the scientific production process obtained superior results in 2020. Correlation analyses between the efficiency scores highlight the fact that the scientific production process significantly impacts the system's results. However, cuts in national budgets earmarked for education and research have been negatively impacting the performance of this activity. Furthermore, it is relevant to map and understand the main difficulties in the formative process because scientific production directly depends on the products generated by it.

The empirical results allow for ranking universities, aid in graduate activities' improvements, and support the development of public policies to enhance Brazilian research results. Despite the relevant results, we must highlight a limitation of the study. A thorough analysis is necessary to investigate more data for both processes to verify whether the superior performance of the scientific production remains. The graduate activities have been resilient throughout a decade of successive budget cuts. However, it is essential to mention that this resilience is not unlimited.

Second, our study did not include quality metrics of graduate activity products. Therefore, more investigations are required to add variables that reflect quality. We can use the classification of publications considering the journal impact factor or quartile to segregate this variable and provide more thorough evaluations.

Third, although federal universities are highly relevant to Brazilian research, the investigation of private, state, and municipal institutions should also be considered to assess the performance of graduate activities. The absence of these HEIs represents the main limitation of this research.

Besides the empirical contributions to the Brazilian HEIs, this paper provides three main methodological contributions. The first relates to a new framework for investigating two-stage systems in a dynamic setting with shared resources between the stages. The second relates to the discussion of efficiency decomposition to verify the uniqueness of the efficiency scores provided by the DNDEA model. The last refers to a simple but effective way to present the results provided by the bi-dimensional representation.

We concluded that extensions of this work are also possible. Initially, the investigations did not consider undergraduate activities, and they represent a significant part of federal universities' operating processes and expenses. It is also relevant to mention that no indicator evaluates undergraduate activities in an aggregate manner to rank the universities. Thus, this extension represents a relevant contribution due to the importance of federal universities to society.

From the methodological point of view, it is important to highlight that the current study evaluates the efficiency decomposition after a cooperative evaluation considering collaboration between the stages. However, analyzing this context from a non-cooperative perspective is interesting for assessing real cases in which cooperation cannot be guaranteed. Modifications of the current model using non-radial measures are extremely valuable in improving the applicability range of the model. Lastly, the model's extension to a multiple-stage and multi-level framework is highly recommended. The extension to multiple stages will allow discussions such as the investigation of the Brazilian university triple helix: teaching, research and extension activities. The multi-level investig ation can aid in investigating how public universities are contributing to obtaining the Ministry of Education goals.

Author Contributions: L.M.L.d.S.T.: conceptualization, methodology, investigation, writing—original draft; F.S.R.: writing—review and editing, supervision. All authors have read and agreed to the published version of the manuscript.

Funding: This research was financed in part by the Coordenação de Aperfeiçoamento de Pessoal de Nível Superior—Brasil (CAPES)—Finance Code 001 and by the Conselho Nacional de Desenvolvimento Científico e Tecnológico–Brasil (CNPq) grant number "PQ 311334/2019-9".

Data Availability Statement: Data are contained within the article.

Acknowledgments: The authors would like to thank the support by the Coordenação de Aperfeiçoamento de Pessoal de Nível Superior—Brasil (CAPES—Coordination for the Improvement of Higher Education Personnel)—Finance Code 001 and Conselho Nacional de Desenvolvimento Científico e Tecnológico—Brasil (CNPq—Conselho Nacional de Desenvolvimento Científico e Tecnológico) grant number "PQ 311334/2019-9"for their financial support for this research.

Conflicts of Interest: The authors declare no conflicts of interest.

Appendix A

In this appendix, we detail a model that does not consider shared resources and allows for exogenous inputs in the second division, as illustrated in Figure A1.

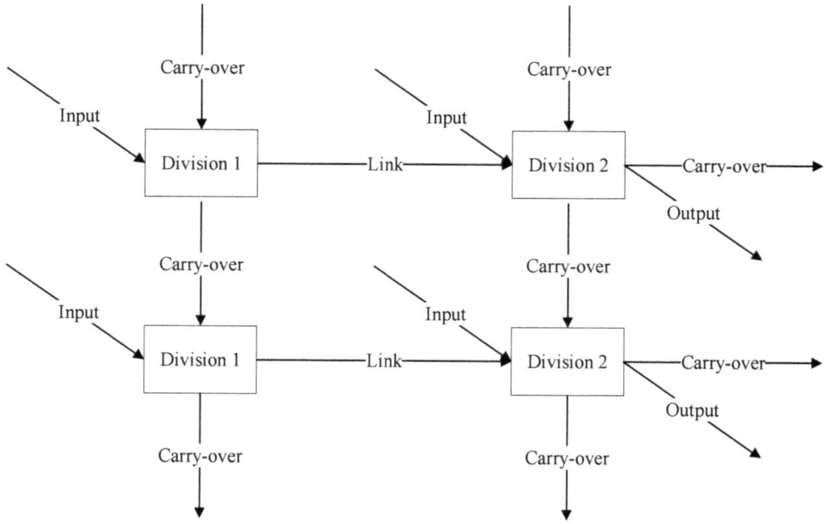

Figure A1. Two-stage dynamic DEA framework.

We consider $x_{ij}^{(t,k)}$ as the ith specific input of DMU j in division k in period t; we follow the same hypothesis discussed in Section 3, and the system's efficiency considers a weighted average of division 1 and 2 for each period, as displayed in (A1).

$$E_j^{(t,\,sys)} = w_1^t * \frac{\sum_{l\in l^1} f_l c_{lj}^{(t,1)} + \sum_{d=1}^{D} w_d z_{dj}^{(t)}}{\sum_{l\in l^1} f_l c_{lj}^{(t-1,1)} + \sum_{m\in m^1} v_i x_{ij}^{(t,1)}} + w_2^t * \frac{\sum_{r=1}^{s} u_r y_{rj}^{(t)} + \sum_{l\in l^2} f_l c_{lj}^{(t,2)}}{\sum_{l\in l^2} f_l c_{lj}^{(t-1,2)} + \sum_{d=1}^{D} w_d z_{dj}^{(t)} + \sum_{m\in m^2} v_i x_{ij}^{(t,2)}} \tag{A1}$$

where $w_1^t + w_2^t = 1$ and are defined as follows:

$$w_1^t = \frac{\sum_{l\in l^1} f_l c_{lj}^{(t-1,1)} + \sum_{m\in m^1} v_i x_{ij}^{(t,1)}}{\sum_{m\in m^1} v_i x_{ij}^{(t,1)} + \sum_{m\in m^2} v_i x_{ij}^{(t,2)} + \sum_{d=1}^{D} w_d z_{dj}^{(t)} + \sum_{k=1}^{K}\sum_{l=1}^{L} f_l c_{lj}^{(t-1,k)}} \quad and$$
$$w_2^t = \frac{\sum_{l\in l^2} f_l c_{lj}^{(t-1,2)} + \sum_{d=1}^{D} w_d z_{dj}^{(t)} + \sum_{m\in m^2} v_i x_{ij}^{(t,2)}}{\sum_{m\in m^1} v_i x_{ij}^{(t,1)} + \sum_{m\in m^2} v_i x_{ij}^{(t,2)} + \sum_{d=1}^{D} w_d z_{dj}^{(t)} + \sum_{k=1}^{K}\sum_{l=1}^{L} f_l c_{lj}^{(t-1,k)}} \tag{A2}$$

Therefore, the system efficiency in each period is detailed in (A3).

$$E_j^{(t,\,sys)} = \frac{\sum_{r=1}^{s} u_r y_{rj}^{(t)} + \sum_{k=1}^{K}\sum_{l=1}^{L} f_l c_{lj}^{(t,k)} + \sum_{d=1}^{D} w_d z_{dj}^{(t)}}{\sum_{k=1}^{K}\sum_{i=1}^{m} v_i\, x_{ij}^{(t,k)} + \sum_{d=1}^{D} w_d z_{dj}^{(t)} + \sum_{k=1}^{K}\sum_{l=1}^{L} f_l c_{lj}^{(t-1,k)}} \tag{A3}$$

We also considered that the overall efficiency is a weighted average of the system efficiency in each period. Therefore, we define w^t in (A4).

$$w^t = \frac{\sum_{k=1}^{K}\sum_{i=1}^{m} v_i x_{ij}^{(t,k)} + \sum_{d=1}^{D} w_d z_{dj}^{(t)} + \sum_{k=1}^{K}\sum_{l=1}^{L} f_l c_{lj}^{(t-1,k)}}{\sum_{t=1}^{T}\sum_{k=1}^{K}\sum_{i=1}^{m} v_i x_{ij}^{(t,k)} + \sum_{t=1}^{T}\sum_{k=1}^{K}\sum_{l=1}^{L} f_l c_{lj}^{(t-1,k)} + \sum_{t=1}^{T}\sum_{d=1}^{D} w_d z_{dj}^{(t)}} \tag{A4}$$

Considering that the overall efficiency is a weighted average of period efficiency, the overall efficiency score of the two-stage process for DMU$_o$ can be evaluated by solving the following fractional program (A5).

$$
\begin{array}{l}
\theta_o^* = Max \frac{\sum_{t=1}^{T}\sum_{r=1}^{s} u_r y_{ro}^{(t)} + \sum_{t=1}^{T}\sum_{d=1}^{D} w_d z_{do}^{(t)} + \sum_{t=1}^{T}\sum_{k=1}^{K}\sum_{l=1}^{L} f_l c_{lo}^{(t,k)}}{\sum_{t=1}^{T}\sum_{k=1}^{K}\sum_{i=1}^{m} v_i x_{io}^{(t,k)} + \sum_{t=1}^{T}\sum_{k=1}^{K}\sum_{l=1}^{L} f_l c_{lo}^{(t-1,k)} + \sum_{t=1}^{T}\sum_{d=1}^{D} w_d z_{do}^{(t)}} \\
\text{s.t } \frac{\sum_{r=1}^{s} u_r y_{rj}^{(t)} + \sum_{k=1}^{K}\sum_{l=1}^{L} f_l c_{lj}^{(t,k)} + \sum_{d=1}^{D} w_d z_{dj}^{(t)}}{\sum_{k=1}^{K}\sum_{i=1}^{m} v_i x_{ij}^{(t,k)} + \sum_{d=1}^{D} w_d z_{dj}^{(t)} + \sum_{k=1}^{K}\sum_{l=1}^{L} f_l c_{lj}^{(t-1,k)}} \le 1 \; (j = 1,\ldots,n; t = 1,\ldots,T) \\
\frac{\sum_{l\in l^1} f_l c_{lj}^{(t,1)} + \sum_{d=1}^{D} w_d z_{dj}^{(t)}}{\sum_{l\in l^1} f_l c_{lj}^{(t-1,1)} + \sum_{m\in m^1} v_i x_{ij}^{(t,1)}} \le 1 \; (j = 1,\ldots,n; t = 1,\ldots,T) \\
\frac{\sum_{r=1}^{s} u_r y_{rj}^{(t)} + \sum_{l\in l^2} f_l c_{lj}^{(t,2)}}{\sum_{l\in l^2} f_l c_{lj}^{(t-1,2)} + \sum_{d=1}^{D} w_d z_{dj}^{(t)} + \sum_{m\in m^2} v_i x_{ij}^{(t,2)}} \le 1 \; (j = 1,\ldots,n; t = 1,\ldots,T) \\
v_i, u_r, w_l, f_d, v_p \ge \varepsilon; \; i = 1,\ldots,m; r = 1,\ldots, s; l = 1,\ldots,L; d = 1,\ldots,D; \; p \\
= 1,\ldots, P
\end{array}
\tag{A5}
$$

With the aid of the Charnes–Cooper transformation, the fractional program proposed in Model (A5) can be converted into Model (A6).

$$
\begin{array}{l}
\theta_o^* = max \sum_{t=1}^{T}\sum_{r=1}^{s} \mu_r y_{ro}^{(t)} + \sum_{t=1}^{T}\sum_{d=1}^{D} \mu_d z_{do}^{(t)} + \sum_{t=1}^{T}\sum_{k=1}^{K}\sum_{l=1}^{L} \gamma_l c_{lo}^{(t,k)} \\
\sum_{t=1}^{T}\sum_{k=1}^{K}\sum_{i=1}^{m} \nu_i x_{io}^{(t,k)} + \sum_{t=1}^{T}\sum_{k=1}^{K}\sum_{l=1}^{L} \gamma_l c_{lo}^{(t-1,k)} + \sum_{t=1}^{T}\sum_{d=1}^{D} \mu_d z_{do}^{(t)} = 1 \\
\sum_{r=1}^{s} \mu_r y_{rj}^{(t)} + \sum_{k=1}^{K}\sum_{l=1}^{L} \gamma_l c_{lj}^{(t,k)} + \sum_{d=1}^{D} \mu_d z_{dj}^{(t)} - \sum_{k=1}^{K}\sum_{i=1}^{m} \nu_i x_{ij}^{(t,k)} - \sum_{d=1}^{D} \mu_d z_{dj}^{(t)} \\
\quad - \sum_{k=1}^{K}\sum_{l=1}^{L} \gamma_l c_{lj}^{(t-1,k)} \le 0 \; (j = 1,\ldots,n; t = 1,\ldots,T) \\
\sum_{l\in l^1} \gamma_l c_{lj}^{(t,1)} + \sum_{d=1}^{D} \mu_d z_{dj}^{(t)} - \sum_{l\in l^1} \gamma_l c_{lj}^{(t-1,1)} - \sum_{m\in m^1} \nu_i x_{ij}^{(t,1)} \\
\quad \le 0 \; (j = 1,\ldots,n; t = 1,\ldots,T) \\
\sum_{r=1}^{s} \mu_r y_{rj}^{(t)} + \sum_{l\in l^2} \gamma_l c_{lj}^{(t,2)} - \sum_{l\in l^2} \gamma_l c_{lj}^{(t-1,2)} - \sum_{d=1}^{D} \mu_d z_{dj}^{(t)} - \sum_{m\in m^2} \nu_i x_{ij}^{(t,2)} \\
\quad \le 0 \; (j = 1,\ldots,n; t = 1,\ldots,T) \\
\nu_i, \nu_p, \mu_r, \gamma_l, \mu_d \ge \varepsilon; \; i = 1,\ldots,m; r = 1,\ldots, s; l = 1,\ldots,L; d = 1,\ldots,D; \; p \\
= 1,\ldots, P
\end{array}
\tag{A6}
$$

After solving Model (A6), it is possible to obtain all efficiency scores discussed previously, namely, process efficiency, system efficiency and overall efficiency. We proceed with efficiency decomposition, similar to the procedure described in Section 3.

The first division has its efficiency maximized, while the overall efficiency is maintained at the level identified with the aid of Model (11). Let $\nu_i^*, \nu_{p,}^*, \mu_r^*, \gamma_l^*, \mu_d^*$ be the optimal weights, while θ_o^*, $\theta_o^{(t,sys)*}$, $\theta_o^{(t,1)*}$ and $\theta_o^{(2,sys)*}$ represent the optimal overall and optimal system efficiency by period and division 1 and division 2 at period t efficiency θ_o^* of an observed DMU$_o$. Suppose we focus on the maximization of the first stage: while maintaining the system by period and overall score, we have:

$$\begin{aligned}
&\theta_o^{(t,1)*} = max \sum_{l\in l^1} \gamma_l c_{lo}^{(t,1)} + \sum_{d=1}^{D} \mu_d z_{do}^{(t)} \\
&\sum_{m\in m^1} v_i x_{io}^{(t,1)} + \sum_{l\in l^1} \gamma_l c_{lo}^{(t-1,1)} = 1 \\
&\sum_{l\in l^1} \gamma_l c_{lj}^{(t,1)} + \sum_{d=1}^{D} \mu_d z_{dj}^{(t)} - \sum_{m\in m^1} v_i x_{ij}^{(t,1)} - \sum_{l\in l^1} \gamma_l c_{lj}^{(t-1,1)} \le 0 \ (j = 1,\ldots,n) \\
&\sum_{r=1}^{s} \mu_r y_{rj}^{(t)} + \sum_{l\in l^2} \gamma_l c_{lj}^{(t,2)} - \sum_{l\in l^2} \gamma_l c_{lj}^{(t-1,2)} - \sum_{m\in m^2} v_i x_{ij}^{(t,2)} - \sum_{d=1}^{D} \mu_d z_{dj}^{(t)} \le 0 \ (j = 1,\ldots,n) \\
&\sum_{r=1}^{s} \mu_r y_{ro}^{(t)} + \sum_{d=1}^{D} \mu_d z_{do}^{(t)} + \sum_{k=1}^{K}\sum_{l=1}^{L} \gamma_l c_{lo}^{(t,k)} \\
&\qquad -\theta_o^{(t,sys)*}\left(\sum_{k=1}^{K}\sum_{i=1}^{m} v_i x_{io}^{(t,k)} - \sum_{d=1}^{D} \mu_d z_{do}^{(t)} - \sum_{k=1}^{K}\sum_{l=1}^{L} \gamma_l c_{lo}^{(t-1,k)}\right) \le 0 \\
&\sum_{t=1}^{T}\sum_{r=1}^{s} \mu_r y_{ro}^{(t)} + \sum_{t=1}^{T}\sum_{k=1}^{K}\sum_{l=1}^{L} \gamma_l c_{lo}^{(t,k)} + \sum_{t=1}^{T}\sum_{d=1}^{D} \mu_d z_{do}^{(t)} \\
&\qquad -\theta_o^{*}\left(\sum_{t=1}^{T}\sum_{k=1}^{K}\sum_{i=1}^{m} v_i x_{io}^{(t,k)} + \sum_{t=1}^{T}\sum_{k=1}^{K}\sum_{l=1}^{L} \gamma_l c_{lo}^{(t-1,k)} + \sum_{t=1}^{T}\sum_{d=1}^{D} \mu_d z_{do}^{(t)}\right) \le 0 \\
&w_1^{t*} * \left(\sum_{l\in l^1} \gamma_l c_{lo}^{(t,1)} + \sum_{d=1}^{D} \mu_d z_{do}^{(t)}\right) \le \theta_o^{(t,sys)*} \\
&\qquad v_i, v_p, \mu_r, \gamma_l, \mu_d \ge \varepsilon;\ i = 1,\ldots,m; r = 1,\ldots, s; l = 1,\ldots,L; d = 1,\ldots,D;\ p = 1,\ldots, P
\end{aligned} \tag{A7}$$

As previously discussed, the system efficiency is a weighted average of the stages; therefore, it is possible to obtain the efficiency of the second stage as $\theta_o^{(t,2)} = \frac{\theta_o^{(t,sys)*} - w_1^{t*}\theta_o^{(t,1)*}}{w_2^{t*}}$. The same hypothesis can be used to investigate Stage 2 efficiency, as shown in Model (A8).

$$\begin{aligned}
&\theta_o^{(t,2)*} = max \sum_{r=1}^{s} \mu_r y_{ro}^{(t)} + \sum_{l\in l^2} \gamma_l c_{lo}^{(t,2)} \\
&\sum_{m\in m^2} v_i x_{io}^{(t,2)} + \sum_{l\in l^2} \gamma_l c_{lo}^{(t-1,2)} + \sum_{d=1}^{D} \mu_d z_{do}^{(t)} = 1 \\
&\sum_{l\in l^1} \gamma_l c_{lj}^{(t,1)} + \sum_{d=1}^{D} \mu_d z_{dj}^{(t)} - \sum_{m\in m^1} v_i x_{ij}^{(t,1)} + \sum_{l\in l^1} \gamma_l c_{lj}^{(t-1,1)} \le 0 \ (j = 1,\ldots,n) \\
&\sum_{r=1}^{s} \mu_r y_{rj}^{(t)} + \sum_{l\in l^2} \gamma_l c_{lj}^{(t,2)} - \sum_{l\in l^2} \gamma_l c_{lj}^{(t-1,2)} - \sum_{m\in m^2} v_i x_{ij}^{(t,2)} - \sum_{d=1}^{D} \mu_d z_{dj}^{(t)} \\
&\le 0 \ (j = 1,\ldots,n) \le 0 \ (j = 1,\ldots,n) \\
&\sum_{r=1}^{s} \mu_r y_{ro}^{(t)} + \sum_{d=1}^{D} \mu_d z_{do}^{(t)} + \sum_{k=1}^{K}\sum_{l=1}^{L} \gamma_l c_{lo}^{(t,k)} \\
&\qquad -\theta_o^{(t,sys)*}\left(\sum_{k=1}^{K}\sum_{i=1}^{m} v_i x_{io}^{(t,k)} - \sum_{d=1}^{D} \mu_d z_{do}^{(t)} - \sum_{k=1}^{K}\sum_{l=1}^{L} \gamma_l c_{lo}^{(t-1,k)}\right) \\
&\qquad \le 0 \\
&\sum_{t=1}^{T}\sum_{r=1}^{s} \mu_r y_{ro}^{(t)} + \sum_{t=1}^{T}\sum_{k=1}^{K}\sum_{l=1}^{L} \gamma_l c_{lo}^{(t,k)} + \sum_{t=1}^{T}\sum_{d=1}^{D} \mu_d z_{do}^{(t)} \\
&\qquad -\theta_o^{*}\left(\sum_{t=1}^{T}\sum_{k=1}^{K}\sum_{i=1}^{m} v_i x_{io}^{(t,k)} + \sum_{t=1}^{T}\sum_{k=1}^{K}\sum_{l=1}^{L} \gamma_l c_{lo}^{(t-1,k)} \right.\\
&\qquad \left. + \sum_{t=1}^{T}\sum_{d=1}^{D} \mu_d z_{do}^{(t)}\right) \le 0 \\
&w_2^{t*} * \left(\sum_{r=1}^{s} \mu_r y_{ro}^{(t)} + \sum_{l\in l^2} \gamma_l c_{lo}^{(t,2)}\right) \le \theta_o^{(t,sys)*} \\
&v_i, v_p, \mu_r, \gamma_l, \mu_d \ge \varepsilon;\ i = 1,\ldots,m; r = 1,\ldots, s; l = 1,\ldots,L; d = 1,\ldots,D;\ p \\
&\qquad = 1,\ldots, P
\end{aligned} \tag{A8}$$

It is possible to obtain the efficiency of the first stage as $\theta_o^{(t,1)} = \frac{\theta_o^{(t,sys)*} - w_2^{t*}\theta_o^{(t,2)*}}{w_1^{t*}}$. It is important to mention that the proposed models and evaluation must be used for each period t under analysis. If $\theta_o^{(t,1)} = \theta_o^{(t,1)*}$ or, $\theta_o^{(t,2)} = \theta_o^{(t,2)*}$, there is a unique decomposition.

Appendix B

This Appendix details the mathematical proof that the modified virtual inputs and outputs proposed in Section 5 maintain the efficiency values obtained with the cooperative DNDEA model. We begin with overall system efficiency. In (17) to (24), the modified virtual inputs and outputs of the all the efficiencies provided by the model are presented.

We obtain any type of efficiency of a given DMU$_o$, $E_O^{(t,k)}$, by dividing the virtual output by the virtual input. In (A9), there is proof that the value obtained by considering the modified virtual inputs and outputs is equal to the value of the unchanged virtual values for the overall system efficiency. It is important to highlight the notations considered in Section 4 and in this appendix. Let $x_{ij} = \sum_{t=1}^{T} x_{ij}^{(t)}$, $x_{pj} = \sum_{t=1}^{T} x_{pj}^{(t)}$, $z_{dj} = \sum_{t=1}^{T} z_{dj}^{(t)}$, $y_{rj} = \sum_{t=1}^{T} y_{rj}^{(t)}$ and $v'_{ij} = \frac{v_{ij}}{S_j}; v'_{pj} = \frac{v_{pj}}{S_j}; \beta'_{pj} = \frac{\beta_{pj}}{S_j} \gamma'_{lj} = \frac{\gamma_{lj}}{S_j}; \mu'_{dj} = \frac{\mu_{dj}}{S_j}; \mu'_{rj} = \frac{\mu_{rj}}{S_j}$.

$$\begin{aligned} E_O^{(sys)} &= \frac{O_o^{(sys)}}{I_o^{(sys)}} = \frac{\sum_{t=1}^{T}\sum_{r=1}^{s}\mu_r y_{ro}^{(t)} + \sum_{t=1}^{T}\sum_{k=1}^{K}\sum_{l=1}^{L}\gamma_l c_{lo}^{(t,k)} + \sum_{t=1}^{T}\sum_{d=1}^{D}\mu_d z_{do}^{(t)}}{\sum_{t=1}^{T}\sum_{i=1}^{m}\nu_i x_{io}^{(t)} + \sum_{t=1}^{T}\sum_{p=1}^{P}\nu_p x_{po}^{(t)} + \sum_{t=1}^{T}\sum_{k=1}^{K}\sum_{l=1}^{L}\gamma_l c_{lo}^{(t-1,k)} + \sum_{t=1}^{T}\sum_{d=1}^{D}\mu_d z_{do}^{(t)}} \\ &= \frac{\sum_{r=1}^{s}\mu_r y_{ro} + \sum_{d=1}^{D}\mu_d z_{do} + \sum_{t=1}^{T}\sum_{k=1}^{K}\sum_{l=1}^{L}\gamma_l c_{lo}^{(t,k)}}{\sum_{i=1}^{m}\nu_i x_{io} + \sum_{p=1}^{P}\nu_p x_{po} + \sum_{d=1}^{D}\mu_d z_{do} + \sum_{t=1}^{T}\sum_{k=1}^{K}\sum_{l=1}^{L}\gamma_l c_{lo}^{(t-1,k)}} \\ &= \frac{\frac{\sum_{r=1}^{s}\mu_r y_{ro} + \sum_{d=1}^{D}\mu_d z_{do} + \sum_{t=1}^{T}\sum_{k=1}^{K}\sum_{l=1}^{L}\gamma_l c_{lo}^{(t,k)}}{S_o}}{\frac{\sum_{i=1}^{m}\nu_i x_{io} + \sum_{p=1}^{P}\nu_p x_{po} + \sum_{d=1}^{D}\mu_d z_{do} + \sum_{t=1}^{T}\sum_{k=1}^{K}\sum_{l=1}^{L}\gamma_l c_{lo}^{(t-1,k)}}{S_o}} \\ &= \frac{\sum_{r=1}^{s}\frac{\mu_r}{S_o} y_{ro} + \sum_{d=1}^{D}\frac{\mu_d}{S_o} z_{do} + \sum_{t=1}^{T}\sum_{k=1}^{K}\sum_{l=1}^{L}\frac{\gamma_l}{S_o} c_{lo}^{(t,k)}}{\sum_{i=1}^{m}\frac{\nu_i}{S_o} x_{io} + \sum_{p=1}^{P}\frac{\nu_p}{S_o} x_{po} + \sum_{d=1}^{D}\frac{\mu_d}{S_o} z_{do} + \sum_{t=1}^{T}\sum_{k=1}^{K}\sum_{l=1}^{L}\frac{\gamma_l}{S_o} c_{lo}^{(t-1,k)}} \\ &= \frac{\sum_{r=1}^{s}\mu'_{ro} y_{ro} + \sum_{d=1}^{D}\mu'_{do} z_{do} + \sum_{t=1}^{T}\sum_{k=1}^{K}\sum_{l=1}^{L}\gamma'_{lo} c_{lo}^{(t,k)}}{\sum_{i=1}^{m}v'_{io} x_{io} + \sum_{p=1}^{P}v'_{po} x_{po} + \sum_{d=1}^{D}\mu'_{do} z_{do} + \sum_{t=1}^{T}\sum_{k=1}^{K}\sum_{l=1}^{L}\gamma'_{lo} c_{lo}^{(t-1,k)}} = \frac{O\prime_o^{(sys)}}{I\prime_o^{(sys)}} \end{aligned} \tag{A9}$$

Also, we prove that the same patterns apply to system efficiency in each period (A10) and to process efficiency in each period. Two different proofs are presented for this last efficiency type because each stage has distinct virtual inputs and outputs. (A11) and (A12) detail the proof for the first and second stages, respectively.

$$\begin{aligned} E_O^{(t,sys)} &= \frac{O_o^{(t,sys)}}{I_o^{(t,sys)}} = \frac{\sum_{r=1}^{s}\mu_r y_{ro}^{(t)} + \sum_{d=1}^{D}\mu_d z_{do}^{(t)} + \sum_{k=1}^{K}\sum_{l=1}^{L}\gamma_l c_{lo}^{(t,k)}}{\sum_{i=1}^{m}\nu_i x_{io}^{(t)} + \sum_{p=1}^{P}\nu_p x_{po}^{(t)} + \sum_{d=1}^{D}\mu_d z_{do}^{(t)} + \sum_{k=1}^{K}\sum_{l=1}^{L}\gamma_l c_{lo}^{(t-1,k)}} \\ &= \frac{\frac{\sum_{r=1}^{s}\mu_r y_{ro}^{(t)} + \sum_{d=1}^{D}\mu_d z_{do}^{(t)} + \sum_{k=1}^{K}\sum_{l=1}^{L}\gamma_l c_{lo}^{(t,k)}}{S_o}}{\frac{\sum_{i=1}^{m}\nu_i x_{io}^{(t)} + \sum_{p=1}^{P}\nu_p x_{po}^{(t)} + \sum_{d=1}^{D}\mu_d z_{do}^{(t)} + \sum_{k=1}^{K}\sum_{l=1}^{L}\gamma_l c_{lo}^{(t-1,k)}}{S_o}} \\ &= \frac{\sum_{r=1}^{s}\frac{\mu_r}{S_o} y_{ro}^{(t)} + \sum_{d=1}^{D}\frac{\mu_d}{S_o} z_{do}^{(t)} + \sum_{k=1}^{K}\sum_{l=1}^{L}\frac{\gamma_l}{S_o} c_{lo}^{(t,k)}}{\sum_{i=1}^{m}\frac{\nu_i}{S_o} x_{io}^{(t)} + \sum_{p=1}^{P}\frac{\nu_p}{S_o} x_{po}^{(t)} + \sum_{d=1}^{D}\frac{\mu_d}{S_o} z_{do}^{(t)} + \sum_{k=1}^{K}\sum_{l=1}^{L}\frac{\gamma_l}{S_o} c_{lo}^{(t-1,k)}} \\ &= \frac{\sum_{r=1}^{s}\mu'_{ro} y_{ro}^{(t)} + \sum_{d=1}^{D}\mu'_{do} z_{do}^{(t)} + \sum_{k=1}^{K}\sum_{l=1}^{L}\gamma'_{lo} c_{lo}^{(t,k)}}{\sum_{i=1}^{m}v'_{io} x_{io}^{(t)} + \sum_{p=1}^{P}v'_{po} x_{po}^{(t)} + \sum_{d=1}^{D}\mu'_{do} z_{do}^{(t)} + \sum_{k=1}^{K}\sum_{l=1}^{L}\gamma'_{lo} c_{lo}^{(t-1,k)}} = \frac{O\prime_o^{(t,sys)}}{I\prime_o^{(t,sys)}} \end{aligned} \tag{A10}$$

$$\begin{aligned} E_O^{(t,1)} &= \frac{O_o^{(t,1)}}{I_o^{(t,1)}} = \frac{\sum_{1\in l^1}\gamma_l c_{lo}^{(t,1)} + \sum_{d=1}^{D}\mu_d z_{do}^{(t)}}{\sum_{i=1}^{m}\nu_i x_{io}^{(t)} + \sum_{p=1}^{P}\beta_p x_{po}^{(t)} + \sum_{1\in l^1}\gamma_l c_{lo}^{(t-1,1)}} = \frac{\frac{\sum_{1\in l^1}\gamma_l c_{lo}^{(t,1)} + \sum_{d=1}^{D}\mu_d z_{do}^{(t)}}{S_o}}{\frac{\sum_{i=1}^{m}\nu_i x_{io}^{(t)} + \sum_{p=1}^{P}\beta_p x_{po}^{(t)} + \sum_{1\in l^1}\gamma_l c_{lo}^{(t-1,1)}}{S_o}} \\ &= \frac{\sum_{1\in l^1}\frac{\gamma_l}{S_o} c_{lo}^{(t,1)} + \sum_{d=1}^{D}\frac{\mu_d}{S_o} z_{do}^{(t)}}{\sum_{i=1}^{m}\frac{\nu_i}{S_o} x_{io}^{(t)} + \sum_{p=1}^{P}\frac{\beta_p}{S_o} x_{pj}^{(t)} + \sum_{1\in l^1}\frac{\gamma_l}{S_o} c_{lj}^{(t-1,1)}} \\ &= \frac{\sum_{1\in l^1}\gamma'_{lo} c_{lo}^{(t,1)} + \sum_{d=1}^{D}\mu'_{do} z_{do}^{(t)}}{\sum_{i=1}^{m}v'_{io} x_{io}^{(t)} + \sum_{p=1}^{P}\beta'_{po} x_{po}^{(t)} + \sum_{1\in l^1}\gamma'_{lo} c_{lo}^{(t-1,1)}} = \frac{O\prime_o^{(t,1)}}{I\prime_o^{(t,1)}} \end{aligned} \tag{A11}$$

$$E_O^{(t,2)} = \frac{O_o^{(t,2)}}{I_o^{(t,2)}} = \frac{\sum_{r=1}^{s} \mu_r y_{ro}^{(t)} + \sum_{1\in 1^2} \gamma_1 c_{lo}^{(t,2)}}{\sum_{1\in 1^2} \gamma_1 c_{lo}^{(t-1,2)} + \sum_{d=1}^{D} \mu_d z_{do}^{(t)} + \sum_{p=1}^{P} \nu_p x_{po}^{(t)} - \sum_{p=1}^{P} \beta_p x_{po}^{(t)}}$$
$$= \frac{\frac{\sum_{r=1}^{s} \mu_r y_{ro}^{(t)} + \sum_{1\in 1^2} \gamma_1 c_{lo}^{(t,2)}}{S_O}}{\frac{\sum_{1\in 1^2} \gamma_1 c_{lo}^{(t-1,2)} + \sum_{d=1}^{D} \mu_d z_{do}^{(t)} + \sum_{p=1}^{P} \nu_p x_{po}^{(t)} - \sum_{p=1}^{P} \beta_p x_{po}^{(t)}}{S_O}}$$
$$= \frac{\sum_{r=1}^{s} \frac{\mu_r}{S_O} y_{ro}^{(t)} + \sum_{1\in 1^2} \frac{\gamma_1}{S_O} c_{lo}^{(t,2)}}{\sum_{1\in 1^2} \frac{\gamma_1}{S_O} c_{lo}^{(t-1,2)} + \sum_{d=1}^{D} \frac{\mu_d}{S_O} z_{do}^{(t)} + \sum_{p=1}^{P} \frac{\nu_p}{S_O} x_{po}^{(t)} - \sum_{p=1}^{P} \frac{\beta_p}{S_O} x_{po}^{(t)}}$$
$$= \frac{\sum_{r=1}^{s} \mu'_{ro} y_{ro}^{(t)} + \sum_{1\in 1^2} \gamma'_{lo} c_{lo}^{(t,2)}}{\sum_{1\in 1^2} \gamma'_{lo} c_{lo}^{(t-1,2)} + \sum_{d=1}^{D} \mu'_{do} z_{do}^{(t)} + \sum_{p=1}^{P} v'_{po} x_{po}^{(t)} - \sum_{p=1}^{P} \beta'_{po} x_{po}^{(t)}} = \frac{O\prime_o^{(t,2)}}{I\prime_o^{(t,2)}} \quad \text{(A12)}$$

Appendix C

This appendix details the sensitivity analysis for process efficiencies.

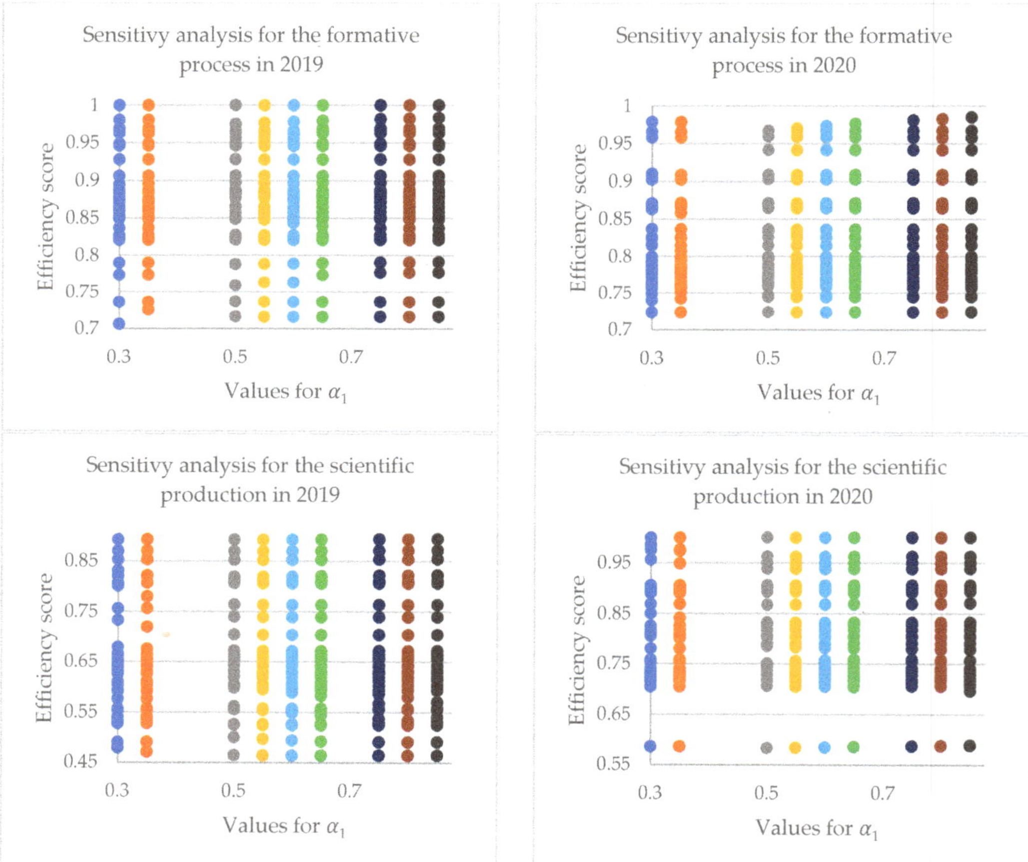

Figure A2. Sensitivity analyses for α_1 values.

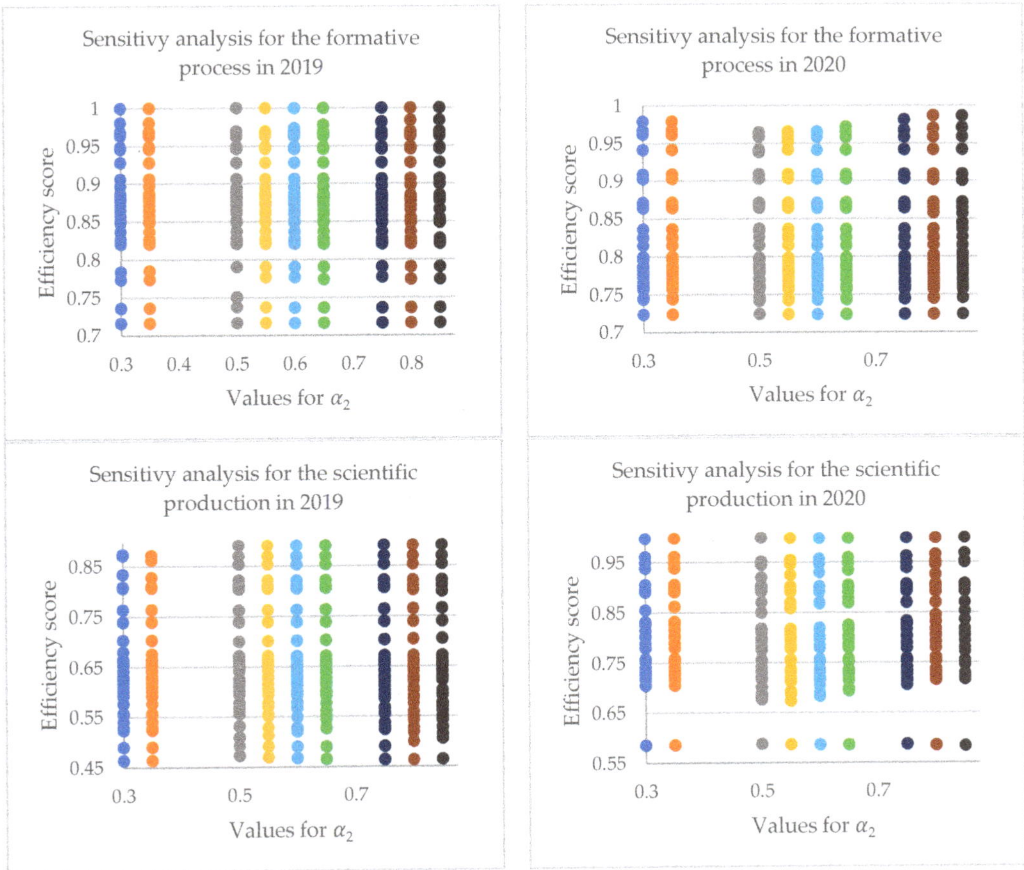

Figure A3. Sensitivity analyses for α_2 values.

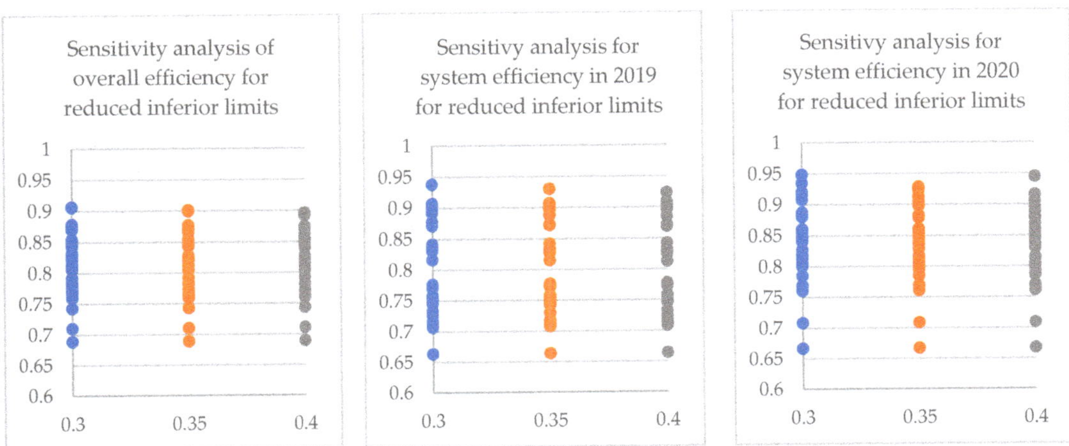

Figure A4. *Cont.*

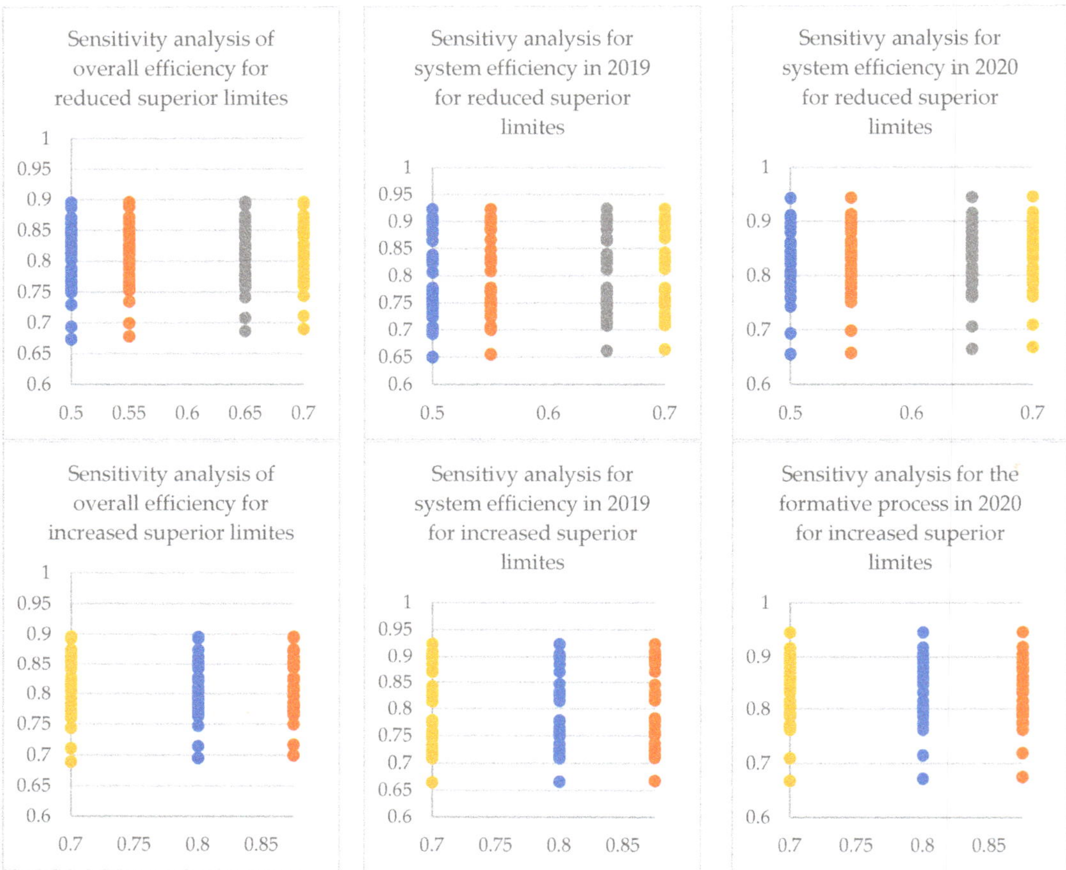

Figure A4. Sensitivity analyses for α_1 and α_2 values.

References

1. Foladi, S.; Solimanpur, M.; Jahangoshai Rezaee, M. Inverse Dynamic Data Envelopment Analysis for Evaluating Faculties of University with Quasi-Fixed Inputs. *Soc. Indic. Res.* **2020**, *148*, 323–347. [CrossRef]
2. Stumbriene, D.; Camanho, A.S.; Jakaitiene, A. The Performance of Education Systems in the Light of Europe 2020 Strategy. *Ann. Oper. Res.* **2020**, *288*, 577–608. [CrossRef]
3. Lee, B.L.; Johnes, J. Using Network DEA to Inform Policy: The Case of the Teaching Quality of Higher Education in England. *High. Educ. Q.* **2022**, *76*, 399–421. [CrossRef]
4. Moncayo-Martínez, L.A.; Ramírez–Nafarrate, A.; Hernández–Balderrama, M.G. Evaluation of Public HEI on Teaching, Research, and Knowledge Dissemination by Data Envelopment Analysis. *Socio-Econ. Plan. Sci.* **2020**, *69*, 100718. [CrossRef]
5. Moore, K.; Coates, H.; Croucher, G. Investigating Applications of University Productivity Measurement Models Using Australian Data. *Stud. High. Educ.* **2019**, *44*, 2148–2162. [CrossRef]
6. ANDIFES. *Universidades Federais: Patrimônio Da Sociedade Brasileira*; ANDIFES: Brasília, Brazil, 2017.
7. UNESCO. *Relatório de Ciências Da Unesco A Corrida Contra O Tempo Por Um Desenvolvimento Mais Inteligente Resumo Executivo E Cenário Brasileiro*; UNESCO: Paris, France, 2021.
8. Thanassoulis, E.; de Witte, K.; Johnes, J.; Johnes, G.; Karagiannis, G.; Portela, C.S. Applications of Data Envelopment Analysis in Education. In *International Series in Operations Research and Management Science*; Springer: New York, NY, USA, 2016; Volume 238, pp. 367–438; ISBN 9781489976840.
9. De Witte, K.; López-Torres, L. Efficiency in Education: A Review of Literature and a Way Forward. *J. Oper. Res. Soc.* **2017**, *68*, 339–363. [CrossRef]
10. Johnes, J. Operational Research in Education. *Eur. J. Oper. Res.* **2015**, *243*, 683–696. [CrossRef]
11. Johnes, J.; Portela, M.; Thanassoulis, E. Efficiency in Education. *J. Oper. Res. Soc.* **2017**, *68*, 331–338. [CrossRef]

12. Ding, T.; Zhang, Y.; Zhang, D.; Li, F. Performance Evaluation of Chinese Research Universities: A Parallel Interactive Network DEA Approach with Shared and Fixed Sum Inputs. *Socioecon. Plann Sci.* **2023**, *87*, 101582. [CrossRef]
13. Parteka, A.; Wolszczak-Derlacz, J. Dynamics of Productivity in Higher Education: Cross-European Evidence Based on Bootstrapped Malmquist Indices. *J. Product. Anal.* **2013**, *40*, 67–82. [CrossRef]
14. Kumar, A.; Thakur, R.R. Objectivity in Performance Ranking of Higher Education Institutions Using Dynamic Data Envelopment Analysis. *Int. J. Product. Perform. Manag.* **2019**, *68*, 774–796. [CrossRef]
15. Xiong, X.; Yang, G.; Liu, K.; Zhou, D. A Proposed Fixed-Sum Carryovers Reallocation DEA Approach for Social Scientific Resources of Chinese Public Universities. *Scientometrics* **2022**, *127*, 4097–4121. [CrossRef]
16. Cossani, G.; Codoceo, L.; Cáceres, H.; Tabilo, J. Technical Efficiency in Chile's Higher Education System: A Comparison of Rankings and Accreditation. *Eval. Program. Plan.* **2022**, *92*, 102058. [CrossRef]
17. Tran, C.-D.; Villano, R. Financial Efficiency of Tertiary Education Institutions: A Second-Stage Dybamic Network Data Envelopment Analysis Method. *Singap. Econ. Rev.* **2021**, *66*, 1421–1442. [CrossRef]
18. Tran, C.-D. Efficiency of the Teaching-Industry Linkage in the Australian Vocational Education and Training. *Empir. Res. Vocat. Educ. Train.* **2021**, *13*, 11. [CrossRef]
19. Fukuyama, H.; Weber, W.L.; Xia, Y. Time Substitution and Network Effects with an Application to Nanobiotechnology Policy for US Universities. *Omega* **2016**, *60*, 34–44. [CrossRef]
20. Tran, C.D.T.T.; Villano, R.A. Measuring Efficiency of Vietnamese Public Colleges: An Application of the DEA-Based Dynamic Network Approach. *Int. Trans. Oper. Res.* **2018**, *25*, 683–703. [CrossRef]
21. Ozcan, Y.A.; Lins, M.E.; Lobo, M.S.C.; da Silva, A.C.M.; Fiszman, R.; Pereira, B.B. Evaluating the Performance of Brazilian University Hospitals. *Ann. Oper. Res.* **2010**, *178*, 247–261. [CrossRef]
22. *Reestruturação e Expansão Das Universidades Federais: Restructuring and Expansion of Federal Universities*; Governo Federal do Brasil: Rio de Janeiro, Brazil, 2007.
23. Pereira, C.; Araújo, J.; Machado, M. de L. Acreditação Do Ensino Superior Na Europa e Brasil: Mecanismos de Garantia Da Qualidade [Accreditation of Higher Education in Europe and Brazil: Quality Assurance Mechanisms]. *Rev. de Políticas Públicas [J. Publ. Policy]* **2015**, *19*, 61–75.
24. Brunstein, J.; Jaime, P.; Curi, D.P.; d'Angelo, M.J.; Mainardes, E.W. Assessment and Evaluation of Higher Education in Business Management: An Analysis of the Brazilian Case in the Light of Social Learning Theory for Sustainability. *Assess. Eval. High. Educ.* **2015**, *40*, 833–854. [CrossRef]
25. Bittencourt, H.; Viali, L.; Rodrigues, A.; Casartelli, A. Mudanças Nos Pesos Do CPC e de Seu Impacto Nos Resultados de Avaliação Em Universidades Federais e Privadas. *Avaliação Rev. Avaliação Educ. Super.* **2010**, *13*, 147–166. [CrossRef]
26. Zanella, A.; Oliveira, R.; de Oliveira, S. Avaliação de Desempenho Na Educação Superior: Uma Abordagem Utilizando a Análise Envoltória de Dados. *Ciência Nat.* **2021**, *43*, e81. [CrossRef]
27. Ikuta, C.Y.S. Sobre o Conceito Preliminar de Curso: Concepção, Aplicação e Mudanças Metodológicas. *Estud. Avaliação Educ.* **2016**, *27*, 938. [CrossRef]
28. BRASIL. *Report CFE No 977/65*; ANPEd: Rio de Janeiro, Brazil, 1965.
29. de França, J.M.F.; de Figueiredo, J.N.; dos Lapa, J.S. A DEA Methodology to Evaluate the Impact of Information Asymmetry on the Efficiency of Not-for-Profit Organizations with an Application to Higher Education in Brazil. *Ann. Oper. Res.* **2010**, *173*, 39–56. [CrossRef]
30. Miranda, R.; Gramani, M.C.; Andrade, E. Technical Efficiency of Business Administration Courses: A Simultaneous Analysis Using DEA and SFA. *Int. Trans. Oper. Res.* **2012**, *19*, 847–862. [CrossRef]
31. Zoghbi, A.C.; Rocha, F.; Mattos, E. Education Production Efficiency: Evidence from Brazilian Universities. *Econ. Model.* **2013**, *31*, 94–103. [CrossRef]
32. Hammes Junior, D.D.; Flach, L.; Mattos, L.K. The Efficiency of Public Expenditure on Higher Education: A Study with Brazilian Federal Universities. *Ensaio* **2020**, *28*, 1076–1097. [CrossRef]
33. Wanke, P.F.; Antunes, J.J.J.; Miano, V.Y.; Couto, C.L.P.d.; Mixon, F.G. Measuring Higher Education Performance in Brazil: Government Indicators of Performance vs Ideal Solution Efficiency Measures. *Int. J. Product. Perform. Manag.* **2022**, *71*, 2479–2495. [CrossRef]
34. Tavares, R.S.; Angulo-Meza, L.; Sant'Anna, A.P. A Proposed Multistage Evaluation Approach for Higher Education Institutions Based on Network Data Envelopment Analysis: A Brazilian Experience. *Eval. Program Plan.* **2021**, *89*, 101984. [CrossRef]
35. Liu, J.; Lu, L.; Lu, W. Research Fronts in Data Envelopment Analysis. *Omega* **2016**, *58*, 33–45. [CrossRef]
36. Fallah-Fini, S.; Triantis, K.; Johnson, A.L. Reviewing the Literature on Non-Parametric Dynamic Efficiency Measurement: State-of-the-Art. *J. Product. Anal.* **2014**, *41*, 51–67. [CrossRef]
37. Mariz, F.B.A.R.; Almeida, M.R.; Aloise, D. A Review of Dynamic Data Envelopment Analysis: State of the Art and Applications. *Int. Trans. Oper. Res.* **2018**, *25*, 469–505. [CrossRef]
38. Kao, C. Network Data Envelopment Analysis: A Review. *Eur. J. Oper. Res.* **2014**, *239*, 1–16. [CrossRef]
39. Halkos, G.E.; Tzeremes, N.G.; Kourtzidis, S.A. A Unified Classification of Two-Stage DEA Models. *Surv. Oper. Res. Manag. Sci.* **2014**, *19*, 1–16. [CrossRef]
40. Tone, K.; Tsutsui, M. Dynamic DEA: A Slacks-Based Measure Approach. *Omega* **2010**, *38*, 145–156. [CrossRef]

41. Nemoto, J.; Goto, M. Measurement of Dynamic Efficiency in Production: An Application of Data Envelopment Analysis to Japanese Electric Utilities. *J. Product. Anal.* **2003**, *19*, 191–210. [CrossRef]
42. Nemoto, J.; Goto, M. Dynamic Data Envelopment Analysis: Modeling Intertemporal Behavior of a Firm in the Presence of Productive Inefficiencies. *Econ. Lett.* **1999**, *64*, 51–56. [CrossRef]
43. Färe, R.; Grosskopf, S. *Intertemporal Production Frontiers: With Dynamic DEA*; Kluwer Academic Publishers: Boston, MA, USA, 1996; ISBN 978-9401073097.
44. Sueyoshi, T.; Sekitani, K. Returns to Scale in Dynamic DEA. *Eur. J. Oper. Res.* **2005**, *161*, 536–544. [CrossRef]
45. Jaenicke, E. Testing for Intermediate Outputs in Dynamic DEA Models: Accounting for Soil Capital in Rotational Crop Production and Productivity Measures. *J. Product. Anal.* **2000**, *14*, 247–266. [CrossRef]
46. Emrouznejad, A.; Thanassoulis, E. A Mathematical Model for Dynamic Efficiency Using Data Envelopment Analysis. *Appl. Math. Comput.* **2005**, *160*, 363–378. [CrossRef]
47. von Geymueller, P. Static versus Dynamic DEA in Electricity Regulation: The Case of US Transmission System Operators. *Cent. Eur. J. Oper. Res.* **2009**, *17*, 397–413. [CrossRef]
48. Tone, K.; Tsutsui, M. Network DEA: A Slacks-Based Measure Approach. *Eur. J. Oper. Res.* **2009**, *197*, 243–252. [CrossRef]
49. Kao, C. Dynamic Data Envelopment Analysis: A Relational Analysis. *Eur. J. Oper. Res.* **2013**, *227*, 325–330. [CrossRef]
50. Färe, R.; Grosskopf, S. Network DEA. *Socioecon. Plan. Sci.* **2000**, *34*, 35–49. [CrossRef]
51. Kao, C.; Hwang, S.N. Efficiency Decomposition in Two-Stage Data Envelopment Analysis: An Application to Non-Life Insurance Companies in Taiwan. *Eur. J. Oper. Res.* **2008**, *185*, 418–429. [CrossRef]
52. Kao, C.; Hwang, S.-N. Efficiency Measurement for Network Systems: IT Impact on Firm Performance. *Decis. Support. Syst.* **2010**, *48*, 437–446. [CrossRef]
53. Kao, C. Efficiency Decomposition in Network Data Envelopment Analysis: A Relational Model. *Eur. J. Oper. Res.* **2009**, *192*, 949–962. [CrossRef]
54. Chen, Y.; Du, J.; David Sherman, H.; Zhu, J. DEA Model with Shared Resources and Efficiency Decomposition. *Eur. J. Oper. Res.* **2010**, *207*, 339–349. [CrossRef]
55. Li, L.; Dai, Q.; Huang, H.; Wang, S. Efficiency Decomposition with Shared Inputs and Outputs in Two-Stage DEA. *J. Syst. Sci. Syst. Eng.* **2016**, *25*, 23–38. [CrossRef]
56. Ma, J. A Two-Stage DEA Model Considering Shared Inputs and Free Intermediate Measures. *Expert. Syst. Appl.* **2015**, *42*, 4339–4347. [CrossRef]
57. Cook, W.D.; Seiford, L.M. Data Envelopment Analysis (DEA)-Thirty Years On. *Eur. J. Oper. Res.* **2009**, *192*, 1–17. [CrossRef]
58. Chen, Y.; Cook, W.D.; Ning, L.; Zhu, J. Additive Efficiency Decomposition in Network DEA. *Eur. J. Oper. Res.* **2009**, *196*, 1170–1176. [CrossRef]
59. Fukuyama, H.; Weber, W.L. A Slacks-Based Inefficiency Measure for a Two-Stage System with Bad Outputs. *Omega* **2010**, *38*, 398–409. [CrossRef]
60. Lozano, S. Technical and Environmental Efficiency of a Two-Stage Production and Abatement System. *Ann. Oper. Res.* **2017**, *255*, 199–219. [CrossRef]
61. Ribeiro, V.M.; Varum, C.; Daniel, A.D. Introducing Microeconomic Foundation in DEA: The Average-Based Approach. *Spat. Econ. Anal.* **2020**, *15*, 145–164. [CrossRef]
62. Ribeiro, V.M.; Varum, C.; Daniel, A.D. Introducing Microeconomic Foundation in Data Envelopment Analysis: Effects of the Ex Ante Regulation Principle on Regional Performance. *J. Knowl. Econ.* **2021**, *12*, 1215–1244. [CrossRef]
63. Carayannis, E.G.; Grigoroudis, E.; Goletsis, Y. A Multilevel and Multistage Efficiency Evaluation of Innovation Systems: A Multiobjective DEA Approach. *Expert. Syst. Appl.* **2016**, *62*, 63–80. [CrossRef]
64. Carayannis, E.G.; Goletsis, Y.; Grigoroudis, E. Multi-Level Multi-Stage Efficiency Measurement: The Case of Innovation Systems. *Oper. Res.* **2015**, *15*, 253–274. [CrossRef]
65. Chao, C.M.; Yu, M.M.; Wu, H.N. An Application of the Dynamic Network Dea Model: The Case of Banks in Taiwan. *Emerg. Mark. Financ. Trade* **2015**, *51*, S133–S151. [CrossRef]
66. Tone, K.; Tsutsui, M. Dynamic DEA with Network Structure: A Slacks-Based Measure Approach. *Omega* **2014**, *42*, 124–131. [CrossRef]
67. Kawaguchi, H.; Tone, K.; Tsutsui, M. Estimation of the Efficiency of Japanese Hospitals Using a Dynamic and Network Data Envelopment Analysis Model. *Health Care Manag. Sci.* **2014**, *17*, 101–112. [CrossRef]
68. Khushalani, J.; Ozcan, Y.A. Are Hospitals Producing Quality Care Efficiently? An Analysis Using Dynamic Network Data Envelopment Analysis (DEA). *Socioecon. Plan. Sci.* **2017**, *60*, 15–23. [CrossRef]
69. Färe, R.; Grosskopf, S.; Whittaker, G. Network DEA II. In *Data Envelopment Analysis: A Handbook on the Modeling of Internal Structures and Networks*; Springer: Boston, MA, USA, 2014; Volume 208, pp. 307–327; ISBN 9780387452814.
70. Soltanzadeh, E.; Omrani, H. Dynamic Network Data Envelopment Analysis Model with Fuzzy Inputs and Outputs: An Application for Iranian Airlines. *Appl. Soft Comput. J.* **2018**, *63*, 268–288. [CrossRef]
71. Yan, Q.; Zhao, F.; Wang, X.; Yang, G.; Baležentis, T.; Streimikiene, D. The Network Data Envelopment Analysis Models for Non-Homogenous Decision Making Units Based on the Sun Network Structure. *Cent. Eur. J. Oper. Res.* **2019**, *27*, 1221–1244. [CrossRef]

72. Moreno, P.; Lozano, S. Super SBI Dynamic Network DEA Approach to Measuring Efficiency in the Provision of Public Services. *Int. Trans. Oper. Res.* **2018**, *25*, 715–735. [CrossRef]
73. Gharakhani, D.; Toloie Eshlaghy, A.; Fathi Hafshejani, K.; Kiani Mavi, R.; Hosseinzadeh Lotfi, F. Common Weights in Dynamic Network DEA with Goal Programming Approach for Performance Assessment of Insurance Companies in Iran. *Manag. Res. Rev.* **2018**, *41*, 920–938. [CrossRef]
74. You, Y.Q.; Jie, T. A Study of the Operation Efficiency and Cost Performance Indices of Power-Supply Companies in China Based on a Dynamic Network Slacks-Based Measure Model. *Omega* **2016**, *60*, 85–97. [CrossRef]
75. Yu, M.M.; Chen, L.H.; Hsiao, B. Dynamic Performance Assessment of Bus Transit with the Multi-Activity Network Structure. *Omega* **2016**, *60*, 15–25. [CrossRef]
76. Omrani, H.; Soltanzadeh, E. Dynamic DEA Models with Network Structure: An Application for Iranian Airlines. *J. Air Transp. Manag.* **2016**, *57*, 52–61. [CrossRef]
77. Kahi, V.S.; Yousefi, S.; Shabanpour, H.; Saen, R.F. How to Evaluate Sustainability of Supply Chains? A Dynamic Network DEA Approach. *Ind. Manag. Data Syst.* **2017**, *117*, 1866–1889. [CrossRef]
78. Chao, C.M.; Yu, M.M.; Lee, Y.T.; Hsiao, B. Measurement of Banking Performance in a Dynamic Multiactivity Network Structure: Evidence from Banks in Taiwan. *Emerg. Mark. Financ. Trade* **2017**, *53*, 786–805. [CrossRef]
79. Tone, K.; Kweh, Q.L.; Lu, W.M.; Ting, I.W.K. Modeling Investments in the Dynamic Network Performance of Insurance Companies. *Omega* **2019**, *88*, 237–247. [CrossRef]
80. Farrell, M. The Measurement of Productive Efficiency. *J. R. Stat. Soc. Ser. A* **1957**, *120*, 253–290. [CrossRef]
81. Bana e Costa, C.A.; Soares De Mello, J.C.C.B.; Angulo Meza, L. A New Approach to the Bi-Dimensional Representation of the DEA Efficient Frontier with Multiple Inputs and Outputs. *Eur. J. Oper. Res.* **2016**, *255*, 175–186. [CrossRef]
82. Charnes, A.; Cooper, W.W.; Rhodes, E. Measuring the Efficiency of Decision Making Units. *Eur. J. Oper. Res.* **1978**, *2*, 429–444. [CrossRef]
83. Belton, V.; Vickers, S. Demystifying DEA-A Visual Interactive Approach Based on Multiple Criteria Analysis. *J. Oper. Res. Soc.* **1993**, *44*, 883–896.
84. El-Mahgary, S.; Lahdelma, R. Data Envelopment Analysis: Visualizing the Results. *Eur. J. Oper. Res.* **1995**, *85*, 700–710. [CrossRef]
85. Porembski, M.; Breitenstein, K.; Alpar, P. Visualizing Efficiency and Reference Relations in Data Envelopment Analysis with an Application to the Branches of a German Bank. *J. Product. Anal.* **2005**, *23*, 203–221. [CrossRef]
86. Adler, N.; Raveh, A. Presenting DEA Graphically. *Omega* **2008**, *36*, 715–729. [CrossRef]
87. Appa, G.; Bana E Costa, C.A.; Chagas, M.; Ferreira, F.; Soares De Mello, J.O. *DEA in X-Factor. Evaluation for the Brazilian Electricity Distribution Industry*; Ratio Analysis View Project Project Appraisal and Portfolio Selection View Project; London School of Economics and Political Science: London, UK, 2010.
88. Akçay, A.E.; Ertek, G.; Büyüközkan, G. Analyzing the Solutions of DEA through Information Visualization and Data Mining Techniques: SmartDEA Framework. *Expert. Syst. Appl.* **2012**, *39*, 7763–7775. [CrossRef]
89. Assunção, M.V. O Estudo Da Eficiência Ambiental Dos Campos Onshore Das Bacias Costeiras Brasileiras Na Perspectiva Da Água Produzida de Petróleo. Ph.D. Thesis, Centro de Ciências Exatas e da Terra, Universidade Federal do Rio Grande do Norte, Natal, Brazil, 2018.
90. Torres, B.G.; Reis, J.D.C.; Mello, J.C.C. Two-Dimensional Representation for Two-Stage Network Dea Models. *Pesqui. Oper.* **2022**, *42*, e258055. [CrossRef]
91. Despotis, D.K.; Koronakos, G.; Sotiros, D. Composition versus Decomposition in Two-Stage Network DEA: A Reverse Approach. *J. Product. Anal.* **2016**, *45*, 71–87. [CrossRef]
92. Liu, J.S.; Lu, L.Y.Y.; Lu, W.M.; Lin, B.J.Y. A Survey of DEA Applications. *Omega* **2013**, *41*, 893–902. [CrossRef]
93. Liang, L.; Cook, W.D.; Zhu, J. DEA Models for Two-Stage Processes: Game Approach and Efficiency Decomposition. *Nav. Res. Logist.* **2008**, *55*, 541–550. [CrossRef]
94. Li, H.; Chen, C.; Cook, W.D.; Zhang, J.; Zhu, J. Two-Stage Network DEA: Who Is the Leader? *Omega* **2018**, *74*, 15–19. [CrossRef]
95. Dyson, R.G.; Allen, R.; Camanho, A.S.; Podinovski, V.V.; Sarrico, C.S.; Shale, E.A. Pitfalls and Protocols in DEA. *Eur. J. Oper. Res.* **2001**, *132*, 245–259. [CrossRef]

mathematics MDPI

Article

Navigating the Complexity of HRM Practice: A Multiple-Criteria Decision-Making Framework

Vuk Mirčetić [1,*], Gabrijela Popović [1], Svetlana Vukotić [1], Marko Mihić [2], Ivana Kovačević [2], Aleksandar Đoković [2] and Marko Slavković [3]

[1] Faculty of Applied Management, Economics and Finance, University Business Academy in Novi Sad, Belgrade 11000, Serbia; gabrijela.popovic@mef.edu.rs (G.P.); svetlana.vukotic@mef.edu.rs (S.V.)
[2] Faculty of Organisational Sciences, University of Belgrade, Belgrade 11000, Serbia; marko.mihic@fon.bg.ac.rs (M.M.); ivana.kovacevic@fon.bg.ac.rs (I.K.); aleksandar.djokovic@fon.bg.ac.rs (A.Đ.)
[3] Faculty of Economics, University of Kragujevac, Kragujevac 34000, Serbia; mslavkovic@kg.ac.rs
* Correspondence: vuk.mircetic@mef.edu.rs

Abstract: A myriad of diverse factors affect the contemporary business environment and all business areas, causing organisations to innovate new business models, or to use innovations to navigate the complexity of contemporary HRM practice successfully. Despite the plenitude of notable studies, a particular theoretical gap exists regarding the innovation's impact on particular HRM practices and on understanding how multiple-criteria decision-making (MCDM) methods can be effectively applied in the context of human resource management (HRM) to address important aspects of successful practices and prioritise the considered alternative solutions. Recognising the potential of the MCDM field highlighted the possibility of involving the MCDM methods in detecting the most influential and innovative HRM practices and defining the rank of companies that are most successful in applying them. The innovative MCDM approach proposed here utilises the CRITIC (CRiteria Importance Through Intercriteria Correlation) method and PIPRECIA-S (Simple PIvot Pairwise RElative Criteria Importance Assessment) method for prioritising innovative HRM practices, and the COBRA (COmprehensive Distance Based RAnking) method for assessing the companies under evaluation. The research, which involved 21 respondent experts from the HRM field and 12 companies from the Republic of Serbia, revealed that employee participation is the most significant innovative HRM practice that yields the best results in the contemporary business environment. Consequently, the first-ranked company most successfully met the requirements of the innovative HRM practices presented.

Keywords: human resource management; innovations; MCDM; CRITIC; PIPRECIA-S; COBRA; HRM; ranking; contemporary business environment

MSC: 90B50

1. Introduction

The contemporary business environment is affected by a plenitude of challenges, changes, and factors. The turbulence of the current business environment means that changes are more numerous and occur rapidly. These modern trends, and an increasingly enunciated uncertainty, impact all business areas, making them more challenging and ambivalent to deal with, and causing organisations to innovate new business models or use innovations. Organisations face intricate challenges and are required to make important decisions and explore diverse methods to make their processes more sustainable (Turskis and Šniokienė, 2024 [1]). Organisations that stagnate in product and business process development are denied the opportunity to prosper. Innovative organisations have additional possibilities to endure in the hypercompetitive business environment (Elshaer, Azazz and Fayyad, 2023 [2]).

Innovation is one of the most important factors contributing to a competitive advantage (Penjišević and Sančanin, 2024 [3]). Today, in many business spheres, companies increasingly focus on innovation and different approaches to previous operating methods. If innovations are observed in the context of human resource management (HRM), it is important to emphasise their varied importance. Shen et al. (2022) [4] underscored that human resource management significantly influences organisations' development and competitiveness. Many scholars have examined the relationship between human resource management and organisational performance. Haque (2023) [5] also points out the disadvantages related to the application of innovations in human resource management, and states that online recruitment has challenges, given that there is no direct interaction involved. Innovations in human resources management involve using information and communication technologies and numerous innovative approaches to human resources management. Certain respectable studies (Corral de Zubielqui, Fryges and Jones, 2019; Hong, Zhao and Stanley Snell, 2019; Papa et al., 2020 [6–8]) aimed to integrate human resource management and open innovation. Engelsberger et al. (2021) [9] define open innovation using mindset, and point out that it represents values, attitudes, and beliefs in the context of an individual's openness to sharing knowledge.

It is impossible to know with certainty which innovations will be created and incorporated into human resource management in the following years, but certain innovations will undoubtedly be present. Given that the contemporary business environment is undoubtedly challenging and uncertain, there is a need for an effective approach to decision-making processes.

An effective tool in various areas with practical and theoretical applications regarding the decision-making process is the employment of multi-criteria decision-making (MCDM) methods (Turskis and Keršulienė, 2024, Zavadskas et al., 2022 [10,11]). Many scholars have demonstrated that MCDM methods can address complex challenges and empower decision-makers to select the best solution for many challenges in an uncertain environment. Multi-criteria decision-making methods strive to assist decision-makers in examining possible decisions and select the most adequate one of the available alternatives (Karamaşa, 2021; Özdağoğlu et al., 2021 [12,13]).

MCDM has many different applications, such as choosing the most suitable alternative, ranking alternatives (partially or completely), sorting a set of alternatives into the categories created earlier, assembling a set of criteria, specifying the performance of alternatives, and elaborating on alternatives (Roy, 1981 [14]). Pinto-DelaCadena, Liern, and Vinueza-Cabezas (2024) [15] point out that mathematical methods are increasingly being utilised to underpin decision-making in human resource management. In the context of human resource management, MCDM methods are used mainly in segments of HRM practices, such as selection, training, and maintaining skills that are necessary for the safe work of personnel; (Gendler, Tumanov and Levin, 2021 [16]); the selection of personnel (Karabašević et al., 2015; Ulutaş et al., 2020; López et al., 2022; Tuğrul, 2022 [17–20]); and the evaluation of human resources (Jakovljević et al., 2021 [21]).

So far, a myriad of respectable studies have been published regarding the employment of different MCDM methods in diverse human resource management contexts. However, despite the surplus of notable studies in this regard, a particular theoretical gap exists, specifically in understanding how MCDM methods can be effectively applied in the context of HRM practices and innovations to successfully navigate the complexity of contemporary HRM practice.

Building on the work of Heidary Dahooie et al. (2022) [22], this paper strives to narrow an existing gap by providing a fine-grained systematic literature review regarding MCDM methods, mainly elaborating on the CRITIC (CRiteria Importance Through Intercriteria Correlation) method, PIPRECIA-S (Simple PIvot Pairwise RElative Criteria Importance Assessment) method, and COBRA (COmprehensive Distance Based RAnking) method, proposing a new and innovative MCDM approach to tackle the aforementioned decision-making challenges. The essential incentive for employing the MCDM approach was its ability to respect all the criteria involved in the decision process. Furthermore, research that utilises MCDM methods does not require the involvement of many respondents, which facilitates the data gathering procedure. Unlike usual statistical methods, incorporating meticulously selected expert groups leads to adequate scientific results. Nevertheless, it is worth mentioning that the issue of innovative HRM practices and their influence on the company's position was not very often perceived through the MCDM prism.

This article is meticulously structured to provide a comprehensive, in-depth analysis of the aforementioned problems. Therefore, the article is organised as follows. The introduction is followed by the first chapter, which explains the materials and methods that were used. The first subchapter of this chapter analyses the empirical research methods, while the next three subchapters explain the three employed MCDM methods, analysing groundbreaking and reputable study papers from the last five years to provide a fine-grained perspective on the significance of these MCDM methods in a myriad of areas and disciplines. The next chapter discusses the results and the numerical illustration of these. It critically evaluates the results and their implications, providing a holistic perspective of the research contributions to human resource management theory. The final chapter consists of conclusions, limitations, and suggestions for future research.

2. Materials and Methods

This chapter is divided into five subchapters. The first segment sets up the research hypotheses, while the second subchapter elaborates on the empirical research methods used in this study. The following three subchapters introduce the selected MCDM method, present its computational procedure, and describe the different fields of application in which it is used, providing the reasoning for the selection of that method in this paper. The CRITIC method is first analysed, followed by the subchapters examining PIPRECIA-S and the COBRA method. The final section of this chapter introduces the Borda rule.

2.1. Empirical Research Methods

A methodological procedure is carefully crafted as a roadmap for this research. To ensure the effective monitoring of the implementation of the relevant activities and paper segments, a detailed research implementation scheme is illustrated in Figure 1.

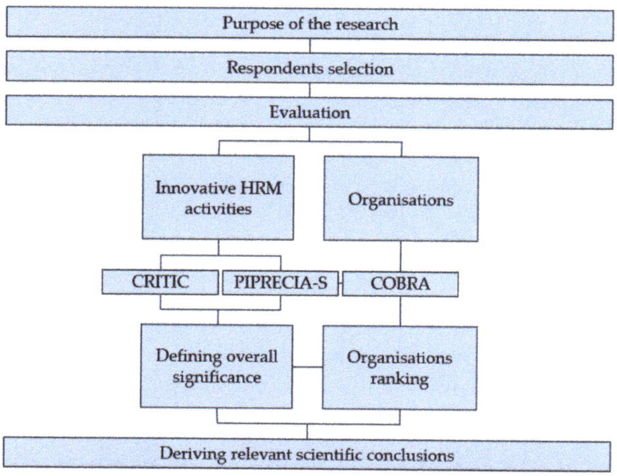

Figure 1. The innovative MCDM approach to HRM practices.

As introduced at the beginning of the article, and as can be observed in the previous Figure, this paper employed three MCDM methods: CRITIC and PIPRECIA-S to determine the significance of the selected innovative HRM practices, and the COBRA method for the evaluation of the selected alternatives. Many MCDM methods could be used to facilitate the decision process in the HRM field. However, besides the well-known CRITIC method, we decided to employ two relatively new methods (PIPRECIA-S and COBRA), because researchers continually create and improve MCDM methods and models, so we believed that these new methods would bring new insights and contribute to the decision-making process, making it more reliable.

Defining the criteria weights represents an essential step in the MCDM analysis. Different weighting methods impact the decision process differently and could result in mutually differing weighting coefficients. Ponhan and Sureeyatanapas (2022) [23] analysed the discrepancy between weighting results that is gained by applying objective and subjective methods. Fourteen experts were assessed using eighteen criteria, with linguistic variables representing the base data, to employ subjective (direct rating, rank sum, and rank-order centroid) and objective (entropy and standard deviation) weighting methods. The final results outlined the volatility in the weighting coefficients depending on the method used.

Furthermore, Paramanik et al. (2022) [24] proposed the objective-subjective weighted method for minimising inconsistency (OSWMI) that involves an improved CRITIC method, BWM, and LINMAP II using a multi-objective non-linear programming (MONLP) model. The leading idea was to propose a model to reduce the possibility of manipulating weighting coefficients. In the present case, we proposed combining the CRITIC method and PIPRECIA-S to craft such an approach, which is sufficiently simple but also reliable. We aimed to define such an approach, that will enable the significance of the HRM practice to be defined while avoiding the extreme or biased weighting coefficients.

In this article, we propose a combination of the CRITIC and PIPRECIA-S methods to define the significance of the HRM practice. The possibilities of such an integrated MCDM approach have yet to be observed, and an explanation of why we chose to use it is given below. The CRITIC method belongs to the group of objective weighting methods that define the significance of the criteria based on the input data regarding the performance ratings of the evaluated alternatives. It is comprehensive and facilitates the process of determining the criteria weights. However, in some cases, the criterion with a high standard deviation and a low correlation with the other criteria may have a high weighting coefficient. As a result, such a dominant criterion relegates other criteria to the background and determines the final result. Therefore, to resolve this issue, we employed the PIPRECIA-S method, which is a subjective method for determining criteria weightings and is very applicable and easy to use. Even the respondents who were unfamiliar with MCDM methods understood the procedure of the PIPRECIA-S more easily, and learned to use it relatively quickly. Besides, the PIPRECIA-S method is very convenient for application in the group decision environment. However, as is the case with every subjective weighting method, the subjective judgements of decision-makers could lead to inadequate weighting coefficients (Paramanik et al., 2022; Mufazzal et al., 2021 [24,25]). Decision-makers could be dishonest or biased, which compromises the evaluation process (Liu et al., 2021 [26]). Because of the abovementioned reasons, we combined the CRITIC and PIPRECIA-S methods to (1) reduce the possibility of dominant weighting values, and (2) manipulate the results of decision-makers. The obtained weighting coefficients represent the input for further assessment using the COBRA method.

The COBRA method is a relatively new method that incorporates three types of distances from possible solutions. By calculating the distance from the positive ideal, negative ideal, and average solution, the reliability of the performed procedure increases, while the possibility of making the wrong decision or choice decreases. Although the procedure is somewhat complex, the reliability of the obtained results is expected to be higher because the distance measurements from different solutions are calculated.

2.2. CRITIC Method: Revolutionising Distance-Based Ranking in Scientific Studies

The CRITIC method (Diakoulaki, Mavrotas and Papayannakis, 1995 [27]) is a correlation method that aims to define the objective weights of relative importance in multi-criteria decision-making problems. Many scholars underlined the efficiency of this method in various multi-criteria issues, particularly when the decision-maker is absent. It facilitates the decision-maker's vocalisation of his argument or belief about the relative importance of the criteria, thereby decreasing the subjective character of the decision-making process. The method also assists in discarding the non-salient attributes in a primary weighting of the evaluation criteria, ensuring a fair and objective process.

The computational procedure of the CRITIC method comprises three steps, which are demonstrated below.

Step 1. Forming the decision-making matrix D as follows:

$$D = \left[x_{ij}\right]_{mxn} \tag{1}$$

where x_{ij} represents the ratings of the alternative i according to criterion j, m indicates the number of alternatives, and j denotes the number of criteria.

Step 2. Constructing the normalised decision-making matrix R as follows:

$$R = \left[r_{ij}\right] \tag{2}$$

where r_{ij} denotes the normalised ratings of the alternative i according to criterion j, and is calculated as follows:

$$r_{ij} = \frac{x_{ij-min_i\, x_{ij}}}{max_i\, x_{ij} - min_i\, x_{ij}} \tag{3}$$

Step 3. Determining the weights of criteria w_j using the following formula:

$$R = \frac{C_j}{\sum_{j=1}^{n} C_j} \tag{4}$$

where C_{ij} represents a quantity of information contained in criterion j, and is calculated in the following way:

$$C_j = \sigma_j \sum_{j=1}^{n} (1 - cr_{jj}) \tag{5}$$

and σ_j indicates the standard deviation of criterion j, cr_{jj} denotes the correlation coefficient between the two criteria.

The CRITIC method is multidisciplinary and applied across various domains, as presented in Table 1, which summarises the reputable and innovative studies that employ this method.

Table 1. Research goal or the field of application of the CRITIC method.

Year	Authors	Research Goal or Field of Application
2024	Chang [28]	Evaluation method for the classroom
2024	Krishnan [29]	Research trends in the CRITIC method
2024	Saensuk, Witchakool and Choompol [30]	Detection of fake news
2024	Shrinivas Balraj et al. [31]	Optimisation of machining parameters
2023	Hassan, Alhamrouni and Azhan [32]	Selection of a solar power plant location
2023	Hosseinzadeh Lotfi et al. [33]	Prioritisation and evaluation of projects based on different criteria
2023	Mishra, Chen and Rani [34]	Proposition of a model established on Fermatian fuzzy numbers
2023	Silva et al. [35]	Selection of investment portfolio
2023	Zhang et al. [36]	Evaluation of the rock burst intensity evaluation
2022	Bhadra, Dhar and Salam [37]	Natural fibres selection
2022	Haktanır and Kahraman [38]	Wearable health applications selection
2022	Kumari and Acherjee [39]	Unconventional processing method selection

Table 1. *Cont.*

Year	Authors	Research Goal or Field of Application
2022	Pamučar, Žižović and Đuričić [40]	CRITIC method modification using fuzzy rough numbers
2021	Mukhametzyanov [41]	Examination and comparison of different methods
2021	Zafar, Alamgir and Rehman [42]	Blockchain system evaluation
2020	Peng and Huang [43]	Financial risks analysis
2020	Peng, Zhang and Luo [44]	5G industry analysis
2019	Tuş and Aytaç Adalı [45]	Software selection

2.3. PIPRECIA-S Method: A New and Simplified Frontier for Assessment in Scientific Research

The PIPRECIA method (Stanujkić et al., 2017 [46]) is a subjective MCDM method for determining the criteria weights that were introduced and established based on the SWARA (Stepwise Weight Assessment Ratio Analysis) method (Keršuliene, Zavadskas and Turskis, 2010 [47]). Unlike in the SWARA method, the criteria are not required to be sorted according to their expected significance before starting the evaluation procedure.

This method was further developed, and one of the originated methods employed in this article is the PIPRECIA-S method (Stanujkić et al., 2021 [48]), which is easier to use for the respondents because they only perform the comparison regarding the first criterion.

The computational procedure of the PIPRECIA-S method includes five steps, as shown below.

Step 1. Determining the set of evaluation criteria.

Step 2. Setting the relative significance s_j of each criterion, except the first, as follows:

$$s_j = \begin{cases} 1 \ if \ c_j > c_1 \\ 1 \ if \ c_j = 1 \\ 1 \ if \ c_j < 1 \end{cases} \tag{6}$$

where $j \neq 1$.

The value of s_1 is set to 1, while the values of s_j belong to the interval (1, 1.9] when $C_j \succ C_1$, that is to the interval [0.1, 1) when $C_j \prec C_1$.

Step 3. Calculating the value of the coefficient k_j in the following way:

$$k_j = \begin{cases} 1 & if\, j = 1 \\ 2 - s_j & if\, j > 1 \end{cases} \tag{7}$$

Step 4. Calculating the recalculated weight q_j as follows:

$$q_j = \begin{cases} 1 & if\, j = 1 \\ \frac{1}{k_j} & if\, j > 1 \end{cases} \tag{8}$$

Step 5. Determining the relative weights of the evaluation criteria in the following way:

$$w_j = \frac{q_j}{\sum_{k=1}^{n} q_k} \tag{9}$$

The PIPRECIA-S method is used in many different domains. Table 2 provides a summary of the reputable studies that employed this method and the methods from the PIPRECIA family across various areas.

Table 2. Research goal or field of application of the methods from PIPRECIA family.

Year	Authors	Research Goal or Field of Application
2024	Mirčetić, Popović and Vukotić [49]	Determining characteristics of the charismatic leaders in the EU
2024	Rizwan, Fizza and Mumtaz [50]	Evaluating strategies for the growth of fibreglass composites industry
2024	Sarbat [51]	Job satisfaction analysis
2024	Setiawansyah et al. [52]	Personnel selection
2024	Stanujkić et al. [53]	Personnel selection in a group decision-making environment
2023	Hadad et al. [54]	Student ranking based on learning assessment
2023	Mladenović, Đukić and Popović [55]	Financial platforms reporting analysis
2023	Setiawansyah and Saputra [56]	Head of the school organisation selection
2023	Stanujkić et al. [57]	Improvement of the decision-making process in the IT industry
2023	Sulistiani et al. [58]	Employees in an educational institution evaluation
2022	Aytekin [59]	Vehicle tracking system
2022	Đukić, Karabašević and Popović [60]	Evaluation of different aspects of cognitive skills
2022	Ulutaş and Topal [61]	Renewable energy sources selection and criteria evaluation
2021	Popović et al. [62]	Identification of key determinants of tourism development
2021	Ulutaş et al. [63]	Transportation company selection
2020	Jauković Jocic, Karabašević and Jocić [64]	Quality of e-learning materials assessment

2.4. COBRA Method: A New Paradigm for Comprehensive Scientific Analysis

The COBRA method (Krstić et al., 2022 [65]) is one of the newer multi-criteria decision-making methods. This method belongs to the multi-criteria decision-making methods based on distance determination. A key advantage of the COBRA method is its comprehensive nature. Alternatives are ranked based on their comprehensive distance from three types of possible solutions: positive ideal, negative ideal, and average. This method implies Euclidean and taxicab distance measures when calculating the distances for all solutions, which contributes to increasing the reliability of the defined solutions.

The computational procedure of the COBRA method incorporates six steps, which are exhibited as follows.

Step 1. Forming an initial decision-making matrix.

Step 2. Normalising the initial decision-making matrix, using the following formula:

$$\alpha_{ij} = \frac{\xi_{ij}}{\max\limits_{i}\xi_{ij}} \tag{10}$$

Step 3. Forming the weight-normalised decision matrix Δ_w:

$$\Delta_w = \left[w_j \times \xi_j\right]_{m\times n} \tag{11}$$

where w_j is the relative weight of criterion j.

Step 4. Defining the positive ideal (PIS_j), negative ideal (NIS_j), and average solution (AS_j) for each criterion function, as presented in the following formulae:

$$PIS_j = \max_i\left(w_j \times \xi_{ij}\right),\quad \forall j = 1,\ldots,m \; for\; j \in B \tag{12a}$$

$$PIS_j = \min_i\left(w_j \times \xi_{ij}\right),\quad \forall j = 1,\ldots,m \; for\; j \in C \tag{12b}$$

$$NIS_j = \min_i\left(w_j \times \xi_{ij}\right),\quad \forall j = 1,\ldots,m \; for\; j \in B \tag{13a}$$

$$NIS_j = \max_i\left(w_j \times \xi_{ij}\right),\quad \forall j = 1,\ldots,m \; for\; j \in C \tag{13b}$$

$$AS_j = \frac{\sum_{i=1}^{n}\left(w_j \times \xi_{ij}\right)}{n},\quad \forall j = 1,\ldots,m \; for\; j \in B, C \tag{14}$$

where B represents the set of benefits, and C denotes the set of cost criteria.

Step 5. Defining the distances from the positive ideal $(\mathrm{d}(\mathrm{PIS_j}))$ and negative ideal $(\mathrm{d}(\mathrm{NIS_j}))$ solutions for each alternative, as well as the positive $\left(\mathrm{d}\left(\mathrm{AS_j^+}\right)\right)$ and negative distances $\left(\mathrm{d}\left(\mathrm{AS_j^-}\right)\right)$ from the average solution, as follows:

$$d(S_j) = dE(S_j) + \beta \times dE(S_j) \times dT(S_j),\ \ \forall_j = 1, \ldots, m \tag{15}$$

where S_j is any solution $(PIS_j,\ NIS_j$ or $AS_j)$, and β is the correction coefficient acquired in the following way:

$$\beta = \max_i dE(S_j)_i - \min_i dE(S_j)_i \tag{16}$$

where $dE(S_j)_i$ and $dT(S_j)_i$ represent Euclidian and Taxicab distances, which are, for the positive ideal solution, acquired as follows:

$$dE(PIS_j)_i = \sqrt{\sum_{j=1}^{m} (PIS_j - w_j \times \xi_{ij})^2},\ \ \forall_i = 1, \ldots, n,\ \ \forall_j = 1, \ldots, m \tag{17}$$

$$dT(PIS_j)_i = \sum_{j=1}^{m} |PIS_j - w_j \times \xi_{ij}|,\ \ \forall i = 1, \ldots, n,\ \ \forall j = 1, \ldots, m \tag{18}$$

For the negative ideal solution, Euclidian and Taxicab distances are determined in the following way:

$$dE(NIS_j)_i = \sqrt{\sum_{j=1}^{m} (NIS_j - w_j \times \xi_{ij})^2},\ \ \forall_i = 1, \ldots, n,\ \ \forall_j = 1, \ldots, m \tag{19}$$

$$dT(NIS_j)_i = \sum_{j=1}^{m} |NIS_j - w_j \times \xi_{ij}|,\ \ \forall i = 1, \ldots, n,\ \ \forall_j = 1, \ldots, m \tag{20}$$

For the positive distance from the average solution, acquired as follows:

$$dE(AS_j)_i^+ = \sqrt{\sum_{j=1}^{m} \tau^+ (AS_j - w_j \times \xi_{ij})^2},\ \ \forall_i = 1, \ldots, n,\ \ \forall_j = 1, \ldots, m \tag{21}$$

$$dT(AS_j)_i^+ = \sum_{j=1}^{m} \tau^+ |AS_j - w_j \times \xi_{ij}|,\ \ \forall i = 1, \ldots, n,\ \ \forall_j = 1, \ldots, m \tag{22}$$

$$\tau^+ = \begin{cases} 1\ if\ AS_j < w_j \times \xi_{ij} \\ 0\ if\ AS_j > w_j \times \xi_{ij} \end{cases} \tag{23}$$

For the negative distance from the average solution, acquired as follows:

$$dE(AS_j)_i^- = \sqrt{\sum_{j=1}^{m} \tau^- (AS_j - w_j \times \xi_{ij})^2},\ \ \forall_i = 1, \ldots, n,\ \ \forall_j = 1, \ldots, m \tag{24}$$

$$dT(AS_j)_i^- = \sum_{j=1}^{m} \tau^- |AS_j - w_j \times \xi_{ij}|,\ \ \forall i = 1, \ldots, n,\ \ \forall_j = 1, \ldots, m \tag{25}$$

$$\tau^- = \begin{cases} 1\ if\ AS_j > w_j \times \xi_{ij} \\ 0\ if\ AS_j < w_j \times \xi_{ij} \end{cases} \tag{26}$$

Step 6. Ranking the alternatives by increasing the values of the comprehensive distances (dC_i) obtained using the following formula:

$$dC_i = \frac{d(PIS_j)_i - d(NIS_j)_i - d(AS_j)_i^+ + d(AS_j)_i^-}{4},\ \forall i = 1, \ldots, n \tag{27}$$

Despite being a new method, the scholars found many different areas where the COBRA method can be employed. The respectable studies that used this method are shown in Table 3.

Table 3. Research goal or the field of application of the COBRA method.

Year	Authors	Research Goal or Field of Application
2024	Asker [66]	Financial performance assessment
2024	Krstić et al. [67]	Risk analysis of the agricultural products supply
2024	Oğuz and Satır [68]	Retail trade enterprises' financial performance assessment
2024	Sahak and Karsli [69]	Environmental degradation in urban conditions analysis
2024	Tadić, Krstić and Radovanović [70]	Strategies for using drones in logistics analysis
2024	Ulutaş et al. [71]	Supplier selection
2024	Verma, Koul and Ajaygopal [72]	Cyber security platforms assessment and selection
2024	Zorlu, Tuncer and Yılmaz [73]	Evaluation of the potential for geo-tourism development
2023	Krstić, Tadić and Agnusdei [74]	Intermodal terminals analysis
2023	Tadić et al. [75]	Decision-making in logistics
2023	Ulutaş, Balo and Topal [76]	Natural stone selection in the construction industry
2022	Krstić et al. [65]	Evaluation of the scenarios for smart reverse logistics development
2022	Popović, Pucar and Smarandache [77]	E-commerce development strategy selection
2022	Verma, Ajaygopal and Koul [78]	Circular supplier selection

2.5. Borda Rule

The Borda rule is usually employed to aggregate the opinions of different decision-makers (Emerson, 2013; Marchant, 2000 [79,80]). For example, if more different attitudes exist because several decision-makers are choosing between numerous alternatives, each of the decision-makers ranks the given alternatives from best to worst. The Borda rule can be used with or without ponders, depending on the decision problem. If the decision-making process is based on a different number of indicators, pondering is applied to include them when forming the final results. The aforementioned rule was proposed to unify the results obtained by employing diverse models. The aforementioned rule implies that, when ranking the m alternative, the best alternative is assigned a score of $m-1$, the following $m-2$, the subsequent $m-3$, and so on until the last. Based on the overall defined score that considers all the positions that the observed alternative took, the final ranking of the evaluated alternatives is determined (Fedajev, Panić and Živković, 2024 [81]). A plenitude of respectable articles study the generalisation of the Borda method to make it more suitable for application in conditions of uncertainty, competition and fuzzy relations. However, there are particular objections to the Borda rule regarding an alternative being considered better than the alternative only if the difference between the Borda scores of the alternatives is greater than zero, while the amplitude of this difference is not considered (Marchant, 2000 [80]).

3. Case Study

Gathering data about the companies considered was necessary to conduct a planned analysis of the importance and impact of innovative HRM practices on the companies' performance. Twelve Serbian companies were chosen for data collection from the following sectors:

- Agriculture, forestry and fishing—three companies;
- Industry and construction—three companies;
- Services—three companies;
- ICT—three companies.

The company selection was based on the quartile report regarding companies' business operations, which is a part of regular "*Quarterly structural research on business operations of companies*" (SORS, 2024 [82]). According to this document, there are four sectors, as mentioned above. The sectors were selected because they face the most challenges for human resource management in the current conditions. One of the key issues is associated with the talent and general workforce deficits in these sectors. The names of the companies are not revealed because of privacy protection, and they are designated as K_1 to K_{12}. The companies' performances regarding innovative HRM practices were estimated by 21 respondents (R_1–R_{21}), who involved 11 experienced HR managers and 10 members of academia. Respondents used the Likert scale (1—the worst to 5—the best) (Likert, 1932 [83]) to assess the companies relative to the chosen innovative activities.

Innovative HRM activities which are involved in the research procedure were determined based on the paper by Heidary Dahooie et al. (2022) [22]. They represent the criteria against which the selected company's performance was evaluated, and all of them are beneficial. The list of the selected HRM practices and their abbreviations are presented in Table 4.

Table 4. Innovative HRM activities.

Selected HRM Practices	Abbreviation
Employee participation	Ep
Hiring process	Hp
Internal promotion	Ip
Job security	Js
Pay and reward	Pr
Performance management	Pm
Sharing information	Si
Teamwork	Tw
Training and development	Td

Source: Heidary Dahooie et al. (2022) [22].

Due to the thoroughness of our data collection process, which involved extensive data, we have chosen not to present all the data in this article.

The CRITIC approach was applied using the initial data about the respondents' estimation of the companies regarding the selected innovative HRM practices. The results obtained, defined based on the initial data from each respondent, are presented in Table 5.

Table 5. Assessment of the innovative HRM practices—CRITIC method.

	R_1	R_2	R_3	R_4	R_5	R_6	R_7	R_8	R_9	R_{10}	R_{11}	R_{12}	R_{13}	R_{14}	R_{15}	R_{16}	R_{17}	R_{18}	R_{19}	R_{20}	R_{21}
Ep	0.13	0.15	0.13	0.08	0.11	0.11	0.13	0.08	0.13	0.15	0.09	0.09	0.16	0.16	0.14	0.23	0.19	0.20	0.21	0.11	0.14
Hp	0.15	0.16	0.11	0.11	0.11	0.11	0.12	0.11	0.08	0.13	0.12	0.06	0.10	0.08	0.11	0.08	0.07	0.12	0.04	0.14	0.10
Ip	0.09	0.10	0.08	0.18	0.15	0.13	0.10	0.10	0.14	0.09	0.09	0.13	0.09	0.17	0.11	0.11	0.12	0.11	0.10	0.12	0.09
Js	0.08	0.09	0.08	0.09	0.09	0.09	0.12	0.18	0.08	0.09	0.08	0.12	0.13	0.09	0.10	0.13	0.10	0.08	0.06	0.10	0.11
Pr	0.12	0.10	0.11	0.07	0.16	0.12	0.10	0.11	0.11	0.12	0.11	0.08	0.08	0.09	0.11	0.06	0.16	0.10	0.08	0.13	0.11
Pm	0.09	0.09	0.10	0.10	0.08	0.17	0.08	0.11	0.14	0.11	0.10	0.10	0.08	0.14	0.12	0.11	0.10	0.07	0.08	0.08	0.08
Si	0.10	0.13	0.10	0.09	0.10	0.08	0.09	0.10	0.10	0.08	0.12	0.22	0.19	0.07	0.07	0.12	0.10	0.12	0.18	0.09	0.16
Tw	0.10	0.10	0.19	0.18	0.10	0.12	0.17	0.14	0.12	0.12	0.10	0.08	0.07	0.09	0.10	0.09	0.09	0.08	0.16	0.15	0.11
Td	0.14	0.08	0.10	0.10	0.09	0.07	0.10	0.07	0.09	0.11	0.19	0.11	0.11	0.11	0.13	0.07	0.07	0.10	0.09	0.09	0.10

Source: authors' calculations.

The results highlight the volatility of the innovative HR practice importance elicited from each respondent separately. Although the CRITIC method belongs to the group of objective methods and involves the initial data in the assessment procedure, when this data is collected from the respondents, it is nevertheless biased and reflects the personal opinion of the particular person. This subjectiveness is expressed indirectly, because the respondents were unaware that their estimation of the companies according to the selected practices led to the estimation of the practices themselves. To gain the results about the importance of the practice where subjectivity is present and intentionally expressed, we used the PIPRECIA-S method.

The PIPRECIA-S method involves the use of special questionnaires, which were distributed via email to the same group of respondents. The responses obtained were then utilised in the computational procedure to determine the importance of the considered practices. Once again, due to the comprehensive nature of the data, only the final results are presented in Table 6.

Table 6. Assessment of the innovative HRM practices—PIPRECIA-S method.

	R_1	R_2	R_3	R_4	R_5	R_6	R_7	R_8	R_9	R_{10}	R_{11}	R_{12}	R_{13}	R_{14}	R_{15}	R_{16}	R_{17}	R_{18}	R_{19}	R_{20}	R_{21}
Ep	0.11	0.11	0.10	0.10	0.10	0.11	0.11	0.11	0.10	0.11	0.11	0.10	0.11	0.10	0.10	0.10	0.11	0.12	0.11	0.11	0.09
Hp	0.10	0.13	0.10	0.09	0.14	0.11	0.12	0.12	0.13	0.11	0.11	0.11	0.13	0.10	0.11	0.10	0.13	0.10	0.12	0.10	0.08
Ip	0.11	0.10	0.13	0.12	0.09	0.11	0.09	0.11	0.10	0.12	0.11	0.09	0.11	0.09	0.11	0.10	0.11	0.10	0.12	0.12	0.11
Js	0.11	0.10	0.13	0.10	0.10	0.10	0.09	0.11	0.13	0.09	0.11	0.12	0.10	0.12	0.10	0.09	0.11	0.10	0.09	0.12	0.13
Pr	0.13	0.11	0.11	0.14	0.12	0.11	0.14	0.11	0.13	0.12	0.15	0.16	0.14	0.15	0.16	0.14	0.11	0.14	0.12	0.12	0.11
Pm	0.11	0.13	0.13	0.13	0.16	0.12	0.12	0.12	0.10	0.10	0.12	0.12	0.11	0.10	0.14	0.14	0.11	0.11	0.12	0.12	0.13
Si	0.12	0.11	0.11	0.12	0.10	0.11	0.12	0.11	0.10	0.14	0.11	0.11	0.10	0.13	0.14	0.10	0.12	0.12	0.14	0.12	0.15
Tw	0.10	0.09	0.09	0.10	0.09	0.11	0.10	0.11	0.09	0.10	0.11	0.11	0.11	0.10	0.08	0.10	0.11	0.10	0.09	0.10	0.10
Td	0.12	0.11	0.10	0.11	0.11	0.12	0.11	0.12	0.11	0.10	0.10	0.08	0.09	0.10	0.09	0.12	0.12	0.10	0.10	0.10	0.09

Source: authors' calculations.

The importance of the considered practices relative to each respondent varies again. We applied the geometric mean to define the final results regarding the CRITIC and PIPRECIA-S methods and the overall results. Table 7 presents the defined significances and their rank order.

Table 7. Overall innovative HRM practices significance.

	CRITIC	Rank	PIPRECIA-S	Rank	Overall Significance	Rank
Ep	0.1386	1	0.1053	7	0.1211	1
Js	0.1068	5	0.1293	1	0.1178	2
Si	0.1130	4	0.1180	3	0.1158	3
Pr	0.1026	7	0.1204	2	0.1114	4
Ip	0.1151	3	0.1066	5	0.1111	5
Hp	0.1167	2	0.0993	9	0.1080	6
Pm	0.1050	6	0.1097	4	0.1076	7
Tw	0.1009	9	0.1065	6	0.1040	8
Td	0.1012	8	0.1049	8	0.1033	9

Source: authors' calculations.

The overall significance obtained using the CRITIC method emphasised the innovative practice Ep—Employee participation, which is extremely important in modern business conditions. This HRM practice strongly dominates the results that amount to 0.1386. The situation was relatively different when the respondents were asked to intentionally evaluate modern HRM practices. Namely, the PIPRECIA-S results placed the practice Js—Job security as the most significant, followed by the practice Pr—Pay and reward (0.1204). The final ranking prioritised Ep—Employee participation as the practice leading to better operations and positioning in the particular company's market. It is not unexpected that respondents consciously give higher priority to performances such as job security and payment. However, the initial estimation of the chosen companies regarding the considered practices revealed that the involvement of the employees is at the core of a successful company.

The obtained results proved that awareness of the criteria evaluation impacts the results regarding the weighting coefficients (Paramanik et al., 2022 [24]). Additionally, the different MCDM methods are grounded on different approaches that also lead to variations in the results (Ponhan and Sureeyatanapas, 2022 [23]). As Table 7 presents, the CRITIC method highlighted employee participation as the priority, while pay and reward are the most important according to the PIPRECIA-S results. The final ranking order, incorporating both approaches, gives a more realistic perspective on the significance of HRM practice in the contemporary business environment.

After defining the objective and subjective significance of the involved HRM practices, we applied the COBRA method to rank the selected companies. The procedure is performed on the data obtained by each respondent separately. We utilise the CRITIC-COBRA and PIPRECIA-S-COBRA approaches to obtain the relevant results. Table 8 presents the results obtained, and Table 9 contains the defined rankings of the alternative companies.

Table 8. The COBRA results.

	R_1	R_2	R_3	R_4	R_5	R_6	R_7	R_8	R_9	R_{10}	R_{11}	R_{12}	R_{13}	R_{14}	R_{15}	R_{16}	R_{17}	R_{18}	R_{19}	R_{20}	R_{21}
CRITIC-COBRA																					
K_1	0.036	−0.003	−0.043	−0.031	−0.034	−0.045	−0.050	−0.055	−0.033	−0.044	−0.042	−0.060	−0.042	−0.062	−0.065	−0.064	−0.049	−0.040	−0.076	−0.034	−0.026
K_2	0.041	0.029	0.000	0.003	−0.021	−0.057	−0.044	−0.032	−0.040	−0.012	−0.028	−0.009	0.023	0.019	−0.032	−0.051	−0.009	−0.029	−0.007	−0.047	−0.028
K_3	0.029	−0.013	0.005	−0.008	−0.020	−0.019	−0.043	−0.037	−0.035	−0.047	−0.033	−0.043	−0.047	−0.012	−0.012	−0.020	−0.021	−0.032	−0.019	−0.088	−0.032
K_4	−0.030	−0.006	0.016	0.012	−0.011	−0.030	−0.023	−0.036	−0.003	−0.015	−0.006	−0.023	−0.024	0.026	−0.004	−0.034	−0.009	−0.016	0.005	−0.016	−0.015
K_5	−0.037	0.007	−0.006	0.003	0.019	0.003	0.005	−0.020	−0.021	−0.012	−0.002	−0.008	0.004	0.002	0.031	0.014	−0.012	0.043	0.042	0.021	−0.001
K_6	−0.014	−0.021	−0.009	0.005	−0.001	−0.002	0.005	0.014	−0.009	−0.012	0.002	−0.016	0.018	0.015	−0.005	−0.004	0.008	−0.025	0.009	0.027	0.018
K_7	−0.025	−0.018	−0.016	0.007	0.001	0.017	0.058	−0.016	0.000	−0.008	0.017	0.004	−0.010	−0.008	0.008	0.001	−0.006	0.004	−0.013	0.050	0.020
K_8	−0.026	0.021	−0.007	−0.006	0.001	0.032	0.035	0.010	0.022	0.011	−0.002	0.024	0.009	0.018	0.012	0.051	0.030	0.013	0.034	0.047	0.026
K_9	−0.025	−0.028	0.001	−0.007	−0.009	0.017	0.015	0.020	0.007	0.001	−0.004	0.019	−0.009	0.015	−0.010	0.021	0.016	0.025	0.014	0.047	0.006
K_{10}	0.024	0.011	0.009	0.023	0.022	0.004	−0.011	−0.017	0.009	0.018	0.028	0.039	0.025	−0.019	0.011	0.019	−0.018	0.024	0.039	0.015	0.019
K_{11}	0.022	0.020	0.032	0.015	0.025	0.036	0.024	0.012	0.025	0.039	0.029	0.047	0.010	0.004	0.020	0.006	0.007	0.033	0.035	0.043	0.017
K_{12}	0.002	0.011	0.013	0.038	0.047	0.067	0.070	0.033	0.059	0.032	0.027	0.039	0.035	0.028	0.042	0.023	0.016	0.007	0.026	0.026	0.035
PIPRECIA-S-COBRA																					
K_1	0.031	−0.002	−0.045	−0.028	−0.036	−0.046	−0.058	−0.055	−0.042	−0.043	−0.042	−0.070	−0.042	−0.070	−0.065	−0.066	−0.059	−0.039	−0.066	−0.025	−0.026
K_2	0.040	0.021	−0.006	−0.005	−0.040	−0.063	−0.050	−0.022	−0.052	−0.011	−0.025	−0.010	0.005	0.008	−0.037	−0.057	−0.023	−0.025	−0.021	−0.057	−0.028
K_3	0.031	0.000	0.003	0.007	−0.002	−0.020	−0.043	−0.047	−0.032	−0.050	−0.037	−0.047	−0.053	0.005	−0.006	−0.030	−0.023	−0.028	−0.013	−0.088	−0.032
K_4	−0.033	−0.003	0.009	0.004	−0.025	−0.029	−0.017	−0.034	−0.003	−0.016	−0.015	−0.008	−0.021	0.050	0.005	−0.047	−0.027	−0.031	0.002	−0.006	−0.015
K_5	−0.038	0.001	0.005	0.017	0.014	−0.001	−0.009	−0.026	−0.027	−0.011	0.002	−0.002	0.008	−0.003	0.042	0.002	0.000	0.056	0.037	0.035	−0.001
K_6	−0.007	−0.026	−0.015	0.004	−0.007	0.007	0.006	−0.001	−0.010	−0.016	−0.008	−0.026	0.010	0.014	−0.010	0.004	−0.001	−0.026	0.021	0.043	0.018
K_7	−0.021	−0.024	−0.024	0.011	−0.002	0.020	0.057	−0.015	0.006	−0.004	0.011	0.007	0.004	−0.006	0.007	0.012	−0.010	−0.009	−0.008	0.042	0.020
K_8	−0.023	0.019	−0.005	−0.014	−0.008	0.035	0.029	0.003	0.020	0.008	0.004	0.020	0.015	0.025	−0.013	0.052	0.028	−0.002	0.031	0.060	0.026
K_9	−0.026	−0.025	−0.004	−0.009	−0.008	0.016	0.013	0.014	0.007	−0.001	0.002	0.031	0.003	0.027	0.006	0.032	0.039	0.020	0.032	0.038	0.006
K_{10}	0.028	0.022	0.028	0.035	0.020	0.000	−0.010	−0.008	0.009	0.022	0.036	0.042	0.024	−0.020	0.005	−0.004	−0.013	0.033	0.028	0.031	0.019
K_{11}	0.032	0.018	0.029	0.017	0.041	0.043	0.046	0.036	0.041	0.039	0.028	0.050	0.026	0.022	0.022	0.020	0.014	0.044	0.037	0.038	0.017
K_{12}	−0.001	0.016	0.029	0.043	0.049	0.067	0.063	0.024	0.066	0.032	0.028	0.050	0.036	0.039	0.057	0.029	0.018	0.014	0.032	0.020	0.035

Source: authors' calculations

Table 9. Final rankings.

	R_1	R_2	R_3	R_4	R_5	R_6	R_7	R_8	R_9	R_{10}	R_{11}	R_{12}	R_{13}	R_{14}	R_{15}	R_{16}	R_{17}	R_{18}	R_{19}	R_{20}	R_{21}	Final Ranking
CRITIC-COBRA																						
K_1	11	6	1	1	1	2	1	1	3	2	1	1	2	1	1	1	1	1	1	3	3	**1**
K_2	12	12	6	6	2	1	2	4	1	5	3	5	10	10	2	2	6	3	4	2	2	**3**
K_3	10	4	8	2	3	4	3	2	2	1	2	2	1	3	3	4	2	2	2	1	1	**2**
K_4	2	5	11	9	4	3	4	3	6	3	4	3	3	11	6	3	5	5	5	4	4	**4**
K_5	1	7	5	5	9	6	6	5	4	5	6	6	6	5	11	8	4	12	12	6	5	**6**
K_6	6	2	3	7	6	5	7	10	5	4	8	4	9	7	5	5	9	4	6	8	8	**5**
K_7	5	3	2	8	7	9	11	7	7	7	9	7	4	4	7	6	7	6	3	12	10	**7**
K_8	3	11	4	4	8	10	10	8	10	9	7	9	7	9	9	12	12	8	9	10	11	**10**
K_9	4	1	7	3	5	8	8	11	8	8	5	8	5	8	4	10	11	10	7	11	6	**8**
K_{10}	9	9	9	11	10	7	5	6	9	10	11	11	11	2	8	9	3	9	11	5	9	**9**
K_{11}	8	10	12	10	11	11	9	9	11	12	12	12	8	6	10	7	8	11	10	9	7	**11**
K_{12}	7	8	10	12	12	12	12	12	12	11	10	10	12	12	12	11	10	7	8	7	12	**12**
PIPRECIA-S-COBRA																						
K_1	10	5	1	1	2	2	1	1	2	2	1	1	2	1	1	1	1	1	1	3	3	**1**
K_2	12	11	4	4	1	1	2	5	1	5	3	4	6	6	2	2	3	5	2	2	2	**3**
K_3	9	6	7	7	8	4	3	2	3	1	2	2	1	5	5	4	4	3	3	1	1	**2**
K_4	2	4	9	5	3	3	4	3	6	3	4	5	3	12	7	3	2	2	5	4	4	**4**
K_5	1	7	8	10	9	5	6	4	4	5	7	6	7	4	11	6	8	12	11	7	5	**7**
K_6	6	1	3	6	6	7	7	8	5	4	5	3	8	7	4	7	7	4	6	11	8	**5**
K_7	5	3	2	8	7	9	11	6	7	7	9	7	5	3	9	8	6	6	4	10	10	**6**
K_8	4	10	5	2	5	10	9	9	10	9	8	8	9	9	3	12	11	7	8	12	11	**10**
K_9	3	2	6	3	4	8	8	10	8	8	6	9	4	10	8	11	12	9	9	8	6	**8**
K_{10}	8	12	10	11	10	6	5	7	9	10	12	10	10	2	6	5	5	10	7	6	9	**9**
K_{11}	11	9	12	9	11	11	10	12	11	12	11	11	11	8	10	9	9	11	12	9	7	**11**
K_{12}	7	8	11	12	12	12	12	11	12	11	10	12	12	11	12	10	10	8	10	5	12	**12**

Source: authors' calculations.

Table 9 shows that the obtained rankings are relatively uniform, which leads to the conclusion that the respondents were familiar with the business performance of the evaluated companies. The Borda rule enabled defining the final ranking results regarding both approaches, CRITIC-COBRA and PIPRECIA-S-COBRA, and the ranking involving both approaches (Table 10 and Figure 2).

Table 10. Overall ranking results.

	CRITIC-COBRA Rank	PIPRECIA-S-COBRA Rank	Overall Rank
K_1	1	1	1
K_2	3	3	3
K_3	2	2	2
K_4	4	4	4
K_5	6	7	6
K_6	5	5	5
K_7	7	6	6
K_8	10	10	10
K_9	8	8	8
K_{10}	9	9	9
K_{11}	11	11	11

Source: authors' calculations.

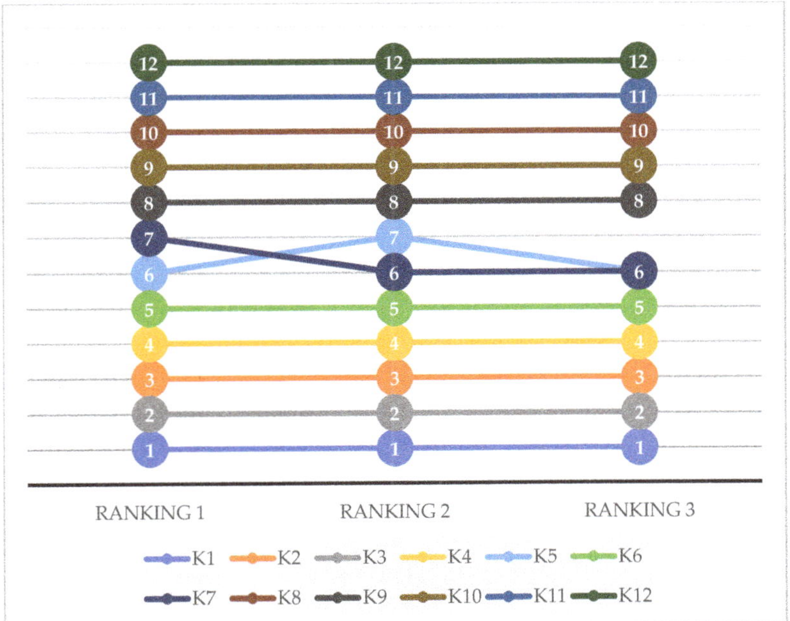

Figure 2. Overall ranking results. Source: authors' calculations.

The final results emphasised company K_1 as the one with the best results and the most successful in applying modern HRM practices.

Heidary Dahooie et al. (2022) [22] based their research on fuzzy DEA and ARAS methods, and highlighted the criteria related to financial results as the most critical HRM practice. Saeidi et al. (2022) [84] assessed sustainable HRM practices using the Pythagorean fuzzy SWARA-TOPSIS method. Because the chosen practices differed, the final results outlined the green work-life balance as the most important practice. In contrast, we applied the crisp model in our case, which outlined employee participation as the most critical practice, followed by pay and reward. These discrepancies are not surprising, because research studies were employed in different landscapes. Furthermore, the crisp MCDM model enables respondents to express their opinions more easily. The resulting estimation of the HRM practices influences the final ranking of the organisations involved.

4. Discussion

The application of the mathematically grounded objective-subjective approach yields adequate scientific results. The CRITIC method defined the objective significance of nine selected innovative HRM practices, while the PIPRECIA-S method helped to find the subjective significance. The geometric mean was used to determine the final significance of the innovative practices. Using the COBRA method, twelve companies from different sectors were evaluated against the mentioned innovative practices. Twenty-one respondents, experts from the HRM field, were involved in gathering data regarding the mentioned companies and fulfilling the special PIPRECIA-S questioners. As is the case with any research study, the methodological approach applied in this article has advantages and disadvantages.

The CRITIC method, which is the objective method intended for calculating the criteria weights, enabled defining innovative HRM practice's significance based on the initial data. However, the method in the current research study is somewhat subjectivised, because the initial data connected to the innovative practices and chosen companies were gathered from twenty-one respondents, reflecting their standpoints. The respondents used the Likert scale to perform previous estimations of the alternative companies regarding the considered innovative practices. However, the fact that the ratings are based on personal views slightly decreases their reliability, which would be higher if this evaluation was based on quantitative and exact data.

The PIPRECIA-S belongs to the category of subjective weighting methods, which are easily understandable and convenient for application when the respondents are unfamiliar with the MCDM techniques. This method has a straightforward computation procedure and is suitable for group decision environments because it enables more accessible group result aggregation. Besides, the estimation procedure in the PIPRECIA-S is much simpler because the respondent constantly compares the criteria with the first one. Nevertheless, this method has shortcomings, too. An essential disadvantage of the PIPRECIA-S is the absence of consistency checking of the gathered estimations from the respondents, which is contrary to the AHP (Analytic Hierarchy Process) (Saaty, 1987 [85]) and PIPRECIA-E (Extended PIPRECIA) (Stanujkić et al., 2017 [46]). This shortcoming of the PIPRECIA-S method makes it difficult to define if the respondents were consistent during the questionnaire filling.

The final analysis and ranking of the alternative companies were meticulously performed using the COBRA method. This method's thoroughness is reflected in its ranking, which is based on the distances from three types of solutions: positive ideal, negative ideal, and average. The distance measures are calculated using the Euclidian and Taxicab distance measures for all possible solutions, thereby increasing the relevance and reliability of the analysis. However, the COBRA method's advantage is also its disadvantage. The method is characterised by a complex and extensive computing procedure, which could be challenging for users who do not frequently use this type of decision support system.

The article expresses the intention of decreasing the research subjectivity level by involving a more significant number of respondents who are familiar with the effects of applying innovative HRM practices and business performances of the considered companies. In that way, the existence of biased estimations is minimised. However, crisp numbers were used to express the respondents' attitudes, which could not transfer the nuances of the respondent's opinions. Applying the fuzzy, grey, or neutrosophic numbers will more accurately reveal the immanent hesitancy and vagueness that characterise every decision-making process. Bearing in mind the previously argued points, it is desirable to use a particular model that involves extensions by applying some of the mentioned logic to observe if the obtained results would be the same. Nevertheless, despite the existing limitations, the applied methodological approach enabled the gaining of relevant scientific results.

The CRITIC method underscored the innovative HRM practice and designated Ep—Employee participation as the most significant and influential factor. This method, while objective, is subjectivised to some degree due to the data types used in the research. The results gained using the PIPRECIA-S method highlighted Jb—Job security as the most important HRM practice. The final results obtained using the geometric mean had employee participation and job security positioned as first and second, respectively. This suggests that employees want to be involved in the company's decision process, but in the current Serbian business environment, job security is nearly equally important to them. It can be concluded that the respondents consciously and unconsciously performed the estimation of the chosen selected innovative HRM practices with the CRITIC method, indirectly by evaluating the companies using the Likert scale, and directly using the questionnaire for the PIPRECIA-S method.

A comparison of the results obtained with those from other research studies revealed some differences. It should be emphasised that these differences originate from variations in the human practices lists and because of the conditions in the countries involved in the research. For example, Heidary Dahooie et al. (2022) [22] discovered that a trained and expert workforce is the most important for promoting innovativeness in Iranian nanotechnology small and medium enterprises (SMEs). The research conducted in Pakistan showed that innovative recruitment practices positively influenced the company's innovativeness (Aslam et al., 2023 [86]). Knowledge acquisition and adequate HRM practices are essential to enhance a company's innovativeness (Papa et al., 2020 [8]). Organisational memory, which represents knowledge acquired and preserved for future needs, is considered a critical HRM practice in India (Soumyaja and Sowmya, 2020 [87]). The fact that we observed the situation in Serbia holistically justifies the obtained results, because the involvement of the employees in the decision-making process still needs to be satisfactory. In addition to this, job security is paramount considering the fragile economic environment.

The question of why companies rank separately for each respondent is raised. The reason is that each respondent estimated the chosen companies himself/herself regarding the chosen innovative HRM performance, so the main idea was that the results obtained in that way would be more realistic and accurate. After defining the ranks of the chosen companies using the objective-subjective approach regarding all respondents, the final rank is defined using the Borda rule. This rule is a beneficial and straightforward approach that enables the calculation of the total score, which defines the final position of the estimated alternative, which in this case was the company. The final results showed that the company performed best in innovative HRM practices, being marked as K_1, while K_{12} had the worst results and was ranked last.

Applying the proposed MCDM model is more comprehensive than just facilitating decision-making in the HRM field; it could also be used to resolve different kinds of business issues. The accuracy of the proposed model could be improved by introducing adequate fuzzy, grey, or neutrosophic extensions. Until now, the authors have introduced the spherical fuzzy COBRA (Zorlu et al., 2024 [73]), fuzzy COBRA (Krstić et al., 2022 [65]), and grey COBRA (Ulutaş et al., 2024 [71]). There are different kinds of CRITIC method extensions (Puška et al., 2022; Wang et al., 2022; Sleem et al., 2023 [88–90]), while for the PIPRECIA-S, the extensions have yet to be introduced. These extensions will increase the model's reliability by incorporating the immanent vagueness into the decision environment. Furthermore, the complexity of the proposed model might be mitigated by developing suitable, more user-friendly tools, such as a software application based on the computational technique proposed by Mandal and Seikh (2023) [91]. This would reduce the time and effort required to perform the procedure, enabling a broader audience to benefit from the software support of the MCDM model during the decision-making process. Developing the specified software would promote the application of the suggested MCDM model, rendering it more accessible for use in various studies across various research fields, HRM practices, or national contexts, hence yielding comparable results and extending its reach beyond the academic community.

5. Conclusions

The main aim of this article was to determine the most successful HRM practices and to rank the selected organisations in Serbia according to them. The evaluation procedure was performed using the hybrid MCDM approach, consisting of the CRITIC, PIPRECIA-S, and COBRA methods. Nine innovative HRM practices and eleven organisations were submitted for evaluation. The results shed light on employee participation as a practice that is a primary one in the current business environment. The organisation designated as K1 is the first-ranked and represents the most successful utilisation of innovative HRM practices. This study aimed to employ the HRM domain as an initial application of this method, and to advocate for adopting the suggested model within HRM and other fields. This model's limitations lie in its complexity, as it incorporates three MCDM approaches, with the COBRA method being particularly difficult and potentially intricate for users. The identified deficiency may be addressed by developing a computer program that enhances the efficacy of applying the suggested MCDM model in facilitating the decision-making process. The suggested perspective would improve accessibility across varied scientific fields, facilitating additional studies and obtaining comparable results. Having user-friendly software would make the proposed MCDM model accessible beyond the academic community and allow practitioners and a broader audience to employ it in real business conditions in the decision-making process. Scientifically, the proposed MCDM model employs relatively new MCDM methods (besides the CRITIC method), the potential of which have yet to be discovered. Nevertheless, the results proved its applicability and usefulness to practitioners. By applying the proposed approach, managers could easily prioritise HRM practices and compare their organisations with others within the selected business field. In that way, they could perform benchmarking to highlight their vital business aspects and the areas they should improve. This MCDM model could be applied to finding solutions or selecting appropriate options in the different business fields such as sustainable development (Hasankhani et al., 2024 [92]), artificial intelligence (Alshahrani et al., 2024 [93]), the economy (Baydaş et al., 2024 [94]), hospitality management (Ayvaz-Çavdaroğlu et al., 2024 [95]), e-learning (Al-Gerafi et al., 2024 [96]), supply chains (Dohale et al., 2024 [97]), etc. Additionally, further research can also use the single-valued (Smarandache, 2020 [98]) or interval-valued (Wang et al., 2005 [99]) neutrosophic sets to make an extension of the PIPRECIA-S method, or involve the extended PIPRECIA method (Stanujkić et al., 2017 [46]) in the procedure of criteria weightings determination.

Author Contributions: Conceptualisation, V.M. and G.P.; methodology, V.M. and G.P.; software, V.M., G.P. and A.Đ.; validation, S.V., I.K. and A.Đ.; formal analysis, V.M. and M.M.; investigation, V.M. and G.P.; resources, V.M. and S.V.; data curation, A.Đ.; writing—original draft preparation, V.M., G.P. and S.V.; writing—review and editing, M.M., I.K. and M.S.; visualisation, V.M.; supervision, S.V. and M.S. All authors have read and agreed to the published version of the manuscript.

Funding: This research received no external funding.

Data Availability Statement: The data supporting this study's findings are available from the corresponding author upon reasonable request.

Conflicts of Interest: The authors declare no conflicts of interest.

References

1. Turskis, Z.; Šniokienė, V. IoT-Driven Transformation of Circular Economy Efficiency: An Overview. *Math. Comput. Appl.* **2024**, *29*, 49. [CrossRef]
2. Elshaer, I.A.; Azazz, A.M.S.; Fayyad, S. Green Human Resources and Innovative Performance in Small- and Medium-Sized Tourism Enterprises: A Mediation Model Using PLS-SEM Data Analysis. *Mathematics* **2023**, *11*, 711. [CrossRef]
3. Penjišević, A.; Sančanin, B. Ispitivanje uticaja ljudskog potencijala na organizacionu efikasnost kompanija. *Revizor* **2024**, *27*, 1–8. [CrossRef]

4. Shen, L.; Liu, K.; Chai, J.; Ma, W.; Guo, X.; Li, Y.; Zhao, P.; Liu, B. Research on the Mathematical Model for Optimal Allocation of Human Resources in the Operation and Maintenance Units of a Heavy Haul Railway. *Mathematics* **2022**, *10*, 3707. [CrossRef]
5. Haque, S.M.S. The Impact of Remote Work on HR Practices: Navigating Challenges, Embracing Opportunities. *Eur. J. Hum. Resour. Manag. Stud.* **2023**, *7*, 56–84. [CrossRef]
6. Corral de Zubielqui, G.; Fryges, H.; Jones, J. Social Media, Open Innovation & HRM: Implications for Performance. *Technol. Forecast. Soc. Change* **2019**, *144*, 334–347. [CrossRef]
7. Hong, J.F.; Zhao, X.; Stanley Snell, R. Collaborative-Based HRM Practices and Open Innovation: A Conceptual Review. *Int. J. Hum. Resour. Manag.* **2019**, *30*, 31–62. [CrossRef]
8. Papa, A.; Dezi, L.; Gregori, G.L.; Mueller, J.; Miglietta, N. Improving Innovation Performance Through Knowledge Acquisition: The Moderating Role of Employee Retention and Human Resource Management Practices. *J. Knowl. Manag.* **2020**, *24*, 589–605. [CrossRef]
9. Engelsberger, A.; Cavanagh, J.; Bartram, T.; Halvorsen, B. Multicultural Skills in Open Innovation: Relational Leadership Enabling Knowledge Sourcing and Sharing. *Pers. Rev.* **2021**, *51*, 980–1002. [CrossRef]
10. Turskis, Z.; Keršulienė, V. Sharda–Aras: A Methodology for Prioritising Project Managers in Sustainable Development. *Mathematics* **2024**, *12*, 219. [CrossRef]
11. Zavadskas, E.K.; Stanujkić, D.; Karabašević, D.; Turskis, Z. Analysis of the Simple WISP Method Results Using Different Normalization Procedures. *Stud. Inform. Control.* **2022**, *31*, 5–12. [CrossRef]
12. Karamaşa, Ç. Ranking Service Quality Using Multi-Criteria Decision-Making Methods: Example of Erzurum Province. *J. Process Manag. New Technol.* **2021**, *9*, 1–12. [CrossRef]
13. Özdağoğlu, A.; Keleş, M.K.; Altınata, A.; Ulutaş, A. Combining Different MCDM Methods with the Copeland Method: An Investigation on Motorcycle Selection. *J. Process Manag. New Technol.* **2021**, *9*, 13–27. [CrossRef]
14. Roy, B. The Optimisation Problem Formulation: Criticism and Overstepping. *J. Oper. Res. Soc.* **1981**, *32*, 427–436. [CrossRef]
15. Pinto-DelaCadena, P.A.; Liern, V.; Vinueza-Cabezas, A. A Comparative Analysis of Multi-Criteria Decision Methods for Personnel Selection: A Practical Approach. *Mathematics* **2024**, *12*, 324. [CrossRef]
16. Gendler, S.G.; Tumanov, M.V.; Levin, L.Y. Principles for Selecting, Training and Maintaining Skills for Safe Work of Personnel for Mining Industry Enterprises. *Nauk. Visnyk Natsionalnoho Hirnychoho Universytetu* **2021**, *2*, 156–162. [CrossRef]
17. Karabašević, D.; Stanujkić, D.; Urošević, S.; Maksimović, M. Selection of Candidates in the Mining Industry Based on the Application of the SWARA and the MULTIMOORA Methods. *Acta Montan. Slovaca* **2015**, *20*, 116–124.
18. López, J.C.; Solano-Noriega, J.J.; Gastelum-Chavira, D.A.; Gaxiola-Valenzuela, T. A Personnel Selection Model for a Software Development Company Based on the ELECTRE III Method and a Variant Of NSGA-II. *Innovar* **2022**, *85*, 117–132. [CrossRef]
19. Tuğrul, F. Personnel Selection Utilizing the Decision Making Mechanism Created With The Intuitionistic Fuzzy Topsis Method. *Mugla J. Sci. Technol.* **2022**, *8*, 16–21. [CrossRef]
20. Ulutaş, A.; Popović, G.; Stanujkić, D.; Karabašević, D.; Zavadskas, E.K.; Turskis, Z. A New Hybrid MCDM Model for Personnel Selection Based on a Novel Grey PIPRECIA and Grey OCRA Methods. *Mathematics* **2020**, *8*, 1698. [CrossRef]
21. Jakovljević, V.; Žižović, M.; Pamučar, D.; Stević, Ž.; Albijanić, M. Evaluation of Human Resources in Transportation Companies Using Multi-Criteria Model for Ranking Alternatives by Defining Relations between Ideal and Anti-Ideal Alternative (RADERIA). *Mathematics* **2021**, *9*, 976. [CrossRef]
22. Heidary Dahooie, J.; Estiri, M.; Zavadskas, E.K.; Xu, Z. A Novel Hybrid Fuzzy DEA-Fuzzy ARAS Method for Prioritizing High-Performance Innovation-Oriented Human Resource Practices in High Tech SME's. *Int. J. Fuzzy Syst.* **2022**, *24*, 883–908. [CrossRef]
23. Ponhan, K.; Sureeyatanapas, P. A comparison between subjective and objective weighting approaches for multi-criteria decision making: A case of industrial location selection. *Eng. Appl. Sci. Res.* **2022**, *49*, 763–771.
24. Paramanik, A.R.; Sarkar, S.; Sarkar, B. OSWMI: An objective-subjective weighted method for minimizing inconsistency in multi-criteria decision making. *Comput. Ind. Eng.* **2022**, *169*, 108138. [CrossRef]
25. Mufazzal, S.; Masood, S.; Khan, N.Z.; Muzakkir, S.M.; Khan, Z.A. Towards minimization of overall inconsistency involved in criteria weights for improved decision making. *Appl. Soft Comput.* **2021**, *100*, 106936. [CrossRef]
26. Liu, Y.; Sun, Z.; Liang, H.; Dong, Y. Ranking range model in multiple attribute decision making: A comparison of selected methods. *Comput. Ind. Eng.* **2021**, *155*, 107–180. [CrossRef]
27. Diakoulaki, D.; Mavrotas, G.; Papayannakis, L. Determining Objective Weights in Multiple Criteria Problems: The CRITIC Method. *Comput. Oper. Res.* **1995**, *22*, 763–770. [CrossRef]
28. Chang, Z. Design and Application of Evaluation Method for Civics Classroom Based on CRITIC Fuzzy Algorithm. *EAI Endorsed Trans. Scalable Inf. Syst.* **2024**, 11, *in press*. [CrossRef]
29. Krishnan, A.R. Research Trends in Criteria Importance Through Intercriteria Correlation (CRITIC) Method: A Visual Analysis of Bibliographic Data Using the Tableau Software. *Inf. Discov. Deliv.* 2024, *in press*. [CrossRef]
30. Saensuk, M.; Witchakool, S.; Choompol, A. A Hybrid Method Based on CRITIC Method and Machine Learning Models for Effective Fake News Detection in Thai Language. *J. Curr. Sci. Technol.* **2024**, *14*, 24. [CrossRef]
31. Shrinivas Balraj, U.; Uyala, S.; Jannu, S.; Gadakh, V.S. Multi-Objective Optimization of Machining Parameters for EDM Of Magnesium Alloy- ZE41 Using Marcos Algorithm AND Different Weighing Methods: SD, CRITIC and MEREC. *Int. J. Mod. Manuf. Technol.* **2024**, *16*, 124–140. [CrossRef]

32. Hassan, I.; Alhamrouni, I.; Azhan, N.H. A CRITIC–TOPSIS Multi-Criteria Decision-Making Approach for Optimum Site Selection for Solar PV Farm. *Energies* **2023**, *16*, 4245. [CrossRef]
33. Hosseinzadeh Lotfi, F.; Allahviranloo, T.; Pedrycz, W.; Shahriari, M.; Sharafi, H.; Razipour GhalehJough, S. The Criteria Importance Through Inter-Criteria Correlation (CRITIC) in Uncertainty Environment. In *Fuzzy Decision Analysis: Multi Attribute Decision Making Approach. Studies in Computational Intelligence*; Lotfi, F.H., Allahviranloo, T., Pedrycz, W., Sharafi, H., GhalehJough, S.R., Eds.; Springer: Berlin/Heidelberg, Germany, 2023; pp. 309–324. [CrossRef]
34. Mishra, A.R.; Chen, S.M.; Rani, P. Multicriteria Decision Making Based on Novel Score Function of Fermatean Fuzzy Numbers, the CRITIC Method, and the GLDS Method. *Inf. Sci.* **2023**, *623*, 915–931. [CrossRef]
35. Silva, N.F.; dos Santos, M.; Gomes, C.F.S.; de Andrade, L.P. An Integrated CRITIC and Grey Relational Analysis Approach for Investment Portfolio Selection. *Decis. Anal. J.* **2023**, *8*, 100285. [CrossRef]
36. Zhang, Q.; Liu, C.; Guo, S.; Wang, W.; Luo, H.; Jiang, Y. Evaluation of the Rock Burst Intensity of a Cloud Model Based on the CRITIC Method and the Order Relation Analysis Method. *Min. Metall. Explor.* **2023**, *40*, 1849–1863. [CrossRef]
37. Bhadra, D.; Dhar, N.R.; Salam, M.A. Sensitivity Analysis of the Integrated AHP-Topsis and Critictopsis Method for Selection of the Natural Fiber. *Mater. Today Proc.* **2022**, *56*, 2618–2629. [CrossRef]
38. Haktanır, E.; Kahraman, C. A Novel Picture Fuzzy CRITIC & REGIME Methodology: Wearable Health Technology Application. *Eng. Appl. Artif. Intell.* **2022**, *113*, 104942. [CrossRef]
39. Kumari, A.; Acherjee, B. Selection of Non-Conventional Machining Process Using CRITIC-CODAS Method. *Mater. Today Proc.* **2022**, *56*, 66–71. [CrossRef]
40. Pamučar, D.; Žižović, M.; Đuričić, D. Modification of the CRITIC Method Using Fuzzy Rough Numbers. *Decis. Mak. Appl. Manag. Eng.* **2022**, *5*, 362–371. [CrossRef]
41. Mukhametzyanov, I. Specific Character of Objective Methods for Determining Weights of Criteria in MCDM Problems: Entropy, CRITIC and SD. *Decis. Mak. Appl. Manag. Eng.* **2021**, *4*, 76–105. [CrossRef]
42. Zafar, S.; Alamgir, Z.; Rehman, M.H. An Effective Blockchain Evaluation System Based on Entropy CRITIC Weight Method and MCDM Techniques. *Peer–Peer Netw. Appl.* **2021**, *14*, 3110–3123. [CrossRef]
43. Peng, X.; Huang, H. Fuzzy Decision Making Method Based on CoCoSo with CRITIC for Financial Risk Evaluation. *Technol. Econ. Dev. Econ.* **2020**, *26*, 695–724. [CrossRef]
44. Peng, X.; Zhang, X.; Luo, Z. Pythagorean Fuzzy MCDM Method Based on CoCoSo and CRITIC with Score Function for 5G Industry Evaluation. *Artif. Intell. Rev.* **2020**, *53*, 3813–3847. [CrossRef]
45. Tuş, A.; Aytaç Adalı, E. The New Combination with CRITIC and WASPAS Methods for the Time and Attendance Software Selection Problem. *Opsearch* **2019**, *56*, 528–538. [CrossRef]
46. Stanujkić, D.; Zavadskas, E.K.; Karabašević, D.; Smarandache, F.; Turskis, Z. The Use of the Pivot Pairwise Relative Criteria Importance Assessment Method for Determining the Weights of Criteria. *Rom. J. Econ. Forecast.* **2017**, *20*, 116–133.
47. Keršuliene, V.; Zavadskas, E.K.; Turskis, Z. Selection of Rational Dispute Resolution Method by Applying New Step-Wise Weight Assessment Ratio Analysis (SWARA). *J. Bus. Econ. Manag.* **2010**, *11*, 243–258. [CrossRef]
48. Stanujkić, D.; Karabašević, D.; Popović, G.; Sava, C. Simplified Pivot Pairwise Relative Criteria Importance Assessment (PIPRECIA-S) Method. *Rom. J. Econ. Forecast.* **2021**, *24*, 141–154.
49. Mirčetić, V.; Popović, G.; Vukotić, S. Unveiling the Characteristics of the EU Charismatic Leaders Using PIPRECIA-S Method. *J. Process Manag. New Technol.* **2024**, *12*, 99–109. [CrossRef]
50. Rizwan, R.; Fizza, T.; Mumtaz, F. Evaluating Future Strategies for Sustainable Growth of Fiberglass Composites Industry in Developing Countries: A Novel Hybrid SWOT-Fuzzy Extended PIPRECIA Approach. *Heliyon* **2024**, *10*, e32137. [CrossRef]
51. Sarbat, I. A MCDM-Based Measurement Proposal of Job Satisfaction Comprising Psychosocial Risks. *Ergonomics* **2024**, 1–16. [CrossRef]
52. Setiawansyah, S.; Sintaro, S.; Saputra, V.H.; Aldino, A.A. Combination of Grey Relational Analysis (GRA) and Simplified Pivot Pairwise Relative Criteria Importance Assessment (PIPRECIA-S) in Determining the Best Staff. *Bull. Inform. Data Sci.* **2024**, *2*, 57–66. [CrossRef]
53. Stanujkić, M.; Popović, G.; Karabašević, D.; Šarčević, M.; Stanujkić, D.; Novaković, S. Approach to the Personnel Selection in a Group Decision-Making Environment Based on the Use of the MULTIMOORA and PIPRECIA-S Methods. *BizInfo (Blace) J. Econ. Manag. Inform.* **2024**, *15*, 19–26. [CrossRef]
54. Hadad, S.H.; Darwis, D.; Qurania, A.; Aldino, A.A.; Mehta, A.R.; Rahmanto, Y.; Setiawansyah, S. Student Ranking Based on Learning Assessment Using the Simplified PIPRECIA Method and CoCoSoMethod. *J. Comput. Syst. Inform.* **2023**, *5*, 30–39. [CrossRef]
55. Mladenović, M.; Đukić, T.; Popović, G. Analysis of Financial Reporting Platforms Based on the PIPRECIAS Method. *J. Process Manag. New Technol.* **2023**, *11*, 95–104. [CrossRef]
56. Setiawansyah, S.; Saputra, V.H. Kombinasi Pembobotan PIPRECIA-S dan Metode SAW dalam Pemilihan Ketua Organisasi Sekolah. *J. Ilm. Inform. Dan Ilmu Komput. (JIMA-ILKOM)* **2023**, *2*, 32–40. [CrossRef]
57. Stanujkić, M.; Popović, G.; Vukotić, S.; Karabašević, D.; Stanujkić, D. Improvement of Business Decision Making in IT Industry Using the MCDM Approach. *Industrija* **2023**, *51*, 73–88. [CrossRef]

58. Sulistiani, H.; Palupiningsih, P.; Hamidy, F.; Sari, P.L.; Khairunnisa, Y. Employee Performance Evaluation Using Multi-Attribute Utility Theory (MAUT) with PIPRECIA-S Weighting: A Case Study in Education Institution. In Proceedings of the 2023 International Conference on Informatics, Multimedia, Cyber and Informations System (ICIMCIS), Jakarta, Indonesia, 7–8 November 2023; IEEE: New York, NY, USA, 2023; pp. 369–373.
59. Aytekin, A. Determining Criteria Weights for Vehicle Tracking System Selection Using PIPRECIA-S. *J. Process Manag. New Technol.* **2022**, *10*, 115–124. [CrossRef]
60. Đukić, T.; Karabašević, D.; Popović, G. Evaluation of Aspects of Cognitive Skills Using the PIPRECIA method. *Ekonomika* **2022**, *68*, 1–14. [CrossRef]
61. Ulutaş, A.; Topal, A. Evaluation of the Criteria Used in the Selection of Renewable Energy Sources with the Plithogenic PIPRECIA Method. In *Optimization and Decision-Making in the Renewable Energy Industry*; Balo, F., Topal, A., Demir, E., Ulutaş, A., Eds.; Engineering Science Reference: Hershey, PA, USA, 2022; pp. 109–125. [CrossRef]
62. Popović, G.; Stanujkić, D.; Mimović, P.; Milovanović, G.; Karabašević, D.; Brzaković, P.; Brzaković, A. An Integrated SWOT–Extended PIPRECIA Model for Identifying Key Determinants of Tourism Development: The Case of Serbia. *Acta Geogr. Slov.* **2021**, *61*, 23–40. [CrossRef]
63. Ulutaş, A.; Popović, G.; Radanov, P.; Stanujkić, D.; Karabašević, D. A New Hybrid Fuzzy PSI-PIPRECIA-CoCoSo MCDM Based Approach to Solving the Transportation Company Selection Problem. *Technol. Econ. Dev. Econ.* **2021**, *27*, 1227–1249. [CrossRef]
64. Jauković Jocić, K.; Karabašević, D.; Jocić, G. The Use of the PIPRECIA Method for Assessing the Quality of E-Learning Materials. *Ekonomika* **2020**, *66*, 37–45. [CrossRef]
65. Krstić, M.; Agnusdei, G.P.; Miglietta, P.P.; Tadić, S. Evaluation of the Smart Reverse Logistics Development Scenarios Using a Novel MCDM Model. *Clean. Environ. Syst.* **2022**, *7*, 100099. [CrossRef]
66. Asker, V. Financial Performance Analysis Using the Merec-Based Cobra Method: An Application to Traditional and Low-Cost Airlines. *Gospodarka Narodowa, Pol. J. Econ.* **2024**, *318*, 35–52. [CrossRef]
67. Krstić, M.; Elia, V.; Agnusdei, G.P.; De Leo, F.; Tadić, S.; Miglietta, P.P. Evaluation of the Agri-Food Supply Chain Risks: The Circular Economy Context. *Br. Food J.* **2024**, *126*, 113–133. [CrossRef]
68. Oğuz, A.; Satır, H. Analyzing Profitability Performance with the Integrated MEREC-COBRA Method: The Case of BIST Retail Companies. *Bus. Econ. Res. J.* **2024**, *15*, 33–50. [CrossRef]
69. Sahak, A.S.; Karsli, F. A New Approach for the Assessment of Urban Eco-Environmental Quality Based on Remote Sensing: A Case Study of Herat City, Afghanistan. *J. Spat. Sci.* **2024**, *69*, 937–962. [CrossRef]
70. Tadić, S.; Krstić, M.; Radovanović, L. Assessing Strategies to Overcome Barriers for Drone Usage in Last-Mile Logistics: A Novel Hybrid Fuzzy MCDM Model. *Mathematics* **2024**, *12*, 367. [CrossRef]
71. Ulutaş, A.; Krstić, M.; Topal, A.; Agnusdei, L.; Tadić, S.; Miglietta, P.P. A Novel Hybrid Gray MCDM Model for Resilient Supplier Selection Problem. *Mathematics* **2024**, *12*, 1444. [CrossRef]
72. Verma, R.; Koul, S.; Ajaygopal, K.V. Evaluation and Selection of a Cybersecurity Platform—Case of the Power Sector in India. *Decis. Mak. Appl. Manag. Eng.* **2024**, *7*, 209–236. [CrossRef]
73. Zorlu, K.; Tuncer, M.; Yılmaz, A. Assessment of Resources for Geotourism Development: Integrated SWARA-COBRA Approach Under Spherical Fuzzy Environments. *Geoheritage* **2024**, *16*, 89. [CrossRef]
74. Krstić, M.; Tadić, S.; Agnusdei, L. Evaluating Governance Models in Intermodal Terminal Operations: A Hybrid Grey MCDM Approach. *J. Intell. Manag. Decis.* **2023**, *2*, 179–191. [CrossRef]
75. Tadić, S.; Krstić, M.; Dabić-Miletić, S.; Božić, M. Smart Material Handling Solutions for City Logistics Systems. *Sustainability* **2023**, *15*, 6693. [CrossRef]
76. Ulutaş, A.; Balo, F.; Topal, A. A New Hybrid MCDM Method for Optimizing Natural Stone Selection for Building Envelopes. *Rev. De La Construcción* **2023**, *22*, 646–660. [CrossRef]
77. Popović, G.; Pucar, Đ.; Smarandache, F. MEREC-COBRA Approach in E-Commerce Development Strategy Selection. *J. Process Manag. New Technol.* **2022**, *10*, 66–74. [CrossRef]
78. Verma, R.; Ajaygopal, K.V.; Koul, S. Circular Supplier Evaluation and Selection Using Hybrid MCDM Methods: Case of the Steel Manufacturing Industry. In Proceedings of the International Symposium of the Analytic Hierarchy Process, Virtual Meeting, 15–18 December 2022; pp. 1–16.
79. Emerson, P. The Original Borda Count and Partial Voting. *Soc. Choice Welf.* **2013**, *40*, 353–358. [CrossRef]
80. Marchant, T. Does the Borda Rule Provide More than a Ranking? *Soc. Choice Welf.* **2000**, *17*, 381–391. [CrossRef]
81. Fedajev, A.; Panić, M.; Živković, Ž. Western Balkan Countries' Innovation as Determinant of Their Future Growth and Development. *Innov. Eur. J. Soc. Sci. Res.* **2024**, 1–29. [CrossRef]
82. SORS. *Kvartalno Poslovanje Privrednih Društava, IV Kvartal 2023. Godine*; Statistical Office of the Republic of Serbia: Belgrade, Serbia, 2024.
83. Likert, R. A Technique for the Measurement of Attitudes. *Arch. Psychol.* **1932**, *140*, 1–55.
84. Saeidi, P.; Mardani, A.; Mishra, A.R.; Cajas, V.E.C.; Carvajal, M.G. Evaluate sustainable human resource management in the manufacturing companies using an extended Pythagorean fuzzy SWARA-TOPSIS method. *J. Clean. Prod.* **2022**, *370*, 133380. [CrossRef]
85. Saaty, R.W. The Analytic Hierarchy Process—What It Is and How It Is Used. *Math. Model.* **1987**, *9*, 161–176. [CrossRef]

86. Aslam, M.; Shafi, I.; Ahmed, J.; de Marin, M.S.G.; Flores, E.S.; Gutiérrez, M.A.R.; Ashraf, I. Impact of Innovation-Oriented Human Resource on Small and Medium Enterprises' Performance. *Sustainability* **2023**, *15*, 6273. [CrossRef]
87. Soumyaja, D.; Sowmya, C.S. Knowledge Management and Innovation Performance in Knowledge Intensive Organisations-The Role of HR Practices. *Int. J. Knowl. Manag. Stud.* **2020**, *11*, 370–392. [CrossRef]
88. Puška, A.; Nedeljković, M.; Prodanović, R.; Vladisavljević, R.; Suzić, R. Market assessment of pear varieties in Serbia using fuzzy CRADIS and CRITIC methods. *Agriculture* **2022**, *12*, 139. [CrossRef]
89. Sleem, A.; Mostafa, N.; Elhenawy, I. Neutrosophic CRITIC MCDM methodology for ranking factors and needs of customers in product's target demographic in virtual reality metaverse. *Neutrosophic Syst. Appl.* **2023**, *2*, 55–65. [CrossRef]
90. Wang, S.; Wei, G.; Lu, J.; Wu, J.; Wei, C.; Chen, X. GRP and CRITIC method for probabilistic uncertain linguistic MAGDM and its application to site selection of hospital constructions. *Soft Comput.* **2022**, *26*, 237–251. [CrossRef]
91. Mandal, U.; Seikh, M.R. Interval-valued spherical fuzzy MABAC method based on Dombi aggregation operators with unknown attribute weights to select plastic waste management process. *Appl. Soft Comput.* **2023**, *145*, 110516. [CrossRef]
92. Hasankhani, Z.; Zandieh, M.; Tirkolaee, E.B. Identification of optimal waste-to-energy strategies for sustainable development in Iran: SWOT analysis, hybrid MCDM methods, and game theory. *Renew. Energy* **2024**, *237*, 121564. [CrossRef]
93. Alshahrani, R.; Yenugula, M.; Algethami, H.; Alharbi, F.; Goswami, S.S.; Naveed, Q.N.; Lasisi, A.; Islam, S.; Khan, N.A.; Zahmatkesh, S. Establishing the fuzzy integrated hybrid MCDM framework to identify the key barriers to implementing artificial intelligence-enabled sustainable cloud system in an IT industry. *Expert Syst. Appl.* **2024**, *238*, 121732. [CrossRef]
94. Baydaş, M.; Yılmaz, M.; Jović, Ž.; Stević, Ž.; Özuyar, S.E.G.; Özçil, A. A comprehensive MCDM assessment for economic data: Success analysis of maximum normalization, CODAS, and fuzzy approaches. *Financ. Innov.* **2024**, *10*, 105. [CrossRef]
95. Ayvaz-Çavdaroğlu, N.; Iyanna, S.; Foster, M. Smart service quality in hospitality–A quantitative assessment using MCDM and clustering methods. *Int. J. Hosp. Manag.* **2024**, *123*, 103931. [CrossRef]
96. Al-Gerafi, M.A.; Goswami, S.S.; Khan, M.A.; Naveed, Q.N.; Lasisi, A.; AlMohimeed, A.; Elaraby, A. Designing of an effective e-learning website using inter-valued fuzzy hybrid MCDM concept: A pedagogical approach. *Alex. Eng. J.* **2024**, *97*, 61–87. [CrossRef]
97. Dohale, V.; Kamble, S.; Ambilkar, P.; Gold, S.; Belhadi, A. An integrated MCDM-ML approach for predicting the carbon neutrality index in manufacturing supply chains. *Technol. Forecast. Soc. Chang.* **2024**, *201*, 123243. [CrossRef]
98. Smarandache, F. The score, accuracy, and certainty functions determine a total order on the set of neutrosophic triplets (T. I. F). *Neutrosophic Sets Syst.* **2020**, *38*, 1–14.
99. Wang, H.; Smarandache, F.; Sunderraman, R.; Zhang, Y.Q. *Interval Neutrosophic Sets and Logic: Theory and Applications in Computing*; Theory and Applications in Computing. Infinite Study; UNM Digital Repository, University of New Mexico: Albuquerque, NM, USA, 2005; Volume 5. Available online: https://digitalrepository.unm.edu/cgi/viewcontent.cgi?article=1049&context=math_fsp (accessed on 1 November 2024).

MDPI AG
Grosspeteranlage 5
4052 Basel
Switzerland
Tel.: +41 61 683 77 34

Mathematics Editorial Office
E-mail: mathematics@mdpi.com
www.mdpi.com/journal/mathematics

www.ingramcontent.com/pod-product-compliance
Lightning Source LLC
Chambersburg PA
CBHW040835250926
51850CB00007BA/1317
9783725857753